DESIGNING & BUILDING ELECTRONIC GADGETS WITH PROJECTS

Other TAB Books by the Author

No. 901 *CET License Handbook—2nd Edition*
No. 1070 *Digital Interfacing With An Analog World*
No. 1194 *How To Troubleshoot & Repair Amateur Radio Equipment*
No. 1224 *Complete Handbook of Radio Transmitters (The)*
No. 1230 *Complete Handbook of Amplifiers, Oscillators and Multivibrators (The)*
No. 1250 *Digital Electronics Troubleshooting*
No. 1271 *Microcomputer Interfacing Handbook: A/D and D/A*
No. 1273 *Amateur Radio Novice Class License Study Guide—3rd Edition*
No. 1290 *IC Timer Handbook . . . with 100 projects & experiments*
No. 1327 *Amateur Radio Advanced Class License Study Guide—3rd Edition*
No. 1351 *Amateur Radio General Class License Study Guide—3rd Edition*
No. 1396 *Microprocessor Interfacing*
No. 1436 *104 Weekend Electronics Projects*
No. 1482 *Commercial FCC License Handbook—3rd Edition*
No. 1550 *Linear IC/Op Amp Handbook—2nd Edition*
No. 1643 *8-Bit and 16-Bit Microprocessor Cookbook*

No. 1690
$19.95

DESIGNING & BUILDING ELECTRONIC GADGETS WITH PROJECTS

BY JOSEPH J. CARR

TAB TAB BOOKS Inc.
BLUE RIDGE SUMMIT, PA. 17214

FIRST EDITION

FIRST PRINTING

Library of Congress Cataloging in Publication Data

Carr, Joseph J.
Designing and building electronic gadgets, with projects.

Includes index.
1. Electronic apparatus and appliances—Design and construction. I. Title.
TK7870.C288 1984 621.381 84-10827
ISBN 0-8306-0690-4
ISBN 0-8306-1690-X (pbk.)

Contents

Introduction

Electronics is truly the universal hobby of our technological society. One of the exciting things about electronics is that there is something for almost everyone who has a technical bent of mind, from the rank novice to the thoroughly experienced professional-level people who have made their entire life's work in the field. Because of the introduction of modern electronic devices, such as operational amplifiers, digital logic ICs, and microcomputers, we find relative amateurs able to design and build amazingly complex electronic devices.

Hobbyists are probably the biggest buyers of books such as this, and their interests were kept close in mind as I wrote this book. Hobbyists come in such a wide range of capabilities, however, that little can be said regarding them that is truly universal. I have assumed, therefore, that the hobbyist has studied basic electronics, but is not yet an old-timer who is no longer in need of instruction. There is no attempt at explaining elementary theory; such knowledge will be assumed. Those readers who are still too new at this hobby to understand the material are referred to any of dozens of books on the subject in the TAB BOOKS Inc. catalog (TAB BOOKS, Inc., Blue Ridge Summit, PA, 17214).

Also in mind as I wrote this book are the professionals in fields other than electronics who have a need for electronic ability, or, at least feel an interest in the subject. There are numerous scientists and engineers who will occasionally use electronic technology, yet cannot readily consult one of their electronic-engineering colleagues. Much of the material in this book was developed during my seven years as first a biomedical equipment technician and then (after earning my degree) a bioelectronics engineer at George Washington University Medical Center in Washington, D.C. During those years I had to deal with many instrumentation problems presented by chemists, physiologists, physicists, and other engineers. Some of the projects in this book were originally designed for professional colleagues in the medical center. Other material was selected for the needs of the scientist and engineer in a laboratory environment.

Professional design of electronic projects will still often require the services of a professional engineer with either microprocessor or instrumentation experience. But for many applications, you will be able to use the material in this book to perform your own design chores with only an occasional "atta boy" from the electronic-engineering community.

My own personal experience during the years I was a student at the GWU School of Engineering indicates that books such as this are often tucked discreetly inside lab notebooks of students working on their senior projects. While I am the last to disparage the technical instruction in our engineering schools, I also recognize the rightful place of the practical as well.

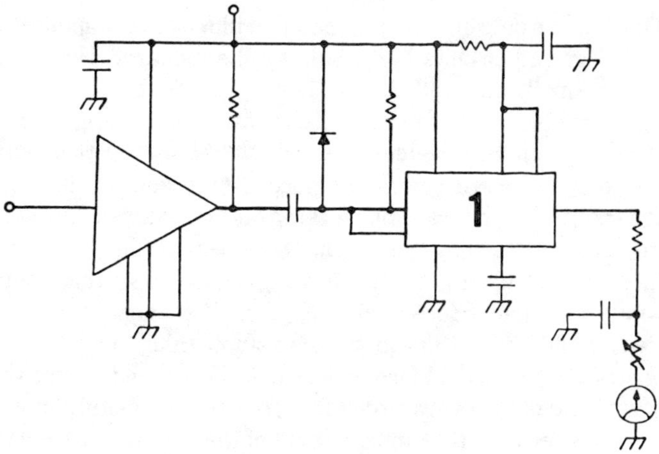

An Art and a Logical Process

The process of electronic project design is both an art and a logical effort in which one can see from past steps where the future steps will lead us. One cannot easily divorce the art from the logical process, although some try. It is especially easy for those who would like to assume that design is a scientific discipline devoid of art, or, alternatively from the other side of the coin, a totally intuitive process like designing a piece of sculpture or a painting. While it is impossible for any author to give you any intuition that you may lack, it is possible to describe the logical process of circuit design. We can lay out the steps of methods that routinely lead more often to a successful design than to failure. You are admonished, however, to not place 100 percent of your faith in the process, and thereby become enslaved to but one more discipline. If you have a hunch on the way that a design should proceed, then it might be fruitful to follow it.

The logical process proves successful most of the time because it is logical and permits one step to follow upon another after previous steps prove successful. Unfortunately, the first thing that most novice and unprofessional designers do is to build circuitry. I can recall from the years that I spent working in the electronics laboratory of a hospital and medical school a certain technician who would *routinely* start every design process with a copy of the *Signetics* or *National Semiconductor* applications manuals propped up on a stack of books on his bench, trying to "design by the seat of his pants" using only circuits presented to illustrate the applications of the integrated circuits being sold by the manuals. His designs only occasionally succeeded, and then by accident.

By contrast, at the same facility we had a design engineer who was really good. His designs worked, they behaved themselves and his projects were completed successfully and on time. This person spent a lot of desk time working out the details of his design before even approaching the bench. When he was finished the desk work, he would have a circuit drawing (or several) that seemed plausibly workable. He would even then breadboard the circuit before committing it to final form. It is a stupid thing to waste money going straight to the final form before breadboarding because there are often little glitches in a circuit that are not immediately apparent on a "desk check" of the circuit. Some of these glitches are due to component imperfections, while others are due to hidden factors not easily seen (e.g., "sneak circuits").

Unfortunately, not even some of our august professors of engineering know the "fact of life" presented above. I recall one fellow who went from high school, through the B.Sc. level, M.Sc. level, to the Ph.D. level in electrical engineering without ever obtaining any practical experience. His sole design experience was his senior project as an undergraduate. Yet, this person taught a circuit design course! To illustrate what this lack of practical experience did to that man, I pass on a little story that undoubtedly repeated itself yearly. The professor stated that the "nice thing about electrical engineering, as opposed to the other forms of engineering, is that *if it works on the blackboard it has to work on the bench!*" Tripe! You could tell in that class who had the practical, technician-level experience before coming to engineering school because about one-third of the class broke into wild, hysterical laughter while the other two-thirds looked on in perplexity wondering what the big joke was all about!

If at all possible, approach a problem in design from a position of low pressure. If you are attempting to design something or another for pay in a professional setting, this ideal may not be possible. Industry is notoriously high pressure for engineering and other designers, while academia can vary from low pressure to extreme pressure (usually from some Ph.D. candidate who needs some do-dad or another designed and built immediately in order to complete his or her doctoral dissertation research before the end of the academic year). I can recall the strategies used in a medical school environment, and some others just discussed. A favorite, at least in coffee-break scuttlebutt, was to go to work immediately on some little, cheap and quickly built circuit that would flash lights, illuminate seven-segment numerical readouts, and possibly "go BEEP" from time to time. Persons who are not into design engineering (and that medical school was overpopulated with such people!) are impressed with such "progress." Hang a voltmeter or two and an oscilloscope off the project and make sure something on the meter/screen moves from time to time; especially when the customer or sponsor or boss comes in. That tactic will either 1) provide you with the time to go off and do a proper job of desk work, or, 2) get you into all kinds of trouble (if the person knows you are faking it).

Do not be afraid to accept the advice of persons whom you consider less qualified than you . . . they may have insight that you lack. I can recall a Master Instrument Maker/Machinist at that university who was at least as qualified as any mechanical engineer I have ever known, at least in the area of practical designing. He used to do wonders in plastic, metal, and other materials for those medical school researchers, despite the fact that few of the intellectual snobs among both students and faculty ever realized what a gem of a guy he was. There was one time when a distinguished member of the faculty (a department chairman, no less) brought the machinist a drawing Xeroxed from a professional journal, and asked him to build it. The machinist examined the drawings for a couple of hours and then went to the professors office to inform him that the device *wouldn't work*. Of course, the distinguished professor exploded in anger over the utter "cheek" of a mere machinist who would *dare* question the design of the great and wonderful Doctor so-and-so, a *well-known expert* in the field! Still, insisted the chastened machinist, the damn thing won't work. A check with the original author revealed, however, that there had been an error in the published version of the drawing. It seems that it is often a long time and a long way from the designers bench to the published version. Very often in academic circles, the designer who should be the one to check the final drawing prior to publication is far down the chain of command so is not even consulted. Pity.

The same problems as presented in the paragraph above, incidentally, also apply to electronic circuits. Very often we find published versions of a circuit do not contain all of the information needed to replicate the design, or, they contain errors that would have been caught in many other publications (for reasons cited above). The point of this little tirade is to warn you against depending on others (even me!), and to use your own mind and your own knowledge. If you know something looks fishy in a circuit, then ask questions . . . do not be swayed by the "argument from intimidation." Experts can be wrong, and slavishly following one can lead to professional disaster.

Besides, published circuits are often *incomplete!* It is frequently the case that published circuits delete for graphics reasons such nice little things as power supply connections to IC devices, offset controls, and others. Unless the purpose of the published design is to allow you to build it (as in magazines such as *Radio-Electronics*), the editor may well delete certain key features. This is extremely common in academic journals, so much so in fact that I sometimes wonder if the *real* purpose is not simplified graphics, but rather, to prevent someone else from performing the same research and thereby exposing how sloppy was the original data (*sigh*). I even know one young engineer who borrowed some computer equipment from me to attempt some interfacing experiments as a learning process. He informed me that one circuit which I gave him (a published circuit) didn't work. After carefully discussing it with him, I noted that the power supply connections were deleted in the drawing. I then asked him if he had made these connections, and was met by a blank stare: " . . . *power supply*?!?!?" Application of +5 volts to the TTL circuits worked wonders for their operation!

THE PROCESS

Begin your design efforts with desk work and a set of drawings. Of course, the complexity of the proposed design will in large measure determine how much time is needed at each step in the process, and also how complex the process must be. Of course, for a little one or two stage project the process is a lot shorter and a lot simpler than for a large project. Obviously, an audio amplifier for headphones on the output of a receiver is a lot less complex than, say, a single-sideband transceiver or a complex medical/physiological instrument. In the case of the latter projects, more rigid discipline to the method is needed for the simple purpose of *maintaining control* of the design over the long period between inception, conception, and delivery of the product.

Before discussing the design procedure in depth, let's pass along one very important admonition: **Document your work!** Even if the project is a small and insignificant little do-dad, keep an accurate record of the circuit and adjustment information. I can recall from bitter experience working on circuits built by other people who did not have the elementary courtesy to make circuit diagrams available. Besides, the poor slob who follows you might be you! It may happen that alignment or repairs are needed a year or two after you finished the work, and it is highly unlikely that you will remember in detail what is in the box. If you failed to document your work earlier, then shame on you . . . but you get what you deserve by having to follow yourself.

Again, the main and most important reason for documenting your work is maintaining control. Without adequate documentation that reflects all changes and tests as they are made, you will lose track of the configuration and become lost in a maze of work done over and over for the lack of prior writing down of progress. Some professional designers keep a bound notebook that is used to record all of their work on a project . . . much akin to the laboratory notebook of the scientist.

There is still another reason why you might want to keep a proper design notebook: patents. Suppose you invent something worthy of being patented. It may someday occur that a challenge to the patent is made, and you will have to prove that you did indeed invent (rather than steal) the invention. A well-documented engineering notebook, showing both successes and failures which lead to the final product, will save your—errrr—case.

One of the complaints heard about the design process is that it seems burdened by needless steps and too much paperwork. Again, control is the reason why that is needed. But don't despair . . . it works, especially for those of us who lack the God-given intuition to jump from intellectual mountaintop to intellectual mountaintop.

Good design work is much like good writing. It involves "writing, rewriting, and more rewriting." Of course, for the circuit and project designer this truism translates into planning, replanning, and more replanning. It also involves plenty of book work and magazine reading. The "literature search" phase is especially important in large projects because it will allow you to draw upon the experience of earlier workers. It will also allow you to home in on the types of features and the specifications required of the particular device or instrument that you are designing. There is no miracle method in designing. It is a bit like music. While the accomplished performer seems to make music with such ease, it is because of many hundreds of hours of learning that backed up his skill. I remember a black woman Gospel singer who had an accompanist who played the piano. The singer was well-known, yet was not too proud to give the young man on the piano some time to display his talent. The woman made a statement in jest to the effect "you white folks say we people have rhythm . . . there it is." Later that evening back at the college dormitory where we were staying, I saw the woman again. She was standing in the lobby while her accompanist entertained the residents. I told her: "Lilly, that's not from 'rhythm,' it comes from lots of sweat at the keyboard." That young man didn't have any kind of genetic magic called rhythm, he had a systematic study of his art . . . so it must be with the proficient designer.

Some of the techniques given in this chapter were derived from, and expanded from, an article titled "A Structured Engineering Approach to the Design and Construction of Electronic Equipment" (*QST*, August 1983, p. 18), by Jerry L. Pittenger. Although that article is directed toward amateur radio operators (who form only a small fraction of the readers of this book) it contains truths that are of use to all people interested in electronic circuit design. In fact, it is an article that ought to be passed around engineering and engineering technology schools. In my own treatment of Pittenger's recommendation, given below, I have modified them according to my own experience (although that does not imply that I found fault with Pittenger).

The steps in the design process are:

a) Define the project
b) Make a block diagram of the project
c) Plan and document the testing of the project
d) Breadboard the project modules
e) Test the modules individually
f) Review the test results and make modifications as needed
g) Build the "final" modules
h) Assemble the final project
i) Test the project
j) Final Acceptance testing
k) Write final documentation
l) Provide user education (if needed)

In the sections to follow we will look at each of these activities in greater detail in order to define just what must be done. I know that the procedure looks long and involved, but it will permit you to design projects of a magnitude that you would have otherwise not thought possible. In fact, relatively large-scale projects become easier and more knowable when we break the big problem into small and easily understood pieces.

DEFINE THE PROJECT

The Pittenger article gives the best definition of this stage, even though it used a different heading: "developing a functional specification." This means that you must think about and write down exactly what you want this project to do. What goal is being sought by this project? Why are you doing it? You must not only decide just what you want the project to do, but also what features and properties of the project are needed in order to accomplish that goal. Is the device a physiological amplifier, for example? Will it drive a strip-chart recorder with a 1-mV input, or an oscilloscope with a 1-volt input? Obviously, the gain required depends much upon the display device (i.e., a difference of 1000:1!). During this phase of the project, it is also appropriate to list alternative means of attempting the goal, and alternative features that "might be nice" if they are affordable or if there is space in the project.

The specification must be *written down*. This document will be the goal that you strive for. In it, you must list minimum acceptable behavior, and also desired optimum behavior. If the project works at a level below the minimum acceptable, then it must either be redesigned or scrapped. It is extremely important that the specifications be written down. This document must be modified anytime the specification changes, or is updated.

MAKE A BLOCK DIAGRAM OF THE PROJECT

Long before you decide what circuits to use, or even which of the several technologies will be used, you must develop a functional block diagram that will implement the specification developed above. The block diagram is a stage-by-stage rendering of how the project will work. Figure 1-1 shows a functional block diagram for a physiological amplifier used to acquire electroencephalogram (brain wave) signals for analysis by computer.

The project of Fig. 1-1 was intended to achieve the following goals:

a) Acquire 5 to 100 microvolt brain-wave signals from a differential source (the scalp) with a 10 kohm impedance
b) Provide rejection of 60-Hz interference on an optional basis
c) Limit bandwidth to 0.05 Hz to 100 Hz
d) Provide *sensitivity* ("span") and *position* controls
e) Provide 1-volt outputs to an oscilloscope
f) Provide a 10-millivolt signal output to a real-time stripchart recorder

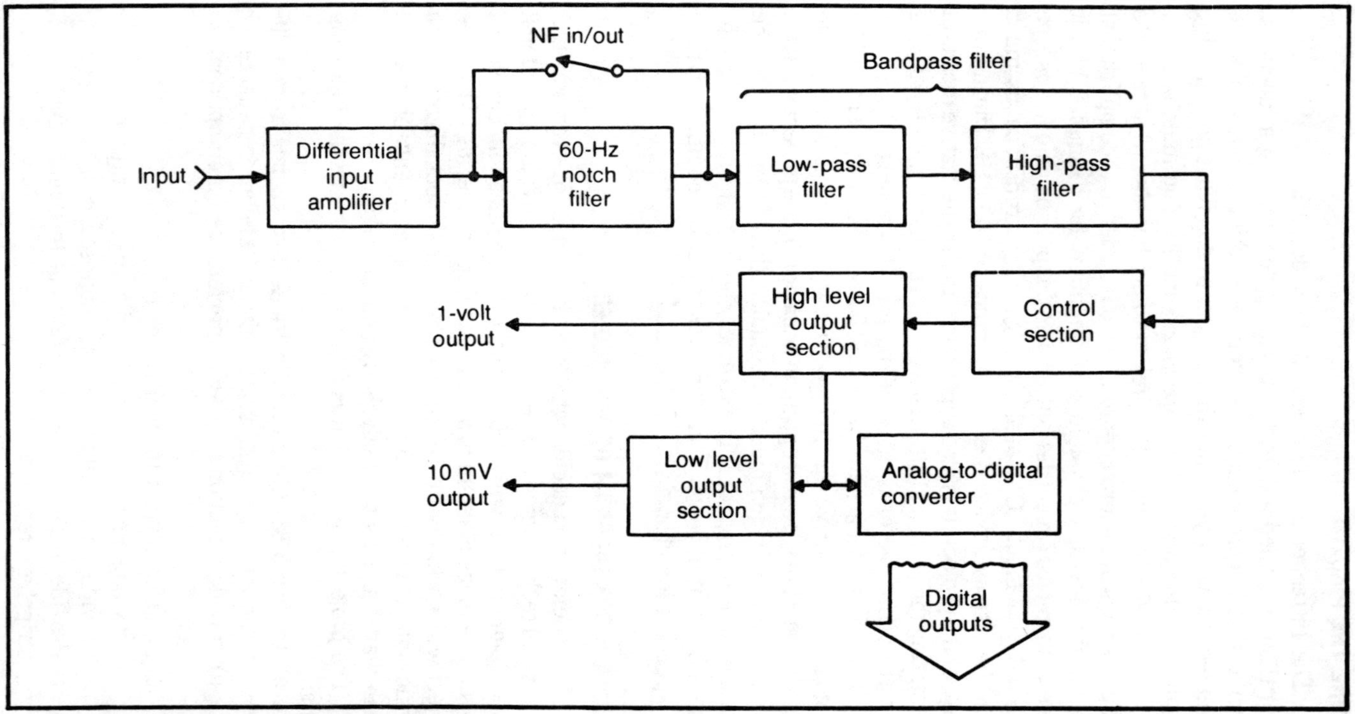

Fig. 1-1. Block diagram of a biophysical amplifier.

g) Provide a 10-bit A/D converter to provide digital data to the computer

Some of the blocks in Fig. 1-1 represent relatively complex circuits, while others (e.g., low-level output section) are simple. The low-level output section could be a 100:1 resistor voltage divider from the 1-volt output. The control section might be three or four operational amplifiers with external potentiometers, resistor voltage dividers/switches, and so forth. The two filters are used for frequency shaping, but could conceivably be replaced with a single bandpass amplifier that has the requisite −3 dB frequency response points.

PLAN AND DOCUMENT THE TESTING OF THE PROJECT

During this phase we will write a test plan, as well as review the final design *prior* to committing anything to so much as a breadboard. Write down the sequence of the tests, the specific steps needed to accomplish each and every test, and the results expected if the circuit does as it is supposed to. Test plans should be written for each subsystem, for integrating the subsystems, and for the final project. These will be used in the breadboard phase, as well as in the final construction.

BREADBOARD THE PROJECT

Finally we are going to build some circuitry! This is the step that most novice designers jump to immediately . . . and fail miserably because of that impetuousness. In this stage, we will actually start building the modules and subassemblies for the project, but not in final form. We will most likely use something like a multi-pin electronic breadboard (e.g., the AP Products, Inc. *Powerace* series, or their equivalent), or, perforated wiring board. Some people like to use the printed circuit patterned stock sold by firms such as *Vector Electronics*.

During the breadboarding phase you will spend a lot of time correcting errors in the design, finding areas which need improvement or compensation (e.g., op-amp output offset) and generally snuffing out problems in the concept. Don't be either frustrated or afraid to make a mistake during this phase. Mistakes are most correctable at this early stage of the game, and it is important that you stay free to make mistakes. Whether from the pressure of others, or from your own idiotic pride, the fear of making mistakes will lead to more problems that it solves. Once the project breadboarding is completed, you will be able to move onto the next stage, module testing.

TEST THE MODULES INDIVIDUALLY

This phase may sometimes be integrated somewhat with the previous phase. Once the breadboards seem completed, then go to full testing of the stages in order to see if they comply with their own specifications and have at least a fighting chance of meeting the project goals. Here is the first place where you implement the test plan.

REVIEW THE TEST RESULTS AND MAKE MODIFICATIONS IF NECESSARY

Hard decisions might have to be made at this phase of the design process. If the project subassemblies do not come up to expectations, and it appears that the project will not "fly" with the present configuration, then it is necessary to decide whether or not the design needs either scrapping or substantial modification. If you have done a good job in the breadboarding and testing phases, then there will be few surprises during the review process. Obviously, if you have been having substantial trouble with the circuitry, then you would have expected negative results.

BUILD THE "FINAL" MODULES

I placed quotation marks around "final" in this section because I know that designers are much like artists in one respect: they like to "diddle" with the finished product! Art gallery owners usually tell clients to never return a painting to an artist because they will want to redo it! The same is true of engineers in design laboratories. Tinkering with the design by the designers will drive the production people absolutely bonkers, so don't allow it. There is an old maxim that I only subscribe to partially: "It's time to shoot the engineers and go into production."

This phase is nearing the end of the project, and is when you build the subassemblies into final form. This is where you will use the best materials, make printed circuit boards (if desired), make the cabinet, and so forth.

ASSEMBLE THE FINAL PROJECT

This is where you "put it all together into the final deliverable form. All of the modules built in the previous phase come together here and become a working system for the first time.

TEST THE PROJECT

Perform all of the final system tests that tell you that the system is working according to the specification. Document the results, which is also a requirement of all other testing.

FINAL ACCEPTANCE TESTING

If you are building the project for someone else, especially if you are being paid to do the job, then it will be necessary to have some criteria on which acceptance is based. You will get paid if the project passes this test or tests. The performance tests done in this phase must have been agreed upon at the start of the project, i.e., when the specification was written. It is at this juncture that most less formal laboratories get into trouble. I can recall customers in the medical school where I worked who were unhappy with custom equipment they had bought or had made because it didn't perform up to expectations. Yet they had no choice except to accept the trash because it was built under a contract which had no acceptance criterion. One such device was a servocontroller that kept the pressure in a monkey's eye constant while measuring the outflow facility of the eye. Three devices were delivered to an opthalmology research laboratory, yet they didn't work. The engineering firm which designed them used an orange to simulate the monkey's eye. But an orange lacks an outflow facility, so would equilibrate shortly after the pressure increment was changed. For all they knew, that was how an eye works. But when the outflow was added, the system became unstable and would oscillate about the correct pressure. The system was also tremendously sensitive to temperature variations. The biophysicists who had to use the equipment had to arrive early in the morning to turn on the power to let it come to thermal stability by noon when her experiment was scheduled.

The above mentioned problem would have been solved if the physician who ordered the equipment had provided a specification that correctly reflected the real-world situation of eyeballs. The engineer who designed the equipment (I eventually met him) was not incompetent, he merely designed according to what he believed was true in the light of no specifications to the contrary. I spent a couple of years, part-time, troubleshooting design problems in that system that could have been ironed out early. If there had been a decent acceptance test for the device, then the firm would have been required to demonstrate that it met the specifications . . . and the university would have saved $20,000.

WRITE THE FINAL DOCUMENTATION

I have nothing but utter contempt for the unscrupulous crud who leaves behind a complex project for someone else to 1) finish, 2) repair, or 3) modify without documentation. Your lab notes are likely to be a mess by this time. During this final phase, write the operations manual, draw final schematics and write a technical description. Give it to the customer and keep a copy yourself.

PROVIDE USER EDUCATION

Unless the user is well-acquainted with the operation of the project, or, the operation is trivial, provide some sort of training for the user. This will cement good relations and, also, provide some coverage in the event of a product liability action. If there is some way that improper operation could cause harm, then you might be liable for that harm if you failed to warn the user. A training session might serve to document the fact that you warned the user of potential pitfalls. Besides, it's just good manners to show the poor slob how to use it.

CONFIGURATION MANAGEMENT

Perhaps no other single subject angers the production people or the field service people more than lack of configuration management. "Configuration" is the design of the circuit and mechanical aspects of the project, and can include software as well if a computer is part of the design. Changes in configuration may be fine as far as the engineer is concerned, but it raises hell with the production department. Also, the field service guy is at a disadvantage if proper configuration management is not employed. During the years when I was working on my engineering degrees I waddled about a major hospital with a tool kit repairing medical instruments. I can't count the number of times that I opened a cabinet only to find that we had *Mark I* manuals and *Mark IV* equipment. One manufacturer sold under the same model number a cardiac monitor with modern integrated circuits and an old tube model. We had drawings on the old tube equipment, yet they had been packed with the IC version (*sigh*—fortunately for the world that manufacturer went out of business—he underbid too much once too often).

Good configuration management really means limiting changes to those that will accomplish some good purpose, and then documenting those changes for all to see. Hewlett-Packard made medical instruments, and we used them in that hospital. Their basic manual was the original edition, but up front there were colored pages (the colors had meaning) that denoted changes, new configurations, warnings, field changes, common problems, and so forth. With that kind of information, we could keep only the newest manual in the shop, and store the others in the archives. The latest manual covered all configurations for that model back to the original. It was merciful and easy to use.

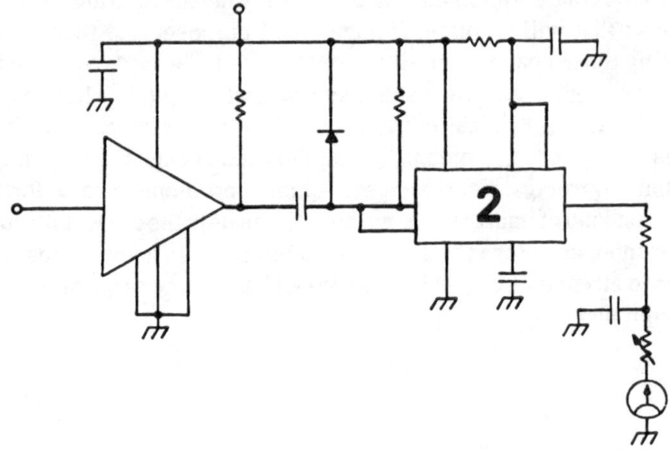

Dc Power Supply Circuits

A scientist in a medical school once invited me into his laboratory in order to proudly show off some new instrumentation he had just obtained. In addition to the costly Tektronix oscilloscope—the pride of his lab in that budget-constrained time—there was a little amplifier with a gain of ×10 and a very high input impedance. The amplifier could be powered with either batteries, which is a pain in the tush, or, a specially-built dc power supply made by the same company who made the amplifier. The amplifier, incidentally, was a relatively ordinary instrumentation amplifier of the sort covered later in this book. My scientist friend paid $295 for that amplifier (we can duplicate it for less than $35 and give it variable gains to ×1000!), and the dc power supply cost *another* $295! That's almost $600 for a pair of "instruments" that could be ginned up in a university electronics shop for less than $50 or $60, including shop time. The dc power supply was a simple affair that offered two voltage outputs, +15 Vdc and −15 Vdc, both rated at 100 milliamperes. In this chapter we will discuss how such a supply can be built for approximately $15 worth of parts and about one hour's worth of shop time.

The dc power supply is literally the heart of almost any electronic project that you can think of . . . but is usually the least considered portion of the design! During a period when I was keeping service records of electronic repairs, I tallied the repair actions according to type of circuit and found that more than *half* of the actions involved the power supply! The power supply makes the project "go," and is critical to its success and proper operation. If we can design an appropriate, reliable, power supply, then we will be a long way home toward having a reliably operating instrument or project.

In this section we will discuss the basic theory behind elementary power supply design, and present some tested projects that cover the vast majority of the needs of readers. But first, let's examine some of the basics of dc power supply design.

POWER SUPPLY BASICS

The purpose of a dc power supply is to convert the bidirectional alternating current (ac) from the power mains to unidirectional direct current (dc) required for the proper operation of electronic circuits. A dc power supply might be relatively crude, like a car battery trickle charger, or, it may be complex. It may produce a "pulsating dc" output that is barely distinguishable as dc, or, it may produce a pure dc that is difficult to differentiate from the pure dc produced by batteries.

The principle components of the dc power supply include: transformer, rectifier, filter and, sometimes, a voltage regulator. Most of the readers of this book will require a relatively pure direct current, so will have to build regulated dc power supplies. For those people, then, the voltage regulator is not optional.

A transformer (Fig. 2-1) is built using coils of wire wound on a common form such that the magnetic field produced when current flows in one will induce a current into the other. There are at least two windings on a true transformer (never mind if you know what an "autoformer" is!) and these are designated *primary* and *secondary*. The primary winding is that which is connected to an ac power source such as the 115 volt ac power line. The secondary is the winding that is connected to some sort of load (e.g., resistor R_L in Fig. 2-1).

Transformers are amazingly efficient devices. Ratings on the order of 97 to 99 percent are not uncommon, although 95 percent efficient is more likely for cheapies. For all transformers, then, we can ignore the losses inherent in any physical system and declare that the following relationship holds true:

$$V_p I_p = V_s I_s \qquad \textbf{(2-1)}$$

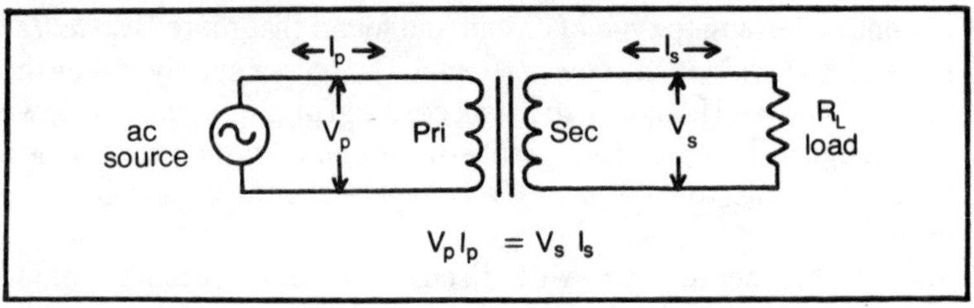

Fig. 2-1. A simple transformer.

Where:

- V_p is the voltage applied to the primary winding
- V_s is the voltage appearing across the secondary when V_p is applied
- I_p is the primary current
- I_s is the secondary current (i.e., V_s/R_L)

It is common practice to use RMS values of voltage and current. Other measures (e.g., peak voltage) are usable, provided that one use the same measure for all four terms, and modify the design equations accordingly.

The purpose of the transformer in the power supply is to scale the ac voltage to a level that will be compatible with the electronic circuits being powered from the supply. For most solid-state projects, this means that the secondary voltage will be lower than the primary voltage; such a transformer is normally called a step-down transformer. Many vacuum-tube projects from past years, and high power vacuum tube radio transmitters (and other devices) use a dc voltage that is higher than the ac power mains potential. The transformer that delivers a higher voltage is called a stepup transformer. For most of our projects, however, stepdown transformers are used. The names of these transformers may well reflect some other application, or some archaic application. In many cases, for example, we will want to use a "6.3 Vac filament transformer" to supply power to +5 volt regulated power supplies. A filament transformer is used in vacuum tube circuits, and the name persists. Other names are signal transformer, rectifier transformer, low-voltage transformer, *et cetera.*

The main ratings of the transformer are the secondary voltage and current, and the primary voltage; we rarely worry about the primary current directly. The secondary voltage and current are determined in part by the application and in part by the values available on the market. For example, we may require a 2.2 ampere transformer, and find that only 3 ampere models are available in the voltages that we need. In general, it is safe to accept a transformer with a higher than needed secondary current rating, but do not use one with a higher secondary voltage unless you either know yourself what is possible to "get away with" or have expert advice. The primary voltage rating is important to plug into the correct ac potential. In the USA, we have 105 to 125 volts ac. (RMS) on the power outlets in homes and labs, with 220 volts ac being available on special installation (for example, air conditioners and electric clothes dryers use 220 volt power).

The work load of the transformer is not given directly by the secondary ratings, but by the primary volt-ampere (VA) rating. The primary and secondary VA product will always be equal, but the primary rating expresses the maximum power (purists, shut up please!) that the device can handle. The reason for this is that, in most transformers, the primary is the winding closest to the core, so will be buried and not exposed to air. In fact, it not only is not exposed to the cooling effects of air but is covered with the heating of the secondary winding. As a result, the primary will burn out at a somewhat lower VA product than will the secondary.

The VA rating is usually predicated upon the maximum secondary current and secondary RMS voltage. If the transformer has a VA rating that is significantly different from the product $V_s I_{s(max)}$, then beware . . . something ain't right!

Be aware also that the VA rating of the transformer is based on the type of service that the transformer was intended to see in actual practice. If the transformer has a center-tapped secondary (about which we will study in a moment), then it is rated for normal full-wave service. The same transformer can be used in bridge rectifier service (also discussed in a moment), but can safely draw only half the rated current without exceeding the primary VA rating.

Figure 2-2 shows some of the different types of transformers that might be found in our projects. Figure 2-2A shows a transformer with multiple secondaries. In some cases, the two secondaries will have different voltage ratings; one might be stepup while the other is stepdown. These transformers are usually called *power transformers*. There are also multiple secondary transformers that have two different stepdown secondaries, or, two equal voltage secondaries.

Figures 2-2B and 2-2C show a transformer with dual primaries that can be connected in either of two manners, one for 110-volt and one for 220-volt applications. In Fig. 2-2B we see the 110-volt case

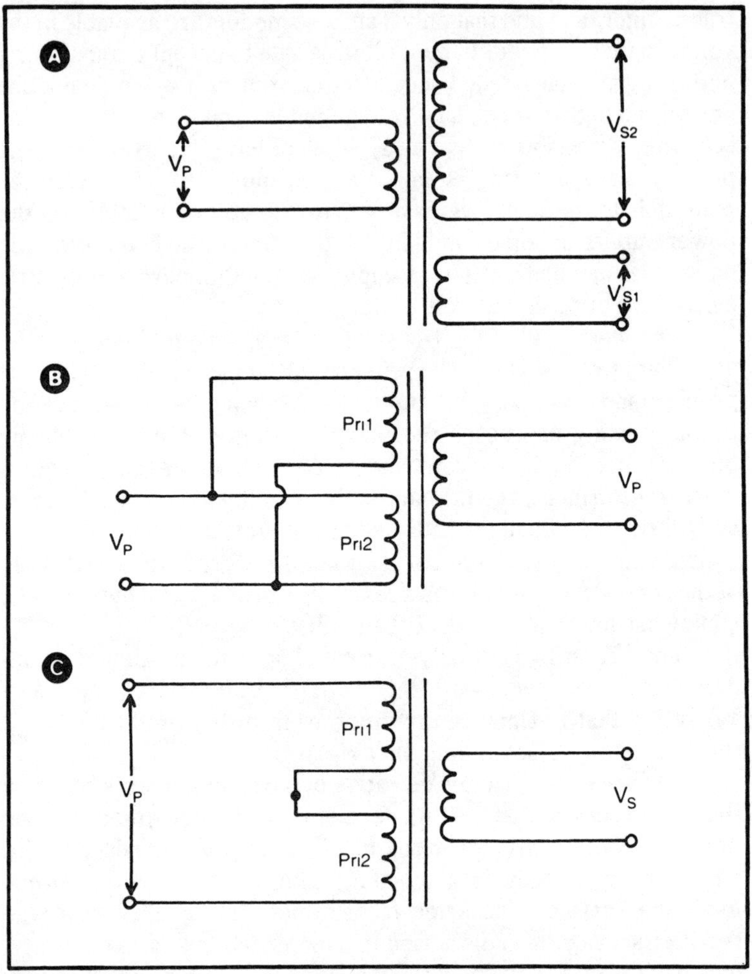

Fig. 2-2. A) Multi-secondary transformer, B) dual primary transformer wired for 110 Vac operation, C) dual primary transformer wired for 220 Vac operation.

in which both primaries are connected in parallel with each other. The configuration shown in Fig. 2-2C has the primaries connected in series for 220-volt ac applications. In some equipment intended for international markets, there is a switch on the rear panel somewhere that allows selection of the correct configuration for local power conditions without the need for rewiring. In still other equipments, the transformer will have a multiply-tapped primary that is switch selected for six or eight different voltage inputs (not all countries of the world are either 110 volt or 220 volt!).

The transformer selected for any given application should have a secondary current rating more than sufficient for the expected load conditions. This criterion is usually taken to mean a 33 percent to 66 percent margin between the maximum load current expected, and the maximum capability of the transformer. Heat kills electronic things, so the idea is to keep the transformer running cool.

RECTIFIERS

The job of the rectifier in a power supply is to convert the bidirectional alternating current to unidirectional direct current. There are a number of different devices which qualify as rectifiers, although only a few are used in modern power supply circuits. All rectifiers have one property in common: they pass electrical current in only one direction. Vacuum tube diodes were once the only available electronic rectifiers, even though devices such as synchronous vibrators and mechanical rectifiers were also used for some very special applications. Today, all common rectifiers are solid-state diodes consisting of a PN junction. When the junction is forward biased by the ac, it will pass current; when the PN junction is reverse biased by the ac the current flow ceases. Figure 2-3 shows what happens when ac is applied to a solid-state rectifier. In Fig. 2-3A the polarity is such that the anode is more positive than the cathode, which is the condition for forward biasing of the PN junction. Under this condition, current will flow in the load, R_L. In Fig. 2-3B we see the condition in which the anode is more negative than the cathode. Since the PN junction is now reverse biased, the diode will not conduct current.

Circuits such as Figs. 2-3A and B are known as *half-wave rectifiers* for reasons that are immediately apparent when examining Fig. 2-3C. Here, we see the effect of a diode that will conduct current only on positive alternations of the applied ac cycle. Current will flow for only half of each cycle, hence the name half-wave rectifier.

Note that the dc output of a rectifier is merely unidirectional, it is not pure dc. The type of direct current which electronic circuits want is the type normally supplied by batteries: it has a constant value over time. If we were to graph true dc it would be represented by a straight horizontal line at some distance up the vertical (voltage) scale, and commences from t/on of the circuit. But the waveform of Fig. 2-3C varies with time in a manner similar to the ac which spawned it. This type of dc is called *pulsating dc,* and must be *filtered* in order to produce nearly pure dc. Filtering will be dis-

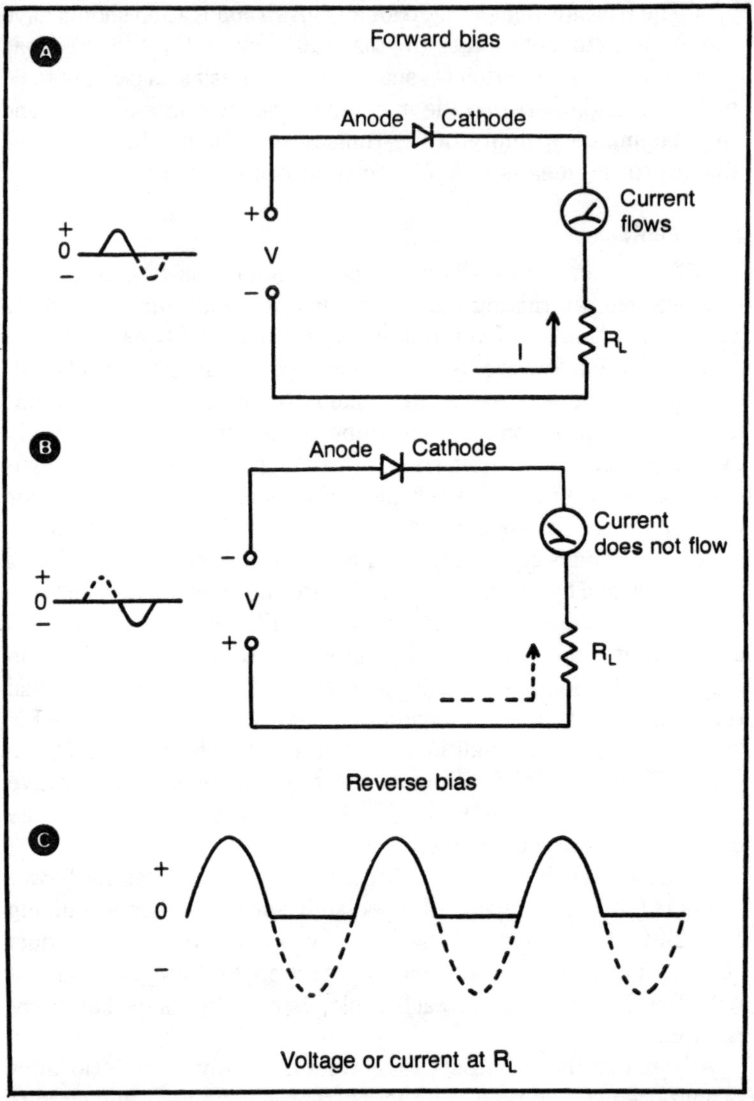

Fig. 2-3. A) Forward biased rectifier diode, B) reverse biased rectifier diode; C) half-wave rectifier output waveform.

cussed in the next section. The departure of pulsating dc from pure dc is called *ripple*, which is measured in percent.

There are several disadvantages to half-wave rectification, and only one advantage. The single advantage is that it uses just one solid-state rectifier as opposed to two or four used by other circuits.

Since popular diodes in 1 to 3 ampere ratings cost around a buck, there is some doubt regarding the usefulness of this "savings." On the disadvantage side is that the half-wave circuit is tremendously wasteful of electrical energy. A transformer used with a half-wave rectifier must have a primary VA rating that is 40 percent higher than would be required to deliver the same load energy in a full-wave situation. Also, the ripple factor of the half-wave circuit is 120 percent, so filtering requires much higher value components. While this factor may be acceptable at small currents, higher current power supplies suffer from much higher costs. The average output potential of a half-wave rectifier is only about 45 percent of the RMS value of the applied ac sine wave.

The basic full-wave rectifier uses two diodes and a center-tapped transformer secondary (Fig. 2-4A), and produces a double-humped pulsating dc waveform (Fig. 2-4B). The transformer secondary uses a center tap that serves as the zero potential common line for the power supply. Since the center tap is at a potential that is one-half the potential appearing across the entire secondary, its use

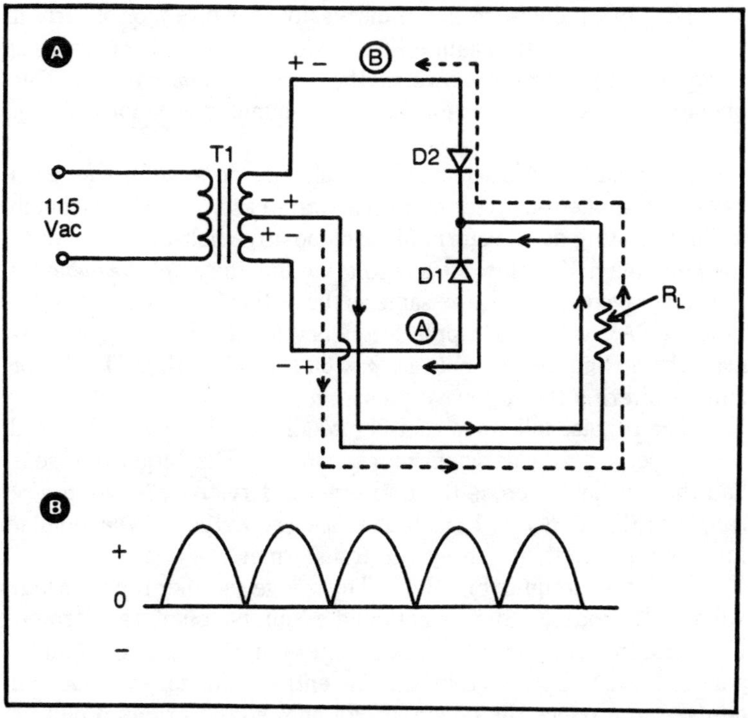

Fig. 2-4. A) Full-wave rectifier circuit. B) full-wave rectifier output waveform.

as a zero-potential reference point means that the "A" end of the secondary will be positive at the same time the "B" end of the secondary is negative, and *vice versa.* As a result, we find that diode D1 will be forward biased when D2 is reverse biased, and D2 will be forward biased when D1 is reverse biased; only one diode conducts at a time.

Let's follow the full-wave rectifier through a single ac cycle. First, we find that the ac cycle is such that point "B" is positive, and point "A" is negative. Under this circumstance, diode D2 is forward biased and will conduct current, while diode D1 is reverse biased and does not conduct current. Current will leave the center tap (follow dotted line), flow through load R_L in the direction shown, through diode D2 and back to the transformer secondary at point "B." When the ac cycle reverses direction, we find that point "B" is negative with respect to the center tap (reverse biasing D2), and point "A" is positive (forward biasing D1). Under this condition (follow solid line), current will leave the center tap, flow through the load resistor R_L in the direction shown, through diode D1 and then back to the transformer secondary at point "A."

Note one point in the above discussion: the direction of current flow in R_L, the load; examine Fig. 2-4A. On both halves of the ac waveform, the current flows in the *same direction* in R_L! This phenomenon is the reason for the double-humped waveform of Fig. 2-4B.

Note that there is a hump in the pulsating dc waveform for each half of the applied ac cycle. In the half-wave case, there is one hump in the pulsating dc waveform for each positive half-cycle, none for the negative half-cycles. As a result, we find the *ripple frequency* of the half-wave rectifier is the same as the ac line frequency (or 60 Hz in the USA), while the ripple frequency for the full-wave case is *twice* the ac line frequency (i.e., 2×60 Hz = 120 Hz). This factor aids in filtering the full-wave pulsating dc waveform.

The regular full-wave rectifier requires only two diodes, and needs a center-tapped transformer secondary. The output voltage is half that available across the entire secondary. A *full-wave bridge rectifier* (Fig. 2-5A) will produce twice the output of the regular full-wave rectifier, for any given transformer.

Note the circuit in Fig. 2-5A. The bridge rectifier requires four solid-state diodes, and does not need a center-tapped transformer. If the transformer you select has a center tap, then ignore it (but be sure to insulate it from ground!). The entire secondary voltage will be used to produce the dc output voltage, so the bridge produces

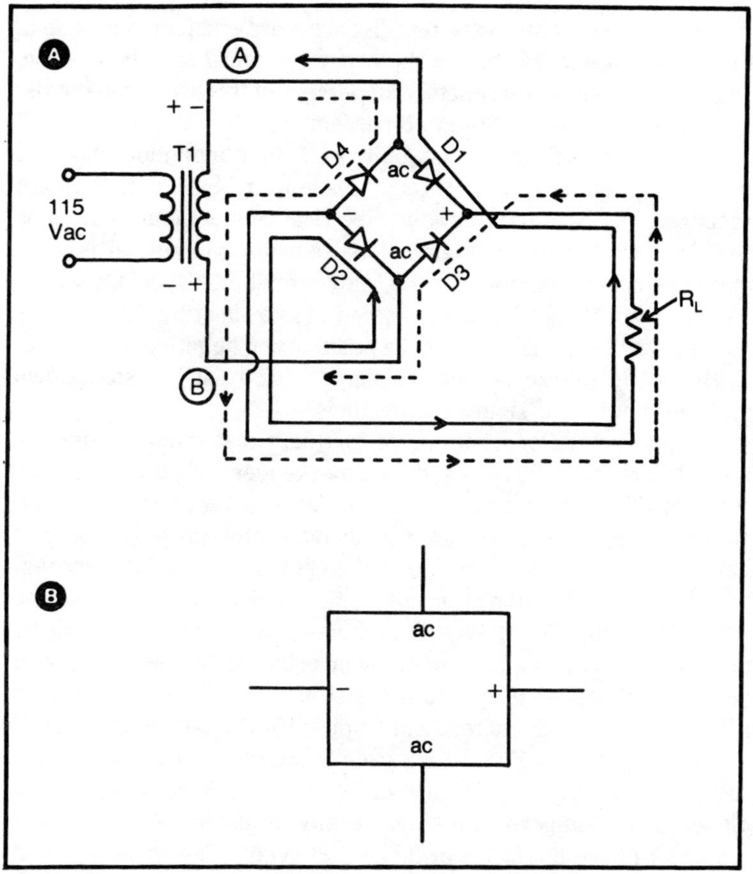

Fig. 2-5. A) Full-wave bridge rectifier circuit, B) circuit symbol.

twice the output of the regular full-wave rectifier.

Let's consider the operation of Fig. 2-5A over the course of a single ac sine wave cycle. First, the polarity of the secondary is such that point "A" is more positive than point "B." Current leaves the transformer at point "B," flows through diode D2, load resistor R_L, diode D1 and back to the transformer secondary at point "B." When the ac polarity reverses direction, we find that point "A" is negative and point "B" is positive. Under this condition, diodes D1 and D2 are reverse biased, and D3/D4 are forward biased. Current leaves the transformer at point "A," flows in diode D4, load resistor R_L, diode D3 and then back to the transformer at point "B." Again, the current flows in R_L in the same direction on both halves of the ac cycle.

Both forms of full-wave rectifier are more efficient in the use of the electrical energy than is the half-wave rectifier. The average output voltage is approximately 90 percent of the RMS value of the applied ac, and the ripple is 48 percent.

Bridge rectifiers can be built from four individual discrete diodes, or, can be bought in ready-made form. Such rectifiers are packaged, and will provide four electrical connections: +dc, −dc and two "ac" terminals. Figure 2-5B shows a symbol that is commonly used in schematics for the bridge rectifier, especially where a ready-made "stack" is used instead of four discrete diodes. The "+" and "−" terminals are for the positive and negative dc, respectively, while the ac is self-explanatory. The "ac" is sometimes replaced with little sine wave symbols.

There are different forms of rectifiers that might be used in projects used by hobbyists and the other readers. Figure 2-6 shows some of the most common forms. Axial lead rectifier diodes are shown in Fig. 2-6A. These are usually made of epoxy or plastic, and have current ratings from 500 milliamperes to about 3 amperes, with 1 ampere being the most common. The popular 1N4000-series are of this type. In all axial lead diodes, the cathode end will be marked with either a band of color or paint, or, by being rounded (see the "bullet" shaped diode). Most of these diodes are from ⅛ inch to ½ inch in diameter, and from 3/16- to ½-inch long.

The rectifier in Fig. 2-6B is called a stud-mounted diode, and is the type used for higher currents (i.e., over 3 amperes). Such diodes in 100-ampere ratings are easily available, although most readers will use 6, 12, 15, or 25 ampere types. The anode on most stud-mounted rectifiers is the terminal end, while the cathode is connected to the mounting stud. An exception to this protocol is the reverse-polarity types, which are usually identifiable by an "R" suffix on the type number, or, by an "R" printed somewhere on the diode body.

Stud-mounted rectifiers are intended for higher current applications. As such, they are expected to heat up quite a bit and must be heatsinked for most reliable operations. The stud-mounted rectifier must be attached via its stud to either a metal chassis, or, preferably, a finned heatsink.

Figures 2-6C and 2-6D are examples of low and high current bridge rectifiers, respectively. The low current type will usually have a 1-ampere rating, although 3-ampere models are known. Most of this type are configured as shown. A modern variation, however, is the "DIP" pack style, in which the bridge diode is inside

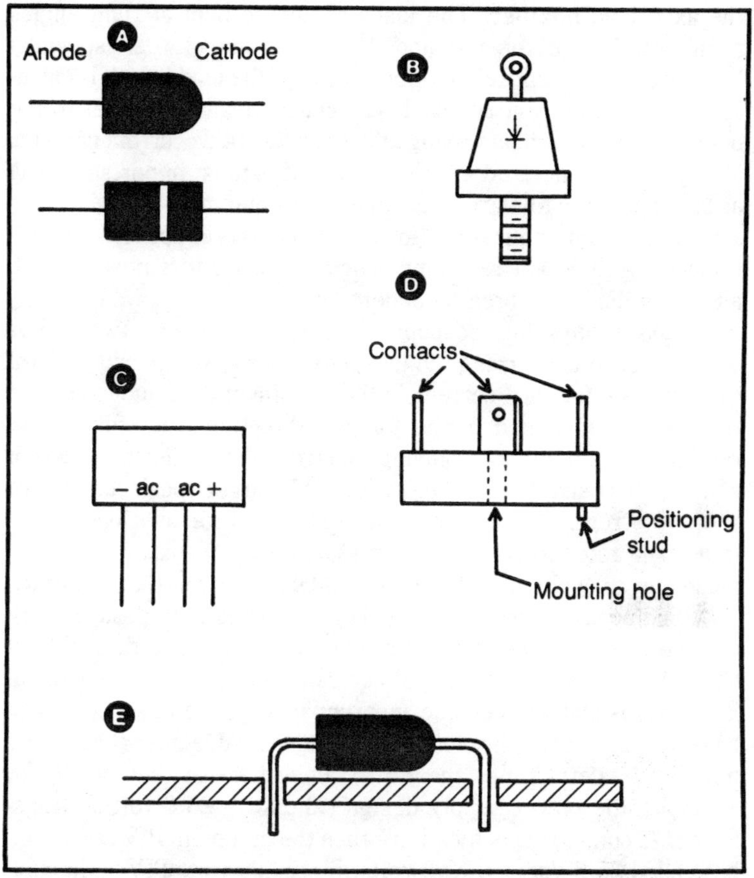

Fig. 2-6. Types of rectifier diodes A) epoxy, B) stud-mounted, C) low-current bridge package, D) high-current bridge package, E) proper mounting to printed circuit board for axial lead epoxy rectifier.

a dual inline package (IC) of the same size as 8-pin DIP integrated circuits. Of course, the bridge rectifier will have only four leads rather than eight.

A higher current bridge pack is shown in Fig. 2-6D. This type will carry ratings to 25 amperes or so, and is mounted with a bolt through a hole in its body to either chassis or heatsink (preferred). The terminals may be solder-only types, or spade/solder types as shown. These terminals will accommodate standard ¼-inch spade lugs. A positioning stud is sometimes provided to serve as a mounting keyway, and to stabilize the rectifier.

Finally, in Fig. 2-6E we see the proper method for mounting

the axial lead rectifier. The leads should be bent at right angles about ¼ to ⅜ inch from the diode body. On some diodes, especially low-cost types of entirely hidden ancestry, the lead length is calculated into the current rating. If you cut the leads off closer to the body, there is less heatsinking effect and the diode current carrying ability must be derated a little bit. Perhaps most important is the airspace underneath the diode. Unless the equipment will be used in a high vibration application (in which case the airspace is usually replaced with a heat resistant *Teflon*® spacer), it is necessary to allow the diode to "breath" underneath.

Diode ratings. For most applications, there are only two ratings of importance concerning diodes: *peak inverse voltage* and *forward current*. The forward current is the maximum continuous current that can be sustained in the forward direction (i.e., the normal rectifier current) without causing damage to the PN junction. Do not confuse this current with the considerably larger "peak current" or "surge current." Such currents are specified for overload conditions, and refer to the maximum *short duration* (usually taken to mean one ac cycle, or 1/60 second in USA) current that can flow.

The peak inverse voltage (PIV), also called the peak reverse voltage (PRV), is the maximum reverse bias voltage that will not harm the PN junction. This voltage can make or break the design. For reasons that will become apparent shortly, the PIV rating must have a value equal to not less than 2.82 (some designers use 3 for a margin of safety) times the RMS value of the applied ac. If, for example, the power supply design calls for a 25.6-volt ac transformer (a common specification), then the minimum PIV rating will be 2.82 × 25.6 Vac, or, 72.2 volts. Since 100 volts PIV is the next higher standard value, select a diode with a PIV rating of 100 volts or more. Personally, I prefer using 1000 volts PIV ratings even for low-voltage power supplies in order to prevent damage due to power line transients.

FILTERING THE DC POWER SUPPLY

The pulsating dc produced by the rectifier is not suitable for use in most electronic circuits. Some means must be found to make the pulsating dc into nearly pure dc similar to the current obtained from batteries. The device that accomplishes this transformation is the *filter*. Various elements are routinely used to make filters, but all circuits involve capacitors; some also involve resistors and inductors in assorted L-section and Pi-section networks. For our present discussion, however, we will consider only the simplest form of filter, the "brute force" or single-capacitor filter.

Figure 2-7A shows the brute force filter, which consists of a single high value capacitor connected in parallel with the load (R_L). Figure 2-7B shows the action of this capacitor. As the output of the rectifier begins to increase, the filter capacitor will begin to charge. After the voltage passes the peak, however, it will begin to decrease. During this period, the voltage across the capacitor will be slightly higher than the voltage across the rectifier output, so the charge stored in the capacitor will be dumped into the circuit. The effect of this charge dumping is to fill in the spaces between the humps (i.e., the shaded area of Fig. 2-7B). If you were to view the waveform on an oscilloscope, the result would be blending of the capacitor and rectifier contributions, so that all you would see is the waveform represented by the heavy line in Fig. 2-7B.

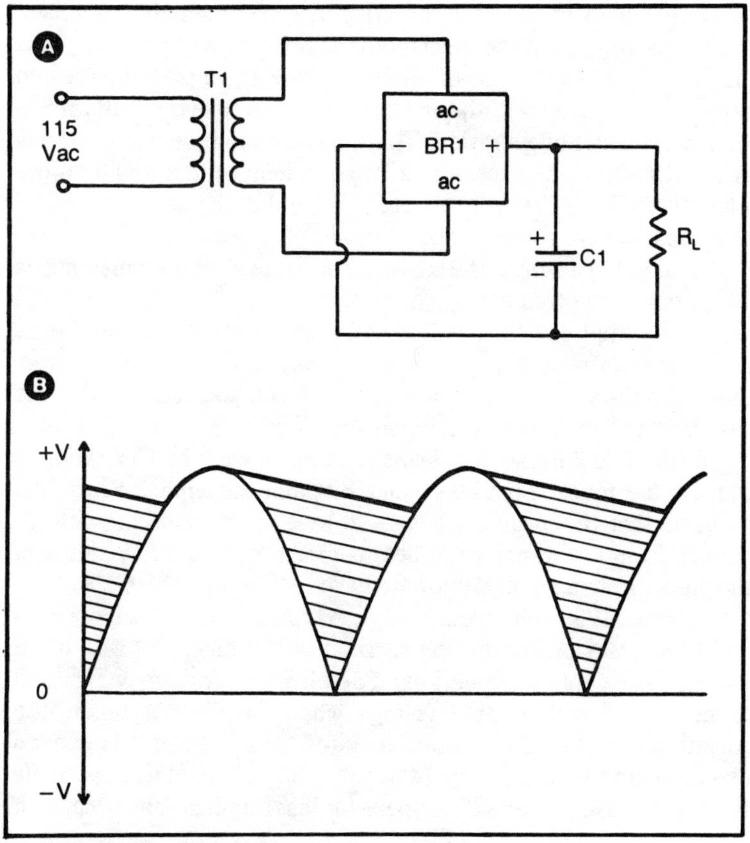

Fig. 2-7. A) Brute force filter circuit, B) waveform.

The capacitor used in the brute force filter typically has a very high value, so must be an electrolytic. Although a few high-voltage power supplies used in some radio transmitter power amplifiers might be found with 4 to 10 microfarad capacitors, these are not typical of low-voltage power supplies. For most of our applications, the power supply filter will have a capacitance of many thousands of microfarads. For this reason, aluminum electrolytics are all but universally used in dc power supplies. These capacitors are the only type in which it is possible to obtain very high capacitance in small packages.

There is a problem with the electrolytic capacitor, however, and that is the polarity. The positive and negative terminals will be marked in most cases, and this polarity *must* be observed. Connecting an electrolytic into a circuit backwards will cause damage to the power supply (blown rectifiers and filter capacitors) and may cause the electrolytic to *explode*. Also, if a power supply is not used for a long time, then the electrolytic may decay and prove useless for filtering when the power supply is once again placed back into service. In many cases, these "dried out" electrolytics will reform after a few hours of operation. This phenomenon is seen sometimes in small radios when there is a large ac hum on the audio output when the radio is first returned to service, but the hum diminishes to normal over a week of operation.

The voltage across the capacitor at the peak of the pulsating dc waveform is approximately the peak voltage, V_p, which is 1.414 times the applied RMS voltage. On the second half of the ac waveform, there will be this V_p *plus* another V_p contribution from the next half-cycle of the applied ac waveform. The result is that the reverse bias applied to the diode rectifiers is 2 V_p, or (2) (1.414 V_{rms}), which is 2.82 V_{rms}. This is the reason why the PIV rating of the rectifier must be not less than 2.82 times the applied RMS. You will find that this problem begins to be a worry once the voltage applied from the transformer begins to approach the PIV rating of the diode. Power supply rectifiers have PIV ratings of 50, 100, 200, 400, 600, or 1000 volts (commonly). A typical 5-volt dc supply uses a 6.3 Vac transformer, so any rectifier will suffice. But a 12 or 15 volt power supply may well use 25-volt transformers, which will produce a 72-volt reverse voltage when the filter is taken into consideration. For this reason, a 100-volt PIV rectifier is needed even though the power supply is only a 15-volt model. I routinely specify the 1N4007 for all 1-ampere (or less) applications because it has a 1000-volt PIV rating, and costs about the same as others of the same series with lower ratings.

The ripple output of a filtered power supply is denoted by a measure called the *ripple factor*. For the brute force filter of Fig. 2-7A, the ripple factor is a function of power line frequency/ripple frequency, the load resistance, and the filter capacitance. Since the ripple frequency (which is determined by power line frequency) is a factor, we need two different equations, one each for half-wave and full-wave cases:

60 Hz ripple frequency (half-wave rectifiers):

$$\Gamma = \frac{1}{208\ R_L\ C_1}$$

and for 120 Hz ripple frequency (i.e., full-wave rectifiers):

$$\Gamma = \frac{1}{416\ R_L\ C_1}$$

Where:

Γ is the ripple factor

R_L is the load resistance minimum value, as determined from Ohm's law: $V_o/I_{o(max)}$

C_1 is the capacitance of C1, in *farads* (*note*: 1 F = 10^6 μF)

Example

Find the ripple factor of a 12 Vdc supply rated to deliver 750 milliamperes if C1 is 2000 μF and full-wave rectification is used. Note: 2000 μF = 0.002 F, and 750 mA is 0.75 amperes. First calculate R_L, or incorporate it into the equation.

$$R_L = V_o/I_{o(max)}$$
$$R_L = (12)/(0.75)$$
$$R_L = 16 \text{ ohms}$$

Next, calculate the ripple factor

$$\Gamma = \frac{1}{(416)\ (16 \text{ ohms})\ (0.002 \text{ F})}$$

$$\Gamma = 1/13.3$$

$$\Gamma = 0.075$$

For most of our applications, we can use a rule of thumb to determine the capacitance. We want to achieve a ripple factor of 0.5 or less, so will use a filter capacitance of not less than 1000 μF per ampere of load current. This is a rule of thumb only, and will work if the ripple factor of 0.5 or so is sufficient. Otherwise, use the equation.

Readers in some countries other than the United States will have 50 Hz ac power mains. These readers will have to scale the constant in the above equations approximately 20 percent lower, or 167 for the half-wave case and 333 for the full-wave case. Alternatively, you can use the equations as is, and then multiply the resultant capacitance by 1.2.

The output voltage from a brute force filter is directly related to the load current, I_o, and the filter capacitance, C_1. This potential is given by:

half-wave case

$$V_o = V_p - (I_o/120C_1)$$

and, for the full-wave case,

$$V_o = V_p - (I_o/240C_1)$$

As in the example above, let's assume that the peak voltage is 12 volts. The output voltage will be (at maximum load current, or 0.75 amperes):

$$V_o = (12) - \left(0.75 /(240)(0.002\text{ F})\right)$$
$$V_o = (12) - (0.75/0.480)$$
$$V_o = 12 - 1.56 = 10.44 \text{ volts}$$

This difference in potential, incidentally, means that unregulated power supplies have only very nominal output voltage ratings. This is one further reason why, now that they are low in cost, regulators should be used on all of our dc power supplies.

VOLTAGE REGULATORS

The main cause of voltage regulation problems in internal resistance in the dc power supply. An ideal dc power supply will have a zero ohms internal resistance, but this ideal is never achieved. There will always be some small internal resistance determined by $V_o/I_{o(groan)}$, as shown in Fig. 2-8. Resistor R_L repre-

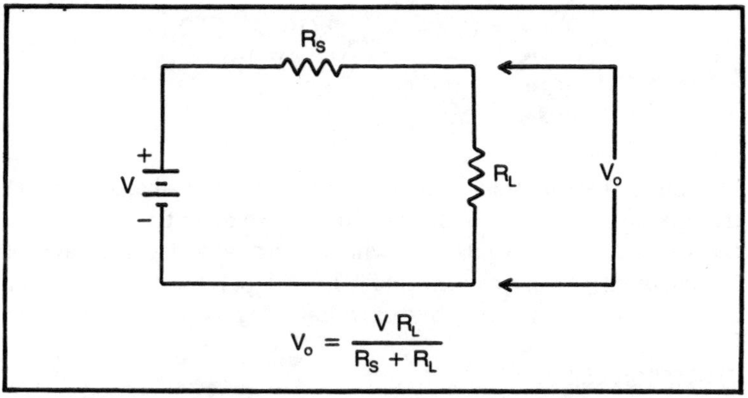

Fig. 2-8. Resistor network model of power supply output circuit.

sents the normal load on the power supply, while R_s is the internal resistance of the power supply. Since these resistors are effectively in series with one another, there will be a voltage drop across R_s which subtracts from the potential available for R_L. Using the ordinary *voltage divider equation* from elementary electrical theory we can calculate V_o:

$$V_o = \frac{V R_L}{R_s + R_L}$$

Obviously, we want to keep R_s as small as possible. Since R_L is determined by the load current, I_o, we find that the output voltage is therefore a function of load current; as I_o increases V_o will decrease. We can find the *percentage of regulation* (PR%) from:

$$PR\% = \frac{V_o' - V_{o(min)}}{V_o'} \times 100\%$$

Where:

PR% is the percentage of regulation

V_o' is the open-terminal, i.e., zero load current, output potential

$V_{o(min)}$ is the output potential when I_o is maximum

Example

Find the percentage of regulation (PR%) when the output voltage drops from 18 to 15 volts as the current increases from zero to full output rating.

$$PR\% = (V_o' - V_{o(min)})\,(100\%)/(V_o')$$
$$PR\% = (18 - 15)\,(100\%)/(15)$$
$$PR\% = (3)\,(100\%)/(15)$$
$$PR\% = 300\%/15 = 20\%$$

Since the variation of the supply voltage will adversely affect the operation of many electronic circuits, we must try to minimize the percentage of regulation (which is actually a percentage of irregulation!); many circuits must have 1 percent regulation, or less. Those circuits are where a voltage regulator is needed.

ZENER DIODES

Perhaps the simplest solid-state voltage regulator is the zener diode shown in Fig. 2-9. The zener diode is a special type of avalanche diode that has a constant reverse bias breakdown potential that is less than the destruction potential. A typical zener diode regulator circuit is shown in Fig. 2-9A, while the *I-vs-V* characteristic curve is shown in Fig. 2-9B. Let's consider Fig. 2-9B first, as it gives us some insight into the operation of the zener diode.

In the forward bias region, the diode operates like any other PN junction diode. Increasing applied voltage (V) will cause increases of forward current in an approximately linear manner. For silicon diodes, the linear operating region begins at a junction voltage V_g of 600 to 700 millivolts.

The region between zero potential and V_g is a transition region in which a forward current will begin to flow, but is not linear with increasing voltage. At low positive values and at negative values of voltage the only current flowing in the diode is a small reverse leakage current, I_L. For silicon diodes of good ancestry this current is minimal.

In the reverse bias region between 0 volts and some special zener potential. V_z, then, the diode also operates like any other PN junction diode. At V_z, however, something happens to make this diode a little different from other diodes: the PN junction avalanches and produces a low resistance. Zener potential V_z is approximately constant, so the zener diode will make a good voltage regulator for a limited range of applications (V_z will wander a little with temperature change). Now let's get back to Fig. 2-9A.

The zener diode will provide voltage regulation in Fig. 2-9A when it is connected in parallel with the load. In that case, V_o will be V_z. Resistor R1 in Fig. 2-9A is used to limit the current in the zener to a safe value. If you will glance again at the curve in Fig. 2-9B you

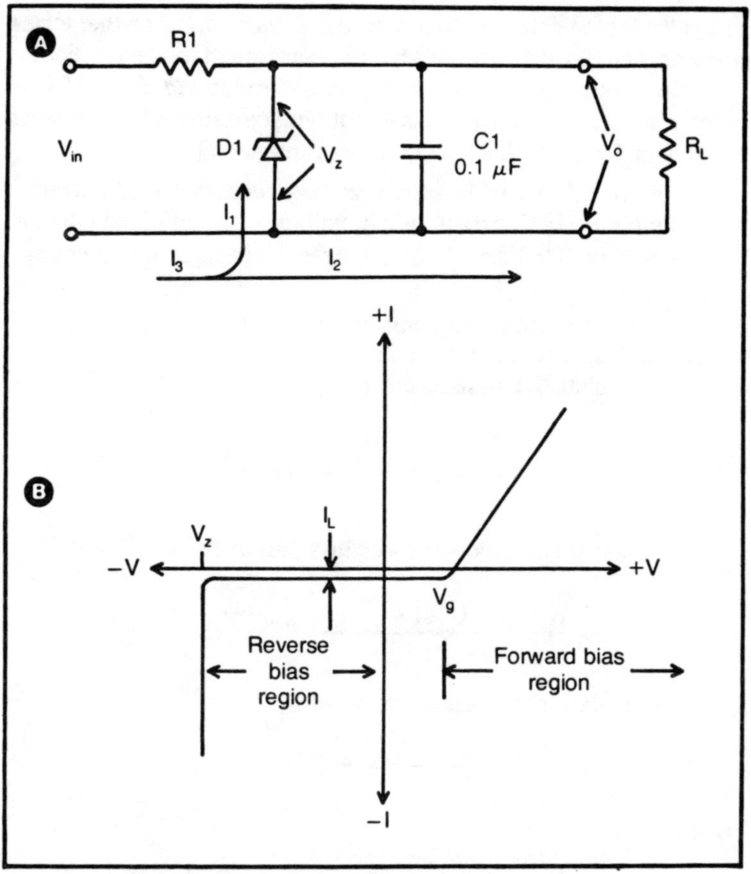

Fig. 2-9. A) Simple zener diode regulator circuit, B) zener diode transfer function.

will see why this is needed. Left unlimited, the current increases extremely rapidly at V_z. Shortly, we will discuss how the resistance and power ratings of R1 are calculated.

The capacitor is optional, and is used to reduce the electrical noise produced by the regulator. Any diode in avalanche mode produces a large amount of "white noise," i.e., "hiss." This noise may or may not be a factor in any given circuit, but must be reduced if it is a factor. The 0.1 μf capacitor is used for this purpose. It is interesting to note that zener diodes are sometimes used to build white noise generators used in testing amplifiers and such rf devices as antennas and networks.

There are three steps in designing a zener diode regulator: 1) select the zener potential required (from the application, normally),

2) select the resistance value and the power rating of the series resistor, and 3) calculate the power dissipation of the zener diode.

The design of the zener regulator circuit *ala* Fig. 2-9A is merely an exercise in Ohm's law, but with certain embellishments to account for three basic sets of operating conditions:

Condition I: Variable supply voltage, constant load current
Condition II: Constant supply voltage, variable load current
Condition III: Variable load current, variable supply voltage

Design for Condition I

1. Select V_z from the application.
2. Calculate load current I2.
3. Calculate R1 resistance.

$$R1 = \frac{V_{min} - V_z}{1.1\ I_2}$$

4. Calculate the diode power dissipation rating.

$$P_{D1} = \frac{(V_{max} - V_z)^2}{R1} - (I_2 V_z)$$

5. Calculate the wattage rating of R_1.

$$P_{R1} = P_{D1} + (I_2 V_z)$$

Design for Condition II

1. Select V_z from the application.
2. Calculate load current $I_{2(max)}$.
3. Calculate the value of R1 from:

$$R1 = \frac{V - V_z}{1.1\ I_{2(max)}}$$

4. Calculate the diode power dissipation (see Condition I above).
5. Calculate the resistor power dissipation (Condition I above).

Design for Condition III

1. Select V_z from the application.
2. Calculate the load current maximum.
3. Determine the minimum expected supply voltage.
4. Calculate R1 from:

$$R1 = \frac{V_{min} - V_z}{1.1\ I_{z(max)}}$$

5. Calculate the diode power dissipation from Condition I preceding.
6. Calculate the resistor power dissipation from Condition I preceding.

Example

A power supply is to produce 6.8 volts dc at a load current of 120 mA maximum. The supply voltage varies from 11 to 16 volts. Find the value and power rating required of R1, and the power rating of the zener diode. Assume that the load current will vary, and therefore design for the worst case.

1. Set the zener potential to 6.8 volts.
2. The load current will be 120 mA, or 0.12 amperes, maximum.
3. Calculate the value of R1:

$$R1 = \frac{V_{min} - V_z}{1.1\ I_{z(max)}}$$

$$R1 = \frac{(11\ V) - (6.8\ V)}{(1.1)(0.12\ amp)}$$

$$R1 = (4.2)/(0.132)$$

$$R1 = 31.2\ ohms$$

Since this application is not extremely critical, we do not need precision in the value of R1, so are able to set R1 equal to the nearest higher value of standard resistance value. In this case the next higher standard value of resistance is 33 ohms, so set R1 = 33 ohms.

4. Calculate the diode power dissipation, and from that calculation select a power dissipation rating.

$$P_{D1} = \frac{(V_{max} - V_z)^2}{R1} - (I_z V_z)$$

$$P_{D1} = \frac{(16 - 6.8)^2}{(33)} - (0.12 \times 6.8)$$

$$P_{D1} = \frac{(9.2)^2}{33} - (0.82)$$

$$P_{D1} = ((85)/33) - (0.82)$$

$$P_{D1} = 2.58 - 0.82 = 1.76 \text{ watts}$$

Zener diode power ratings tend to be 400 mW, 500 mW, 750 mW, 1 watt and 5 watts. For this application, since P_{D1} is greater than 1 watt, we will want to go to the 5-watt rating. Even if there were 2- to 5-watt types available, I would recommend the 5 watt for this application because of the reliability gained by using a diode at a stress a lot less than its rating.

5. Calculate the power dissipation of R1 and from this computation select a power rating for the resistor.

$$P_{R1} = P_{D1} + (I_2 V_z)$$
$$P_{R1} = 1.76 \text{ watts} + (0.12 \times 6.8)$$
$$P_{R1} = 1.76 \text{ watts} + 0.82$$
$$P_{R1} = 2.58 \text{ watts}$$

From this calculation we find that none of the carbon composition resistors will suffice, because they come in ratings of 2 watts and under. The 3-watt wirewound resistor is sufficient, but a 5-watt resistor is more desirable.

HIGHER CURRENT REGULATORS

The zener diode is generally regarded as a voltage regulator only at relatively low current levels. When current levels increase to more than a few hundred milliamperes, it might be wise to consider a circuit similar to Fig. 2-10. In this circuit, the zener diode is used to set a reference voltage V_z at the base of the power transistor (Q1). The heavy current is passed along the collector-emitter path of the transistor, while the zener operates in the relatively low-current base circuit. The output voltage will be approximately equal to the zener voltage (V_z) *less* the base-emitter bias (V_{b-e}) of transistor Q1. This circuit is known as a *series-pass voltage regulator*, and Q1 is generally called a series-pass transistor.

The maximum output current is determined by the base current supplied by the zener circuit and the *beta* gain of the transistor. In general, $I_o = I_b H_{fe}$. Set resistor R1 according to the rules for condition I, above.

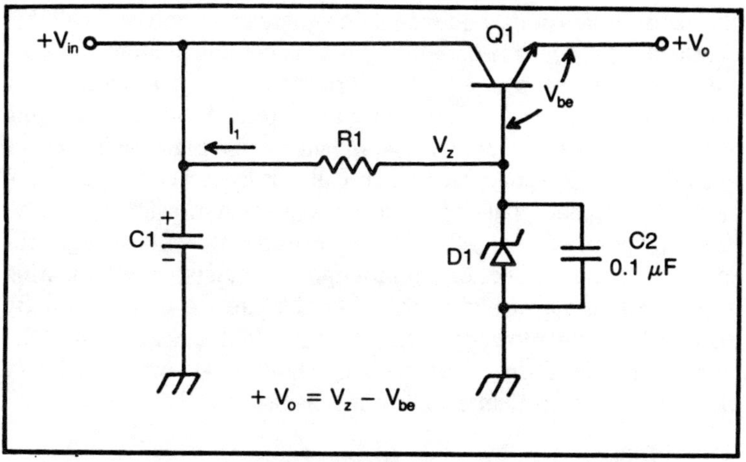

Fig. 2-10. Series-pass voltage regulator.

Capacitor C1 is the normal filter capacitor for the rectifier. Although you can calculate the value of C1 from the considerations given earlier, there is a rule of thumb for the minimum capacitor value. Most authorities agree that C1 should have a capacitance of 1000 microfarads per ampere of output current. Some will increase this to 2000 μF/ampere. If the lesser advice is heeded, and we have a 3-ampere power supply, then we will need a filter capacitance of (3 amperes)(1000 μF/ampere), or 3000 μF. Capacitor C2 is almost optional, and is used to suppress "white noise" produced by the zener diode. Any value from about 0.1 μF to 0.5 μF is sufficient.

The transistor selected for Q1 should have sufficient collector current rating to sustain the maximum value of output current at the maximum applied input voltage ($+V_{in}$). Note that the power dissipation must also be considered. It is likely that any transistor selected will meet the voltage and current requirements, but not the power dissipation requirement. The product of voltage and current (maximum in both cases) should be less than the power dissipation (P_d) of the transistor. Also, the *beta* gain (H_{fe}) should be sufficient to produce the maximum rated output current at the value of base current which can be produced by the zener diode circuit.

It may also be necessary to heatsink transistor Q1. Heat is the great killer of electronic circuitry, and seems especially harmful to power devices which generate their own heat. If the transistor runs so hot that the case cannot be comfortably touched with the bare hand, then it is likely that the heatsinking is insufficient.

The role of heat in electronic equipment cannot be overemphasized. Professional equipment designers know only too well that overheating reduces reliability by a considerable amount. It takes only a few degrees rise in temperature to affect a decrease in reliability. It will pay rich dividends if you provide adequate cooling of electronic circuits, even if that means employing a fan or blower to cool off the components. I can recall servicing a certain brand of medical equipment that had a terribly poor reputation for reliability. From my point of view, all of those power supply voltage regulator and oscilloscope deflection amplifier power transistors overheating meant getting out of bed in the dead of winter to go to work. By installing a pair of blowers in the main console, the heat inside of the equipment was reduced to a point where it all but completely eliminated that particular failure mechanism.

THREE-TERMINAL IC VOLTAGE REGULATORS

For most readers of this book the most practical voltage regulator is the three-terminal IC voltage regulator. These devices reduce the voltage regulator circuit to a minimum of connections: *input, output,* and *common.* These voltage regulators come in standard voltages from 2 volts to 24 volts, and in currents from 100 milliamperes to 8 amperes; the most common forms are the 750 milliampere and 1 ampere models.

Figures 2-11A through 2-11C show several of the more popular regulator package styles. Note that these are the same as for various power transistors. The "H-package" is shown in Fig.

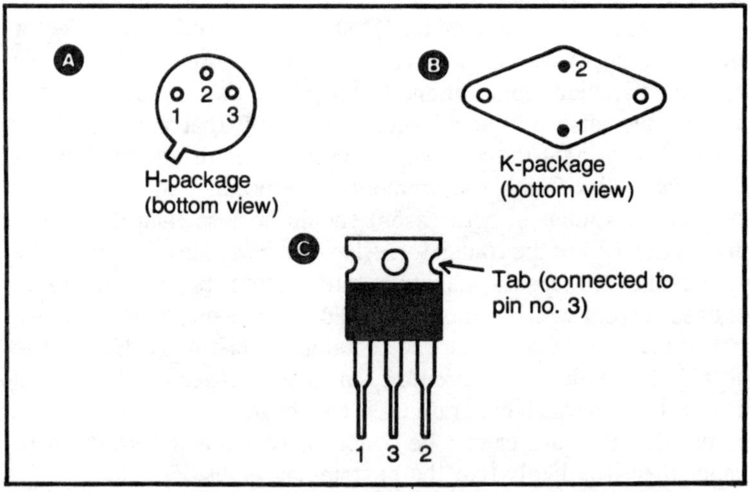

Fig. 2-11A, B, C. Voltage regulator packages (three-terminal).

2-11A. This package is the same as the TO-5 transistor package. There is also a plastic version of this package that is equivalent to the TO-92 transistor package. That device also carries the "H" designation. In most cases, the H-package regulators will pass up to 100 mA of current without heatsinking (or, at least, minimal heatsinking such as connection to a chassis). Note that pin no. 3 is often connected to the case.

The "K-package" is shown in Fig. 2-11B, and is the same as the TO-3 power transistor package. This "diamond-shaped" package is found on 1 ampere and up voltage regulators (up to 8 amperes are available). The third connection for this voltage regulator is the case, which on positive regulators is usually ground. The 1-ampere voltage regulator will usually be capable of 1.5 amperes operation when properly heatsinked. That exceeds the ratings, however, so be careful.

Finally, we have the plastic TO-220 "T-package." This regulator style is becoming more and more popular. Although advertised as a 1-ampere regulator, it is more like a 750 mA regulator unless properly heatsinked. For all heatsinking, incidentally, use silicone heat-transfer grease between the package and the heatsink surface. Tighten the TO-3 and TO-220 ("K" and "T") packages sufficient to mate the package with the surface, but not enough to cause damage.

A typical positive-voltage three-terminal IC voltage regulator circuit is shown in Fig. 2-11D. The three terminals of the voltage regulator are almost universally designated as follows:

1	Input
2	Output
3*	Common

* Also case

Beware of negative regulators, however, because pins 1 and 3 are interchanged on those devices.

Of the four capacitors used in Fig. 2-11D, only C1 is mandatory; the other three (C2 through C4) are optional but highly recommended. Capacitor C1 is the normal filter capacitor used with the rectifier to convert pulsating dc to nearly pure dc (of which, more later). The value of C1 should be 1000 microfarads per ampere (1000 μF/amp) of load current. For a 1-ampere supply, therefore, C1 should be not less than 1000 amperes; some say use 2000 μF/ampere.

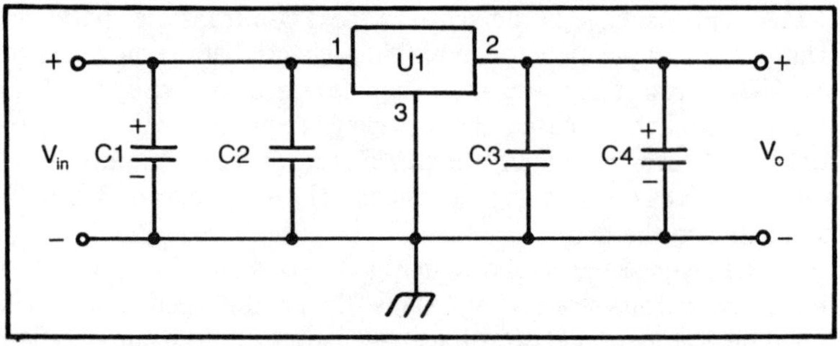

Fig. 2-11D. Voltage regulator circuit (three-terminal).

The voltage rating of capacitor C1 must be sufficient to withstand the input voltage ($+V_{in}$), not the rated output voltage of the regulator. There will always be a 2 to 3 volt differential between input and output voltage (with input voltage being higher), and in some cases the differential might be 30 volts or more. If we used +35 volts input to produce a +5-volt regulated output, then the filter capacitor would have to withstand 35 volts, not 5 volts. In that case, the next higher standard value is 50 WVdc, so we would have plenty of margin. Or would we? It is possible to underrate the capacitor quite a bit because of normal circuit tolerances.

I can remember a system of 12 oscilloscopes that our group serviced in a hospital. There was a ±200-volt power supply in each oscilloscope which seemed to go through one helluva lot of filter capacitors (60 μF/350 WVdc). It seems that someone forgot the basics of tolerances in electronic components. The two 220-volt power supplies (one each positive and negative) had a pre-regulator voltage of 270 volts. Normally, there is a ±15 percent tolerance on unregulated circuit voltage, and a ±20 percent tolerance on component values. Thus, the worst case situation gives us 311 volts across the filter capacitor, and a real rating of (350 × 0.8 volts), or 280 volts for the filter capacitor. Thus, we have the unnerving situation of placing as much as 311-volts across a capacitor which will only withstand 280-volts! Of course, we rarely if ever encountered the "worst case" situation, but the real stresses on those filter capacitors were high enough to cause massive failures. Replacing all 24 60 μF/350 WVdc capacitors with 60 μF/450 WVdc capacitors solved that reliability problem.

Returning to Fig. 2-11D, capacitors C2 and C3 are used to improve the immunity of the voltage regulator to transient noise pulses coming in on the power line from the rectifier. These transients are usually born on the ac power line, and can have quite high amplitudes. In most cases, these capacitors have a value of 0.1 μF to 0.47 μF and are mounted as close to the body of the regulator as possible. Capacitor C4 is used to improve the output transient response of the regulator. In the event of a sudden increase in load current, the regulator might require a few microseconds to readjust to deliver the required current. If we keep capacitor C4 charged, then it will be in a position to supply the needed transient current; C4 acts as a local reservoir to supply current during the brief instant when U1 is "catching up." In most cases, the value of C4 will be approximately 100 μF per ampere of maximum load current.

The voltage regulator serves one other function besides overcoming raw regulation problems. The voltage regulator will also act to substantially reduce the ripple of the pulsating dc from the regulator. The output of the regulator is nearly pure dc, much more "pure" than would be possible with a 2000 μF/ampere filter capacitor. During the early 60s, a manufacturer of automobile radios offered "battery eliminators" that supplied 0 to 18 volts to service 6 Vdc and 12 Vdc car radios. The power supply was offered at first only to factory-authorized service shops for that brand of radio, and were among the first to feature a solid-state voltage regulator. The literature selling that supply claimed (not altogether erroneously!) that it had the equivalent of 1.5 *farads* of filter capacitance from an "electronic filter capacitor." What they meant was the voltage regulator smoothed out the ripple as much as a 1.5 farad (1,500,000 microfarads!) capacitor would. Previous battery eliminators were used as mere battery chargers because the ripple was too great to operate car radios without severe audio hum in the loudspeaker.

REGULATOR TYPE NUMBERS

There are several families of voltage regulators, and several special types which fit into no family but stand alone on the market. The LM-309 is an example of the latter form. The LM-309 device is, perhaps, one of the oldest three-terminal IC regulators on the market, and it is still popular. It produces a fixed output voltage of +5-volts, so is popular in digital circuits. TTL digital IC devices operate from a +5-volt regulated power supply. There are two different versions of the LM-309 device. The LM-309H is a 100-mA regulator and is housed in a TO-3 "H-package." The LM-309K, on the other hand, is a 1-ampere model housed in a TO-3 transistor package (designated "K-package"). The LM-323K is related to the 309, but will produce up to 3 amperes at +5 volts dc output.

The two main families of positive output voltage three-terminal IC regulators are the 78xx and the LM-340n-xx devices. The 78xx type numbers replace the "xx" with the voltage rating of the device. Hence, a −5-volt regulator in the 78xx line will be the 7805, while a +12-volt regulator is the 7812. The LM-340n-xx series follows a similar protocol. The "n" designates the case style (hence the current rating in most cases), so can be H, K, or T. The "xx" indicates the voltage. Thus, we find a regulator bearing the type number LM-340T-05, we know that it is a +5-volt regulator in a "T-package" (i.e., TO-220 plastic package) and will probably be rated at 750 mA in free air and 1 ampere when heatsinked. If, on the other hand, the type number is LM-340K-05, then it is a −5-volt regulator in a TO-3 package.

Sometimes, the regulator type numbers seem a little confused. The 78xx type, for example, should be applied to Motorola branded types with a plastic "hole-in-the-center" package. We often see identical regulators with the same ratings bearing either 78xx or LM-340-xx numbers (sigh). The negative voltage versions of these families are the 79xx and LM-320n-xx, respectively, and follow the same type number protocols.

PRIMARY CIRCUITS OF POWER SUPPLIES

The primary circuit for most of our projects is not shown in detail. There are, after all, several which might be used. Figure 2-12 shows a power supply primary circuit that accomplishes several things: on-off switching, fusing and current in-rush protection.

The main power switch is S1. Some people prefer to use only an SPST (single-pole-single-throw) toggle switch for this purpose. That is reasonable if the circuit and construction are designed such

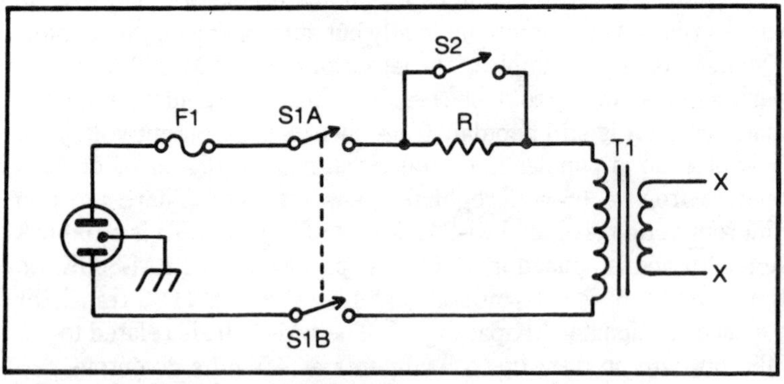

Fig. 2-12. Primary circuit with slow turn-on.

that the switch breaks the hot side of the ac power line. It is unwise to place an SPST switch in the neutral side of the circuit as this can lead to possibly dangerous situations. A better solution is to break *both* sides of the ac power line. Switch S1 is a double-pole-single-throw (DPST), or, alternatively a DPDT used as a DPST. Both the voltage and current ratings of the switch must be sufficient for the intended use.

The fuse or circuit breaker should have a value of approximately 1.5 to 2 times the expected normal current. If heavily reactive loads are anticipated, there will be a surge at first, so a slow-blow fuse might be appropriate.

Switch S2 and resistor *R* are used for current in-rush protection. The extremely high charging current required for high capacitance filters will sometimes cause excessive current flow for the regulator and/or fusing selected. By placing *R* in series with the transformer primary, we can reduce the voltage across the capacitor a little bit until there is some charge in the capacitor. The resistor is then shorted out and the power will operate at full power. This process often takes as much as 10 seconds. Some designers make S2 a timer/switch so that it operates automatically.

KILLING THE "GLITCH WITCH"

If you own a personal computer, or, if you work on or with a professional microcomputer, you are undoubtedly aware of the fact that those machines occasionally "bomb" for no apparent reason. A program that has run successfully before suddenly refuses to run at all until reloaded from the tape or disc. Alternatively, execution will be proceeding, there is a blink on the CRT screen and then the program fouls up. In these cases, we might be seeing the activity of a vicious little electrical gremlin called the "glitch witch."

What is a "glitch?" They are high-voltage electrical transients on the 110 Vac power line. These transients have a short duration (typically less than 30 microseconds), fast rise times and high voltages (3000 volts is not uncommon). Because of their high-voltage, high-frequency nature, these transients will propagate through a power supply, including the regulator and get into the sensitive electronic circuitry. In the analog circuit, the result is soon forgotten as disequilibrated circuits return to their normal states. But in digital electronic devices, gates are opened inappropriately, registers are set or reset incorrectly, and counters are incremented or decremented wrongly. The damage is semipermanent.

Of course, if the pulse is high enough in voltage or long enough in duration, then the damage might be permanent . . . semiconductors don't like to be fried. Fortunately, there is something we can do about it. There are a series of nonlinear voltage-dependent resistors on the market called *varistors* (General Electric's metal-oxide varistors—MOV—or Panasonic's ZNR devices). These devices act like a pair of back-to-back 180 volt zener diodes, so will clip high-voltage transients while not affecting 110 Vac at all.

Figure 2-13 shows the basic manner of connecting a MOV transient suppressor. The MOV device is connected in parallel with the ac line across the primary winding of the power transformer. This is the simplest use of the MOV device. Many of the commercial "static killers" or "transient suppressors" sold to microcomputerists are nothing more than a standard outlet strip with one of these devices across the first socket! For $50 you get a $15 outlet strip and a $3 MOV device.

A more sophisticated version is shown in Fig. 2-14. Here we have an electromagnetic interference (EMI) filter in the circuit, as well as several MOV devices. The EMI filter is an LC low-pass filter that will 1) keep rf generated in the computer (of which, there is plenty) from being radiated via the power lines, and 2) keep rf from the outside from interfering with the operation of the computer. These filters are shielded, and should be located as close to the power line entrance point (on the power supply chassis) as possible. If we leave a few inches of ac power cord before the filter, then it can radiate into the circuitry. Some manufacturers offer EMI filters built into standard chassis-mounted line cord receptacles (e.g., P1 in Fig. 2-14). In any event, try locating the fuse on the line side of the filter in case one of the filter's internal capacitors shorts to ground.

The MOV devices should be connected at least between hot

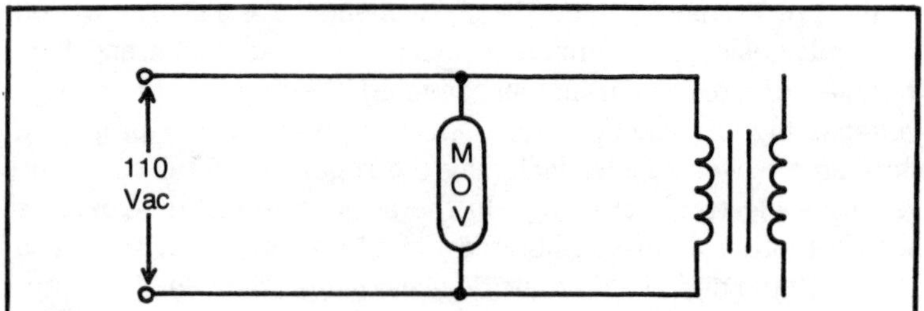

Fig. 2-13. Use of a MOV device for transient suppression.

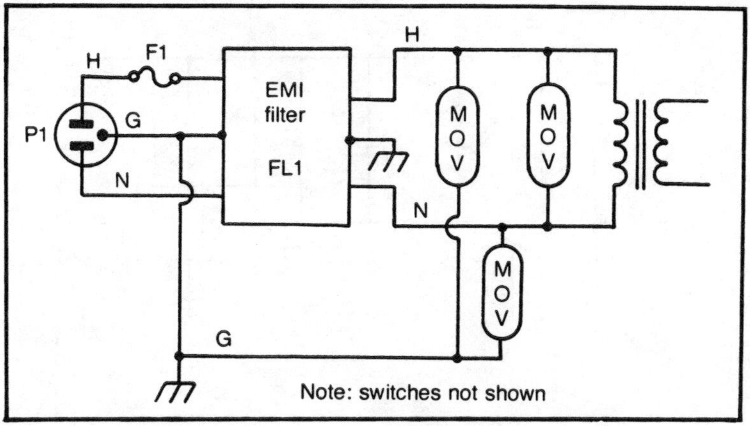

Fig. 2-14. More complex circuit for especially noisy environments.

and neutral, and also between hot and ground. Additional, but an optional MOV device should also be used from ground to neutral.

If the MOV arrangement of Fig. 2-14 doesn't adequately suppress transient glitches, then it may be necessary to go to either an isolation transformer (such as the *Sola* or *Topaz* models) that is designed for this service, or, use a clean special-purpose ac power supply that is guaranteed glitch free.

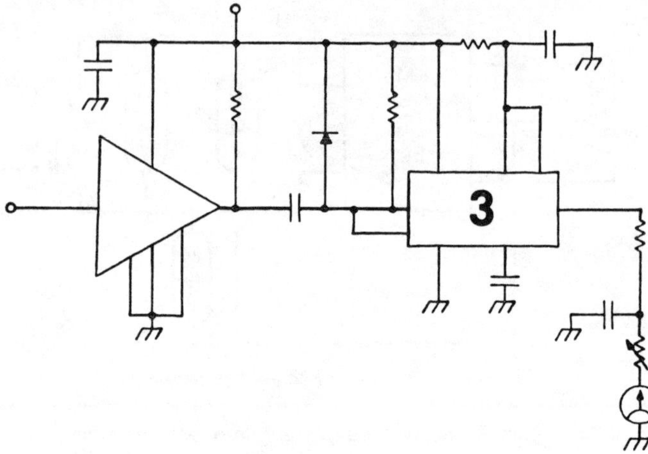

Some Power Supply Projects

The pages to follow contain some elementary projects that are based on the principles given earlier in this book. All of them use standard components that are generally easy to obtain. If you do not have any local supply of electronic components, then almost any of the mail-order parts firms that advertise in hobbyist magazines (e.g., *Radio-Electronics*) can supply these parts. Also, *Radio Shack* stores carry most of these components, including the transformers.

In most cases, the components specified are rated conservatively for the actual application. This should make a more reliable power supply design. In all cases, however, the specified components are at least sufficient if not spectacular. All circuits have been tested.

Since all of these projects are intended to be "stand alone" designs, it is intended that the reader skip those which do not fit his or her needs and go on to those which are applicable. For this reason, elements of each discussion will repeat themselves in each project section. While this will make for boring reading if your intent is to read linearly through the book, it will ensure that some critical bit of information is not missed. This section is intended as reference, so such an approach is reasonable.

±12 VOLTS AT 100 mA

Figure 3-1 shows the circuit for a dual polarity power supply that will provide two different outputs: one negative with respect to

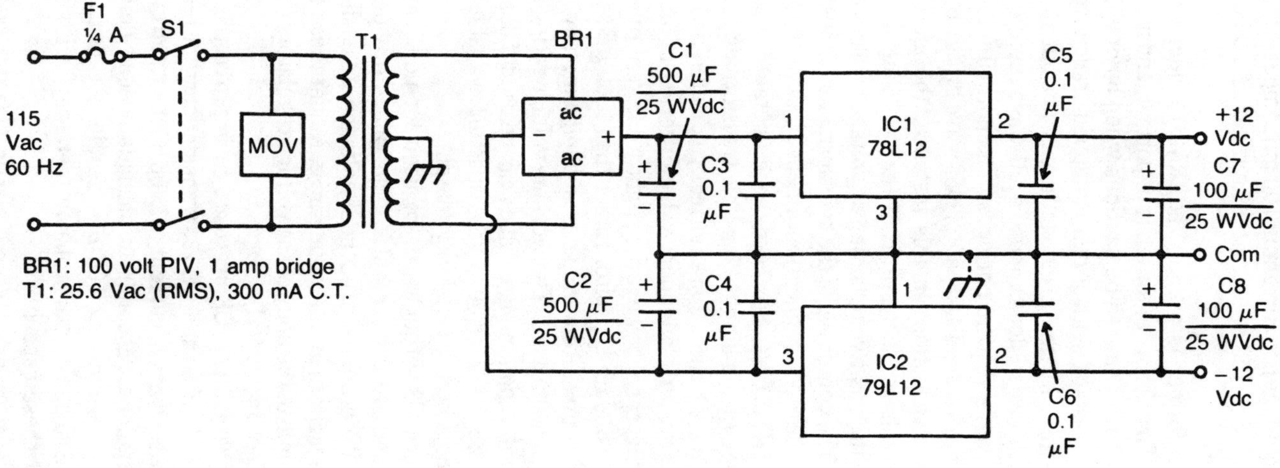

Fig. 3-1. ±12 volts at 100 mA.

common (COM) or ground and the other positive with respect to COM. The rated output current is 100 milliamperes, although that can be increased to 150 milliamperes per side if a different voltage regulator is used.

The transformer is a 25.6 Vac (RMS) model with a center-tapped secondary. That center tap is important, and not all 25.6-volt transformers have one—so be cautious when buying! The use of a bridge rectifier stack with a center-tapped transformer may not seem to square with the previous discussion but is quite easily explained. This circuit does not use a full-wave rectified bridge, but a pair of half-wave rectified bridges. The normal full-wave bridge stack (BR1) may be used as a pair of half-wave bridge rectifiers with a common ac connection. In this instance, therefore, the minus pulsating dc voltage is taken from the "−" side of the rectifier (BR1), while the positive pulsating dc voltage is taken from the "+" side of BR1. The center tap of the transformer supplies the common or "ground" reference connection. In effect, this circuit sees a pair of 12.8 Vac (RMS) transformers, each operated with respect to ground, rather than a single 25.6-volt transformer.

Capacitors C1 and C2 are normally selected according to the 1000 μF/ampere rule minimum value criterion in most regulated supplies. In this case, however, a slightly different rule is followed. Here we use a 500 μF capacitor rather than 1000 μF as specified by the rule. Of course, this value is a minimum; any larger value capacitor can also be used. The working voltage rating (WVdc) is 25 volts minimum. Again, higher values can be used if desired. In fact, if you have a 35 WVdc or 50 WVdc capacitor available, they would probably offer a better reliability margin.

Capacitors C3 through C6 are used for noise immunity improvement, and are 0.1 μF units. The values of working voltage should be 35 WVdc for each, as these voltages are easily obtained in small-size capacitors in this value range. Mount these capacitors as close as possible to the body of the voltage regulators (in fact, *on* the regulator is not incorrect!).

Capacitors C7 and C8 are used to improve the response to output current load changes (i.e., transient current response), and act as little reservoirs of current to keep up the rated output voltage while the regulator responds to changes of load demand. These capacitors are specified at 100 μF, despite the fact that 100 μF/ampere is the rule. You can probably "get away" with values as low as 15 μF for these capacitors.

The regulators are 100 mA plastic (TO-92) case types that

produce a fixed output voltage of 12 Vdc. The 78L12 is a +12-volt model, while the 79L12 is a −12-volt model. It is difficult to heatsink plastic case regulators of this type, but that is probably not needed. If 78H12 and 79H12 devices are substituted for the "L-series" cousins, and if the H-series devices are heatsinked with a "top-hat" or clip-on heatsink, then up to 150 mA can be drawn from this supply.

±15 VOLTS AT 100 mA

The circuit of Fig. 3-2 is very similar to that of Fig. 3-1, except for the voltage regulators used. If 78L15 and 79L15 are substituted for 78L12 and 79L12, respectively, then we obtain a ±15 volt (at 100 mA per voltage) power supply using the same components as before. Thus, you are advised to read the description given above for the ±12 volt @ 100 mA power supply.

In both cases, there is a device called a MOV in the transformer primary circuit. This is a *General Electric* metal-oxide varistor, and is used to clamp high-voltage spikes traveling into the power supply from the ac line. This component is especially needed where the power supply serves digital circuits. Either 12-volt or 15-volt versions may well wind up serving CMOS or HNIL logic devices, so should be protected with a MOV, or similar device.

±12-VOLT DC POWER SUPPLY AT 1 AMPERE

Figure 3-3 shows the circuit for a dual polarity power supply that will produce two different output voltages: one is 12 volts positive with respect to ground (common), while the other is 12 volts negative with respect to ground. The rated output current is 1 ampere for each voltage (i.e., the +12 Vdc side will supply 1 ampere at the same time the −12 Vdc is supplying 1 ampere).

The transformer is a 25.6 Vac (RMS) model with a center-tapped secondary winding. That center tap is important, and not all 25.6 Vac transformers have one, so be cautious when buying. The use of a bridge rectifier with a center-tapped transformer may not seem to square with the previous discussion but is quite easily explained. This circuit does not use a full-wave bridge rectifier, as it might seem, but consists of a *pair* of half-wave bridge rectifiers. The normal full-wave bridge stack (BR1) may be used as a pair of half-wave bridges with common ac connections if the transformer secondary is center tapped. In this instance, therefore, the minus pulsating dc voltage is taken from the "−" terminal of the rectifier (BR1), while the positive pulsating dc is taken from the "+" side of

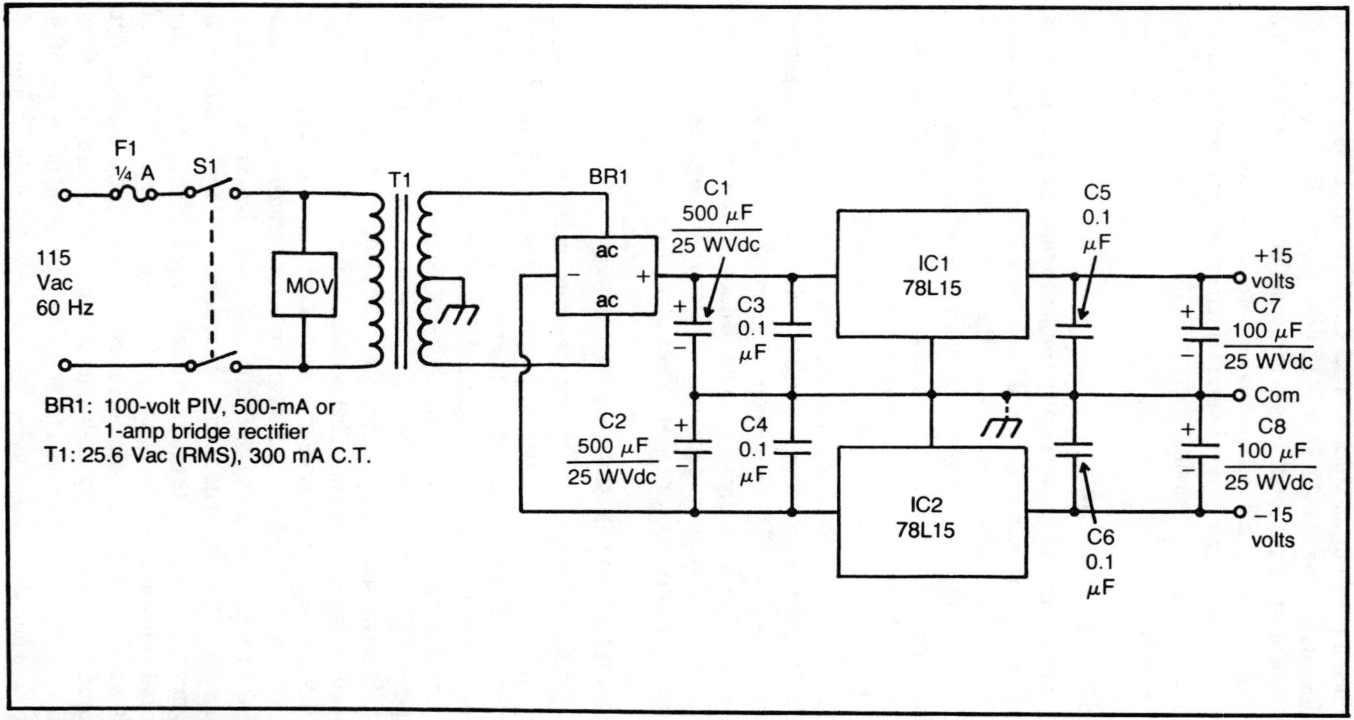

Fig. 3-2. ±15 volts at 100 mA.

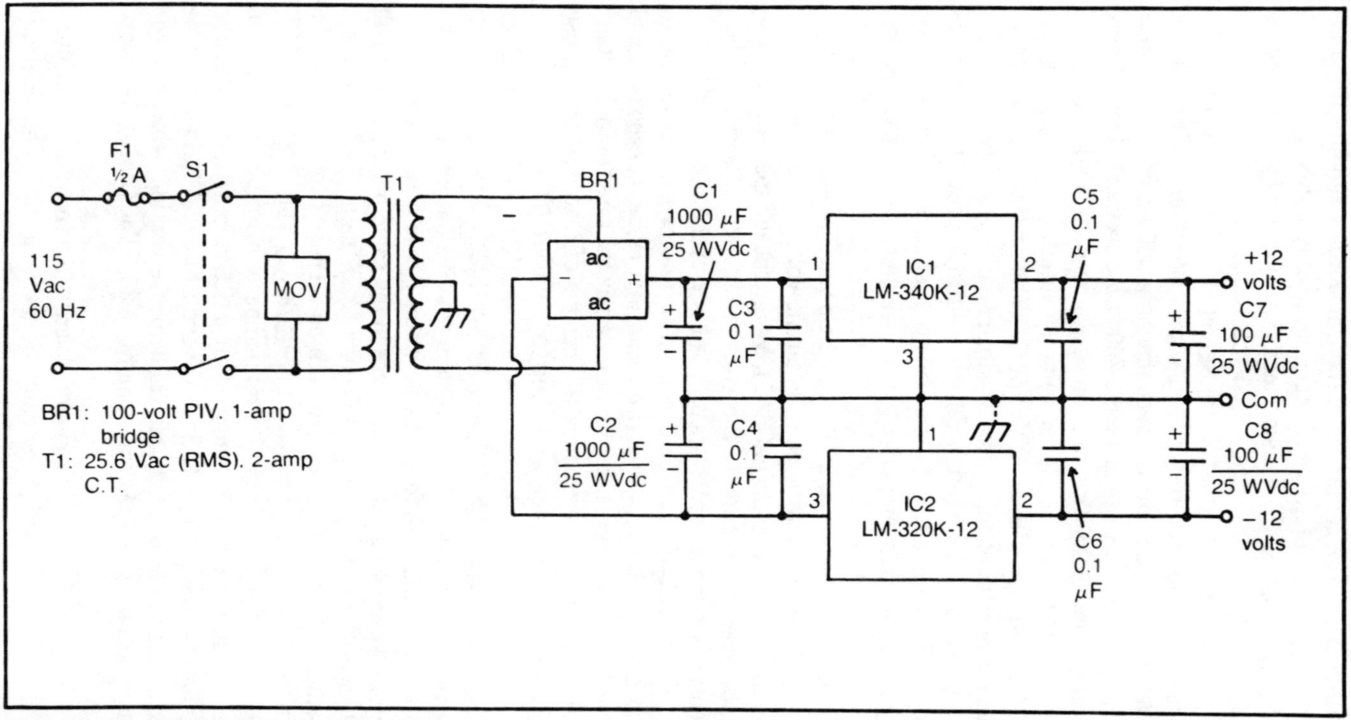

Fig. 3-3. ±12-volt dc power supply at 1 ampere.

rectifier BR1. The center tap of the transformer supplies the common or "ground" reference connection. In effect, this circuit sees a pair of 12.8 Vac (RMS) transformers, each operated with respect to ground, rather than a single 25.6-volt transformer.

Capacitors C1 and C2 are selected according to the 1000 μF/ampere rule. The 1000 μF capacitors, however, are a minimum value and some authorities recommend a 2000 μF/ampere value (which, in this case, would mean a 2000 μF capacitor). The working voltage rating of the capacitor can be as low as 25 WVdc, but again I recommend the use of higher voltage capacitors if they are available. Capacitors with 35 WVdc or 50 WVdc ratings are preferred where a margin of reliability is demanded.

Capacitors C3 through C6 are used to improve the noise immunity of the circuit, and a normally 0.1 μF to 0.5 μF units (some higher current regulators like to see 1 μF units for this application). The values of these capacitors should be 35 WVdc or 50 WVdc for C3 and C4, and as low as 25 WVdc for C5 and C6. Since it is unlikely that one will want to specify different voltages for these units, specify C5/C6 according to C3/C4. Mount these capacitors as close as possible to the body of the voltage regulators. Some builders even mount these capacitors *on* the body of the regulator!

Capacitors C7 and C8 are used to improve the response to output current load changes (i.e., transient current response), and act as little reservoirs of current to keep up the rated output voltage while the regulator responds to changes of load demand. These capacitors are specified according to a 100 μF/amp rule.

The voltage regulators in this circuit (IC1 and IC2) are of the LM-340K-12 and LM-320K-12. The "320" is a negative output voltage model, while the "340" is a positive output voltage type. Note the terminal differences between IC1 and IC2! These regulators will look the same (both are in the same package), but will not have the same pinouts. Wiring a "320" as a "340" (the usual mistake) will cause immediate burn out. These regulators may be replaced with 7812 (for the LM-340K-12) and 7912 (for LM-320K-12). If you need only 750 milliamperes of output current, then the transformer can be reduced to a 1.5-ampere type, and the regulators changed to LM-340T-12 and LM-320T-12. These same regulators can be used in a 500 milliampere version, which is made simply by changing the transformer to a 25.6 Vac (RMS), 1-ampere type.

The "MOV" device connected across the transformer primary is a metal oxide varistor, made by General Electric. This device is used to clamp high-voltage spikes coming in from the ac power line. This component is especially needed where the power supply serves digital electronic circuits, such as when the ±12-volt supply is used for CMOS and HNIL devices.

Like the 100-milliampere project that precedes this one, the 1-ampere power supply can be converted to ±15 Vdc with the simple expedient of using 15-volt regulators instead of 12 volt. For that application, use LM-340K-15 (or 7815), and LM-320K-15 (or 7915).

±24 VOLTS DC AT 1 AMPERE

Figure 3-4 shows the circuit for a dual polarity power supply that will provide two different outputs: one is 12 volts positive with respect to ground, while the other is 12 volts negative with respect to ground (or COM if the center tap of T1 is not grounded). The rated output is 1 ampere for each side of the supply (i.e., the +24 volt side will produce 1 ampere at the same time the −24-volt side is producing 1 ampere).

There are transformers available from certain industrial distributors that will provide sufficient dc output from the rectifier for 25-volt regulators. Select a model with a rating of 48 volts ac (RMS), at 2 amperes of secondary current. Most readers, I suspect, will not have such distributors locally available, and will be unable to meet the $50 to $100 minimum order most such firms charge their mail-order customers. In the event that you cannot locate the "ideal" single transformer, then use the alternate circuit shown in the inset of Fig. 3-4. Here we see a pair of 25.6 Vac (RMS), 1-ampere transformers connected such that their primaries are in parallel with each other, and the secondaries are connected in series-aiding (*note*: If zero output voltage is obtained, then it is likely that you accidentally connected the secondaries in series-opposing. The problem can be solved by reversing *either* the primary or secondary wires of *one* of the transformers. Do not switch both wires of both transformers, or, any one single transformer). The common junction between the two secondaries is the common (COM) for the power supply, and is grounded to chassis in most projects.

The use of a bridge rectifier stack with a center-tapped transformer may not seem to square with the previous discussion of power supply theory, but is actually quite easily explained. This circuit is not actually a full-wave rectified bridge, but is instead a pair of half-wave bridges. The normal full-wave bridge rectifier

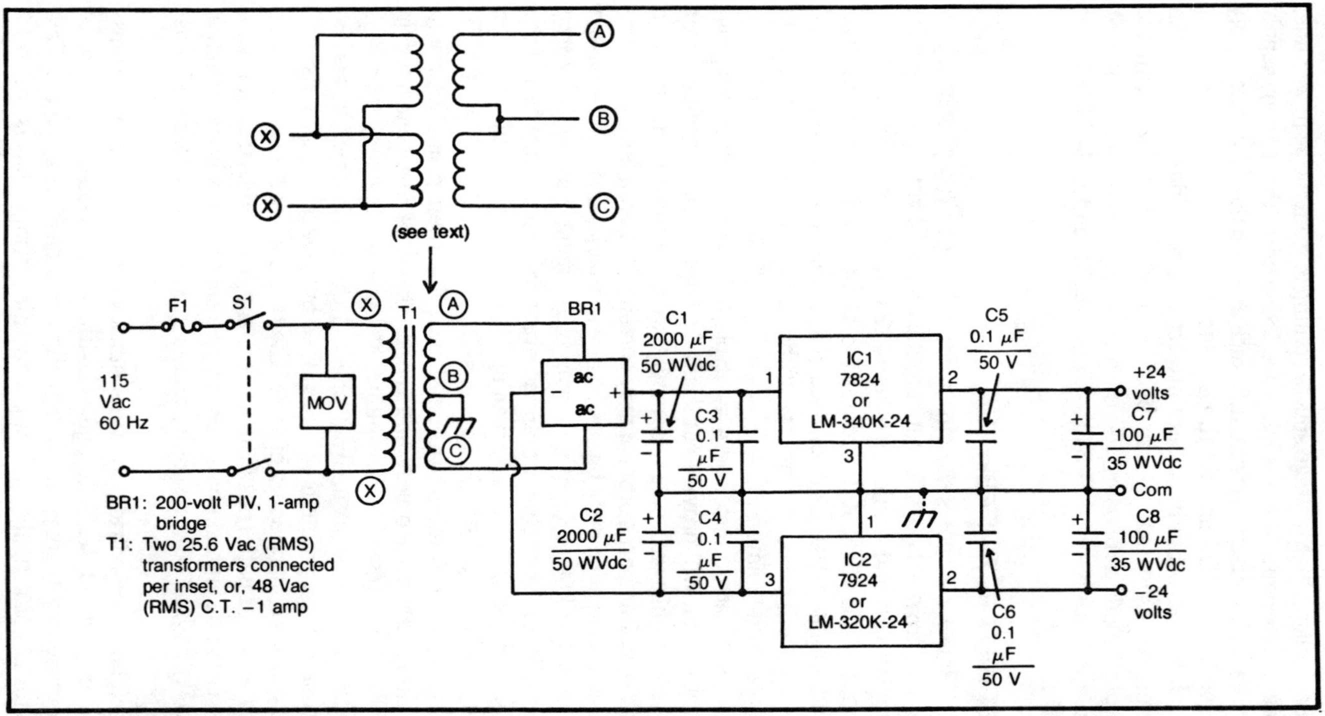

Fig. 3-4. ±24 volts dc at 1 ampere.

stack (BR1) may be used as a pair of half-wave bridges in a common case and with common ac terminals. The positive pulsating dc output is taken from the "+" terminal of BR1, while the negative pulsating dc is taken from the "−" terminal of BR1.

The rectifier is a normal packaged bridge stack (or, it can be made using four independent diodes). The rating of this rectifier is at least 100 volts PIV @ 1 ampere forward current.

Capacitors C1 and C2 are selected according to the 1000 μF/ampere rule normally, but in this case a 2000 μF capacitor is used in these slots. The working voltage rating should be 50 WVdc. These capacitors will be aluminum electrolytics.

Capacitors C3 through C6 are used to improve the noise immunity of the voltage regulators, and are 0.1 μF units. You will see these capacitors also specified up to approximately 0.47 μF for 1-ampere regulators and 1 μF for higher current regulators. The voltage rating for all four of these capacitors is 50 WVdc, and the units should be of high quality. Mount these capacitors as close as possible to the body of the regulators (in fact, *on* the regulator is not incorrect!).

Capacitor C7 and C8 are used to improve the response to output current load changes (i.e., transient current response), and act as little reservoirs of current to keep up the rated output voltage while the regulator responds to changes of load demand. These capacitors are specified according to the 100 μF/ampere rule.

The voltage regulators are 7824 or LM-340K-24 for the positive side of the power supply, and 7924 or LM-320K-24 for the negative side of the supply. These TO-3 package devices are capable of delivering up to 1 ampere of current. Similarly rated (voltage wise) devices are available in 750 mA and 100 mA current ratings. Those can be used whenever you need a smaller power supply.

It is always good practice to use an adequate heatsink on any voltage regulator (or any other power device, for that matter!). When heatsinking the voltage regulator one must keep in mind the output current rating and the voltage differential between output and input (i.e., $V_{in} - V_o$). In all cases, use silicone heat-transfer grease between the heatsink and the regulator package.

+12 VOLTS DC AT 1 AMPERE

Figure 3-5 shows a single-polarity dc power supply that will produce 12 Vdc (regulated) at currents up to 1 ampere. Smaller rated outputs can be achieved with other regulators and/or smaller transformers, but the 1-ampere size is also practical for most lesser-current applications.

Three-terminal IC voltage regulators tend to require an input-output differential voltage of 2.5 to 3 volts. For a 12-volt power supply, therefore, the pulsating dc input must be 15 Vdc or more. The full-wave rectified and filtered output from a 12.6 Vac (RMS) transformer will provide the required voltage with ease. The transformer used for this application, therefore, must be rated at 12.6 Vac (RMS) at a current of 1 ampere or more. Because of the nature of bridge rectifiers, however, it must be regarded as an absolute minimum current rating, and is offered only because of easy availability. It would be a lot smarter to use a 2-ampere transformer, if such is available. No center tap is needed. If the transformer which you select is equipped with a center tap, then insulate it from ground and ignore it.

Capacitor C1 is the normal filter capacitor, and may be specified according to the 1000 μF/ampere rule. In this case, a 2200 μF is specified, but that may be a little high as a minimum. The higher value capacitor won't hurt, and may help. The working voltage rating of this capacitor may be as low as 35 WVdc, but I recommend a 50 WVdc unit for improved reliability.

Capacitors C2 and C3 are used to improve the noise immunity of the voltage regulator. The capacitance may be anything between 0.1 μF and 0.47 μF, and the working voltage is 50 WVdc. These capacitors should be mounted physically as close as possible to the body of IC1. Many builders quite properly place the capacitors *on* the body of the regulator.

The function of capacitor C4 is to improve the transient response of the regulator to changes in output load current. This capacitor serves as a little local current reservoir to supply the needs of the circuit while the voltage regulator "catches up." C4 is specified according to a 100 μF/ampere rule.

The MOV device across the primary winding of the transformer is a metal-oxide varistor made by General Electric. The MOV is used to clamp high-voltage transients that appear on the ac line, and is especially needed when the power supply serves digital circuits such as CMOS and HNIL.

+15 VOLTS AT 1 AMPERE

We can make 15-volt and 24-volt power supplies using the same basic circuit as Fig. 3-5, with suitable changes of components. The regulators, for example, will have to be 7815 (7824) or LM-

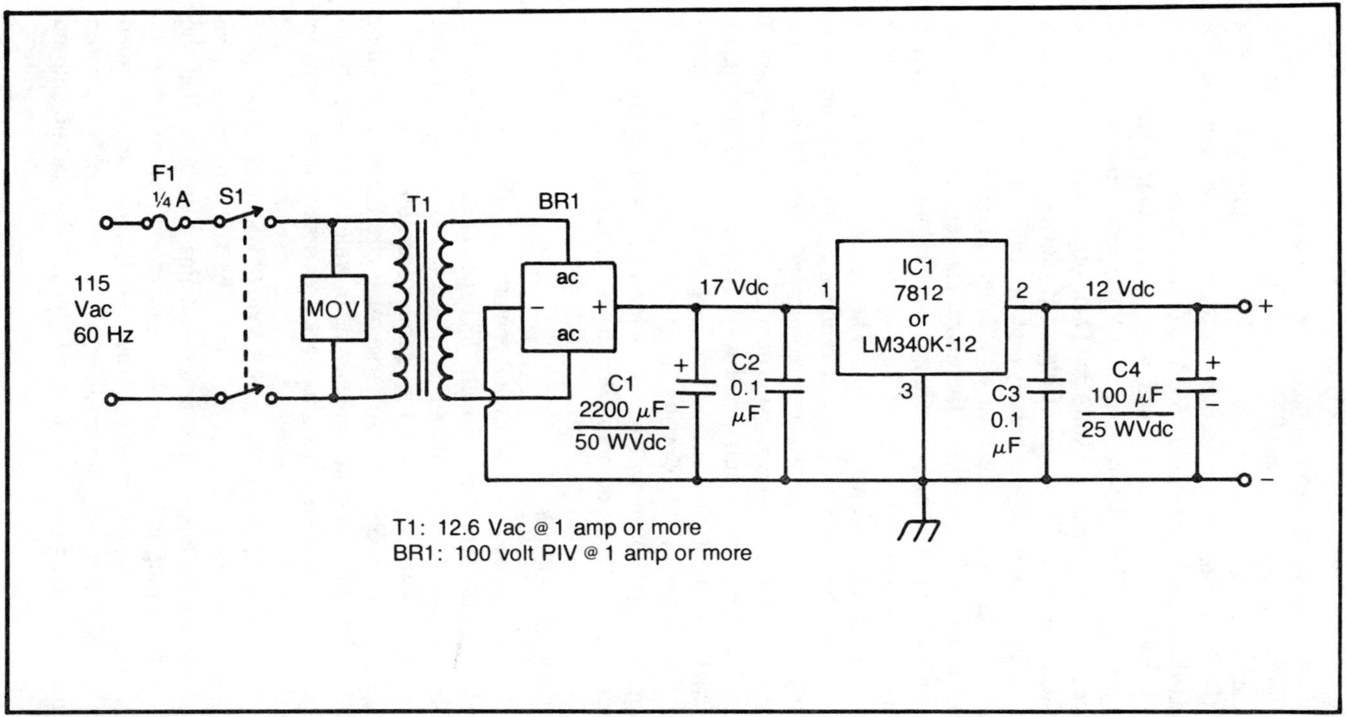

Fig. 3-5. +12 volts dc at 1 ampere.

340K-15 (−24), and 7915 (7924) or LM-320K-15 (−24), and the transformers will have to be changed. For the 15-volt supply, the transformer must be 15 Vac (RMS) or higher (18 Vac is a standard popular value) or 25.6 Vac (RMS); the 24-volt supply may use a 22 Vac (RMS) or higher transformer. Both, of course, may use the standard and easily obtained 25.6 Vac (RMS) types.

Although digital electronics projects are found with other voltages (CMOS can operate to ±15 Vdc), most common digital devices are operated at +5 volts dc. This potential is the standard for the ubiquitous TTL-style of digital IC devices. TTL devices have a requirement for very closely regulated +5 volts, and require substantial currents (as much as 40 mA per chip!). A large digital circuit made of TTL devices, therefore, can require several amperes of current. The output voltage must be (officially) within ± 250 millivolts (i.e., +4.75 to +5.25 volts dc), but that range is a little too wide for all applications. The upper end of the range (+5.25 volts) sometimes leads to reduced reliability from overheating, especially if the project is not properly ventilated. On the other hand, some complex function devices will not always work at voltages around +4.75 volts, and want to see a slightly higher potential. For most applications, the range quoted above is sufficient, but +4.9 to +5.05 volts dc is more reasonable. In the paragraphs to follow we will discuss several dc power supplies that will produce +5 volts dc output for digital projects.

+5 VOLTS AT 100 mA

Figure 3-6 shows a small digital power supply that will produce only 100 mA of current at +5 volts. This power supply will serve small-scale devices that contain only a few (no more than five) digital devices whose current requirements are less than 100 milliamperes. The supply will also serve other circuits which require less than 100 mA at 5 volts, an example of which is a Wheatstone-bridge transducer with a "looking-back" resistance of 50 ohms or more.

The transformer can be a 6.3 Vac (RMS) type rated at a secondary current of 100 mA or more. The standard in this size seems to be 150 mA, so the figure contains a notation that the transformer rating is "6.3 Vac, 150 mA." Such transformers are very small compared with some of the other transformers specified in this book. In fact, many are small enough to be available in printed circuit. There are quite a few different types on the market, and you

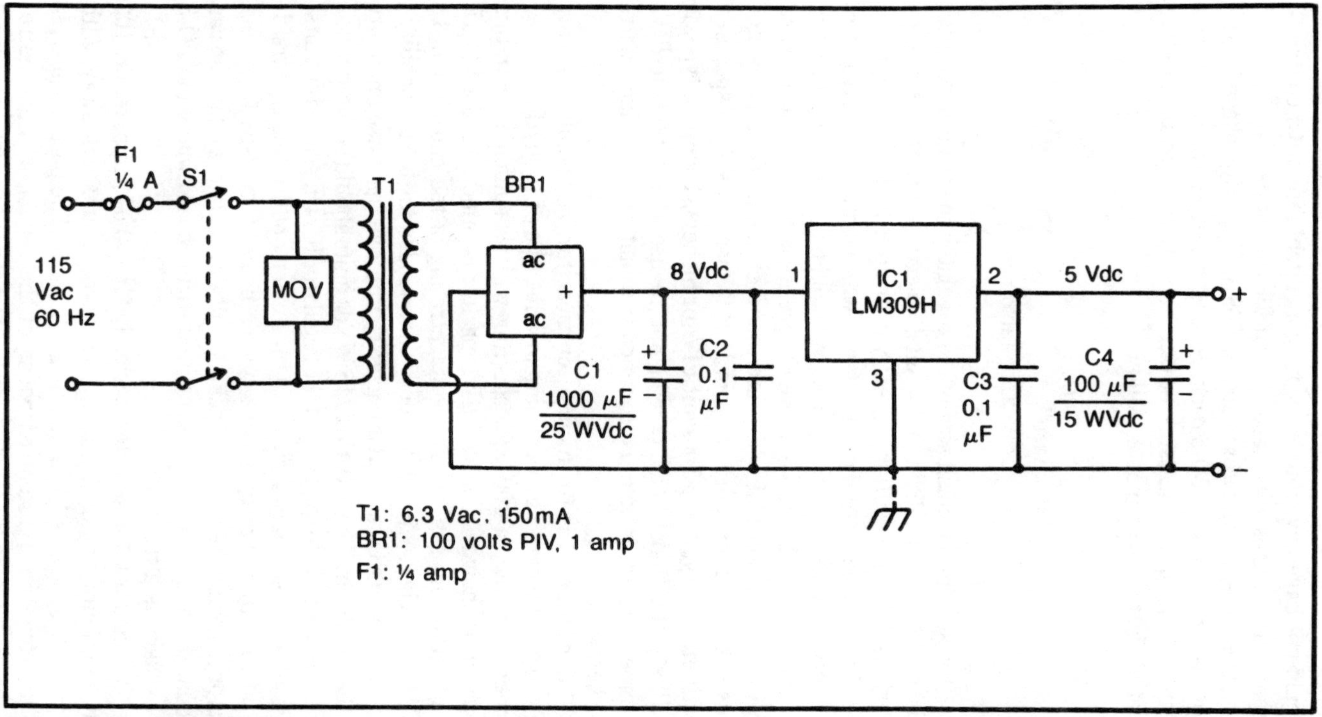

Fig. 3-6. +5 volts at 100 mA.

should not have any trouble locating a suitable item. Of course, a larger current capacity is also useful for this project.

The voltage regulator is an LM-309H. The "H-package" is the same as the transistor TO-5 package, and should be equipped with a "top-hat" or similar heatsink device. The H-package pinouts are related to transistor pins as follows:

1. "Emitter"
2. "Base"
3. "Collector" (also case)

Because of the small size of the regulator and the transformer, this project can be built on a small piece of printed circuit or "perfboard." The rectifier is greatly over-specified for this application, but only because there are few lower current rectifiers easily obtainable at local parts places. We could theoretically use any rectifier rated at 100 mA or more. There are some available in these lower ranges, some of which are found either in TO-5 transistor cans (like the regulator) or 3-pin DIP IC packages. The specified rectifier, however, is a 100-volt PIV (or higher), 1-ampere type. It is likely that most readers will opt to use the 1000-volt PIV, 1-ampere rectifier that is commonly found on the piece parts market.

Capacitor C1 is the main filter capacitor, and would normally be specified according to a minimum value rule of 1000 μF/ampere. Under that rule, a 100 μF capacitor will be sufficient. For this application, however, a 1000 μF capacitor is selected and a 470 μF would probably suffice. The voltage rating of the capacitor is 15 WVdc or higher, with 25 WVdc recommended for best reliability.

Capacitors C2 and C3 are used to improve the noise immunity of the regulator. These 0.1 μF/15 WVdc units should be mounted as close as possible to the body of the regulator IC. On printed circuits, this job is most easily achieved with small monolithic capacitors. It is difficult to mount capacitors *on* the body of an H-package regulator (as would be recommended for other regulators), but one can mount them to the pins on the printed wiring board where the TO-5 H-package is attached.

Capacitor C4 is selected normally for 100 μF/ampere, or, 100 μF minimum value. I suspect that you could get away with capacitors as low as 15 μF in this application. The purpose of C4 is to improve the transient response of the regulator, i.e., the response to large shifts of load current. The capacitor serves as a local current reservoir to keep up the output voltage while the regulator catches up with the current demands.

The MOV device in the primary circuit is a metal oxide varistor manufactured by General Electric (similar devices are also made by other companies). The purpose of the MOV is to clamp high-voltage transients that arrive via the ac power line. Such transients have the nauseating habit of adversely affecting digital electronic circuits. In fact, any time a digital electronic circuit behaves intermittently in a bizarre manner, suspect such power line transients.

+5 VOLTS AT 500 OR 750 mA

The 500- to 750-milliampere power supply shown in Fig. 3-7 is probably the smallest size that should be built in +5 volts for most readers. This size will serve many small digital projects, and can be strained to produce up to 1000 mA (i.e., 1 ampere) in a pinch (and *if* proper heatsinking and ventilation is provided!!!!!).

This circuit is almost identical to that of Fig. 3-6, so one is asked to read the preceding description as well. The differences are the transformer and the regulator used. The transformer is a standard 6.3 Vac (RMS) "filament" transformer (a term harking back to vacuum tube technology) rated at a secondary current of 1 ampere or more. The regulator device is an LM-340T-05, and will pass up to 750 mA freestanding, or 1000 mA if heavily heatsinked.

The capacitor used for the main filter (C1) can be anything from 500 μF (for 500 mA) on up. Use the 1000 μF/ampere rule for the minimum capacitance. The capacitor shown in the figure is according to the 2000 μF/ampere (assuming 750 mA current load) alternate rule. The voltage rating of the filter is not less than 15 WVdc, with 25 WVdc recommended for improved reliability.

Digital power supplies of the +5-volt type are usually found in currents somewhat higher than we have shown here. This supply is for lightweight duty, and should not be used for projects with large numbers of chips.

We can use a simple LED (i.e., light-emitting diode) to make a *power-on indicator.* Simply connect the LED in series with a 330 ohm resistor and then connect the series combination across the output of the power supply in such a polarity as will light the LED when +5 volts is present.

A blown fuse indicator is obtainable by connecting a small neon glow lamp (e.g., NE-2) in series with about 470 kohms, and then connect the pair across the fuse. When the fuse is OK, this combination is shorted out so the lamp is turned off. If, however, the fuse

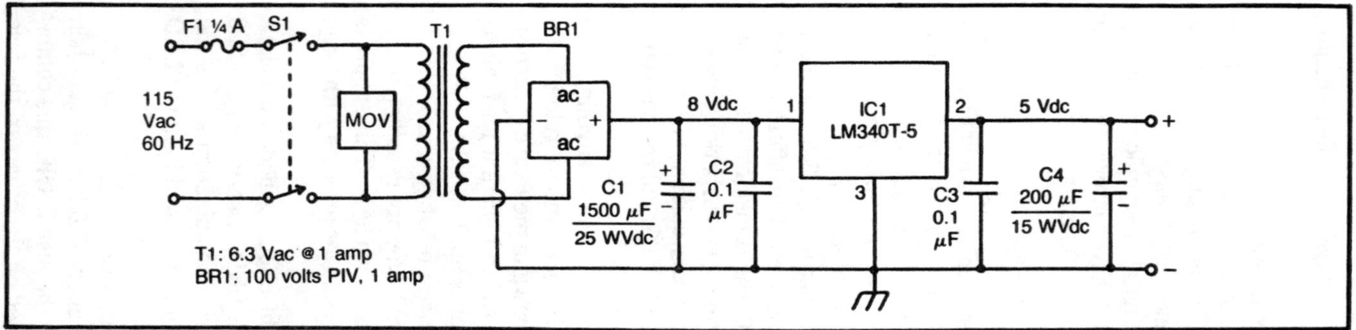

Fig. 3-7. +5 volts at 500 or 750 mA.

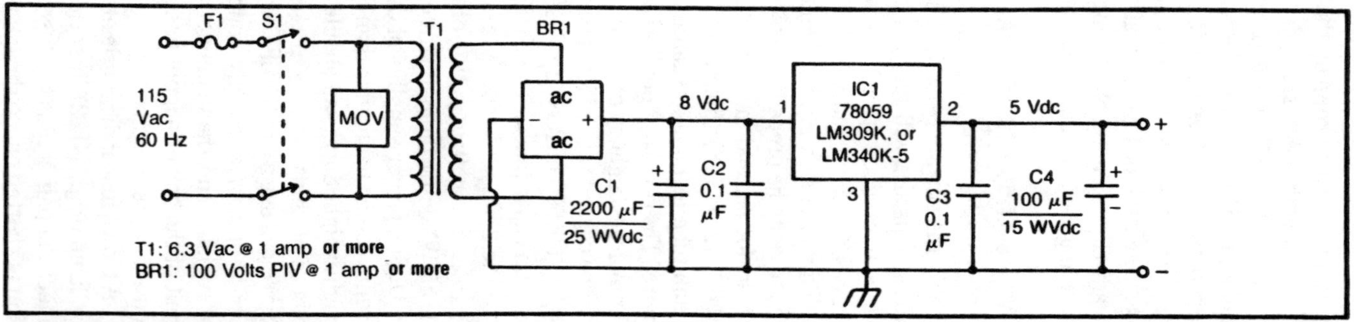

Fig. 3-8. +5 volts dc at 1 ampere.

opens for any reason, there will be 115 volts ac across the combination so the lamp will light. Fuse/resistor combinations sealed in plastic housings that fit a front panel are available.

+5 VOLTS DC AT 1 AMPERE

Perhaps *the* standard dc power supply for hobbyist and other builders of small digital electronic projects is the 1-ampere, 5-volt dc supply. Figure 3-8 shows an example of such a power supply. Note that most of the early three-terminal IC voltage regulators on the market were intended for this class of operation. The basic circuit is very similar to others of this series.

The transformer is a 6.3 Vac (RMS) or more filament transformer. When rectified in a full-wave rectifier and filtered, this size transformer will produce about 8 volts dc at the input of the voltage regulator. This is sufficient because most regulators will operate with 2 to 3 volts input/output differential voltage. We could, of course, use a higher voltage (e.g., a 12.6 Vac filament transformer), but that would increase the input/output differential voltage across the regulator, hence it increases the power dissipation. When power dissipation is increased, then the reliability of the power supply suffers due to increased junction temperatures inside of the IC.

An alternate transformer, made by *Triad,* uses a tapped primary to offer the user a selection of 6.3 Vac or 7.5 Vac output across the secondary. While this feature is probably unneeded at 1 ampere, it is useful at higher currents. Use either the 6.3 or 7.5 Vac taps, as you please. The increase in input/output differential occasioned by using the 7.5 Vac tap is insignificant in this case.

The rectifier can be any with a rating of 50 volts PIV (or higher) at 1 ampere, or higher. In most cases, I suspect that the user will want to use the 1000-volt PIV bridge stack which is so very popular in the 1-ampere size. That rating is a little overkill, but is more than sufficient. It might offer a little bit of protection against high-voltage transients on the power line.

Capacitor C1 is the main filter capacitor. Note the polarity markings. This capacitor is an electrolytic, and those markings must be observed. On most capacitors, there will be either "+" and/or "−" signs, or, there will be a black band to mark the negative terminal. This black band is often found on the side of the capacitor, rather than right at the terminal. It is necessary to observe this polarity, otherwise damage (and danger) could result. The capacitance can be chosen according to the 1000 μF/ampere rule, but in this case the author selected a 2000 μF unit (for convenience, as it turned out).

Capacitors C2 and C3 are intended to improve the immunity of the circuit to noise transients. These capacitors can have a value of 0.1 μF to 0.5 μF, and must be mounted as close as possible to the body of the regulator. In fact, many builders will mount these capacitors physically on the body of the regulator, using the pins as tie-lugs (with a solder lug to ground through one of the TO-3 mounting holes for the other ends).

The last capacitor in this circuit is C4, which is used to improve the response of the power supply to transient demands of current from the load. When the current demand suddenly increases, the voltage output may momentarily drop, thereby causing a "glitch" in the circuit. This capacitor can be anything from 100 μF to about 400 μF, although the standard rule of thumb is 100 μF/ampere. Like its big brother, C1, this capacitor is a polarized electrolytic.

The working voltages for all capacitors in this project can be as low as 15 WVdc, but I recommend that (at least) C1 and C2 have a 25 WVdc rating for improved reliability.

The voltage regulator can be any of several standard three-terminal IC voltage regulators that supply 1 ampere at a rate output potential of +5 volts dc. Any of the following are suitable if they come in a TO-3 package: LM-309K, 7805, LM-340K-05 and so forth. Some 7805 devices are in TO-220 plastic power transistor cases, and these will work if properly heatsinked. All in all, however, I prefer the TO-3 "K-package" for best reliability.

Heatsinking to the chassis is permissible, if you use a chassis, but it is not recommended. I personally favor using a finned heatsink on the regulator. The mounting screws holding the regulator should be torqued down adequately to firmly seat the regulator, but not too much lest damage occur. It is also recommended that heatsinking silicone grease be used between the TO-3 package bottom and the heatsink surface. Since the case is grounded on this project, you can mount the regulator directly to the heatsink without any insulation, or, directly to the metal chassis of the cabinet.

The +5-volt, 1-ampere, dc power supply is, perhaps, the most popular because it allows up to 40 or 50 TTL devices, depending upon type. This is the size which most hobbyists will require. If you need a heavier +5-volt dc supply, then you might want to consider one of the projects to follow.

+5 VOLTS DC AT 3 AMPERES

Most small single-board microcomputers require a dc power

supply of +5 volts at 3 amperes (or more in many cases). This project, therefore, is the smallest which can be successfully used for most small microcomputers. Many others will require the +5-volt, 5-ampere model shown as project no. 11; consult the specifications sheet of the machine which you purchase. *Note*: most microcomputers will also require a source of ±12 volts dc at 500 mA, or so. Such power supplies are discussed earlier in this section.

The transformer used in this project (Fig. 3-9) is a 6.3 Vac (RMS) type, rated at 5 amperes. The reason for using a 5-ampere transformer in a 3-ampere project is two-fold. First, there is the fact that 5-ampere transformers seem easier to obtain locally than 3-ampere transformers (don't ask me why, I've looked!). Second, there is improved reliability in using a transformer at a current that is only 60 percent of its full-rated current will allow the transformer to run cooler, hence more reliably. Keep in mind that some current between 1 and 5 amperes is something of a heating breakpoint. Transformers in the 5-ampere range seem to overheat more than 1-ampere transformers, when each is operated at its full capacity.

The rectifier in this project is rated at a PIV of 50 volts or higher, with 100 or 200 volts PIV being most common. In this current range, there are few low-cost 1000-volt PIV units, so I suspect that most readers will prefer to use the device specified rather than a higher-voltage type. The rectifier is a bridge stack rather than individual diodes. These are typically available in 6 amperes, 12 amperes, 15 amperes, or 25 amperes. I suspect that most electronics parts stores frequented by hobbyists will stock only the 25-ampere size. That is the experience in my locality. Keep in mind that rectifiers in this range tend to be somewhat overrated by some hobbyist suppliers. This is not a sin, but one must keep in mind what is being done. The supplier is merely using a rating method analogous to the old "Intermittent Commercial and Amateur Service" (ICAS) rating of vacuum tubes in the "good ol' days." If you use the power supply heavily, as in a microcomputer that is turned on all day, day in and day out, then opt for a higher rating. It has been my experience that some suppliers will overrate rectifiers as much as 100 percent in the hobbyist market. This is not a problem if you avoid off-brand parts and stick with those you know. There is not much in the way of real problems with off-brands, as long as you take precautions such as derating anything you don't recognize.

The rectifier will more than likely come in a chassis-mounted package. In the most usual configuration, the rectifier is a small,

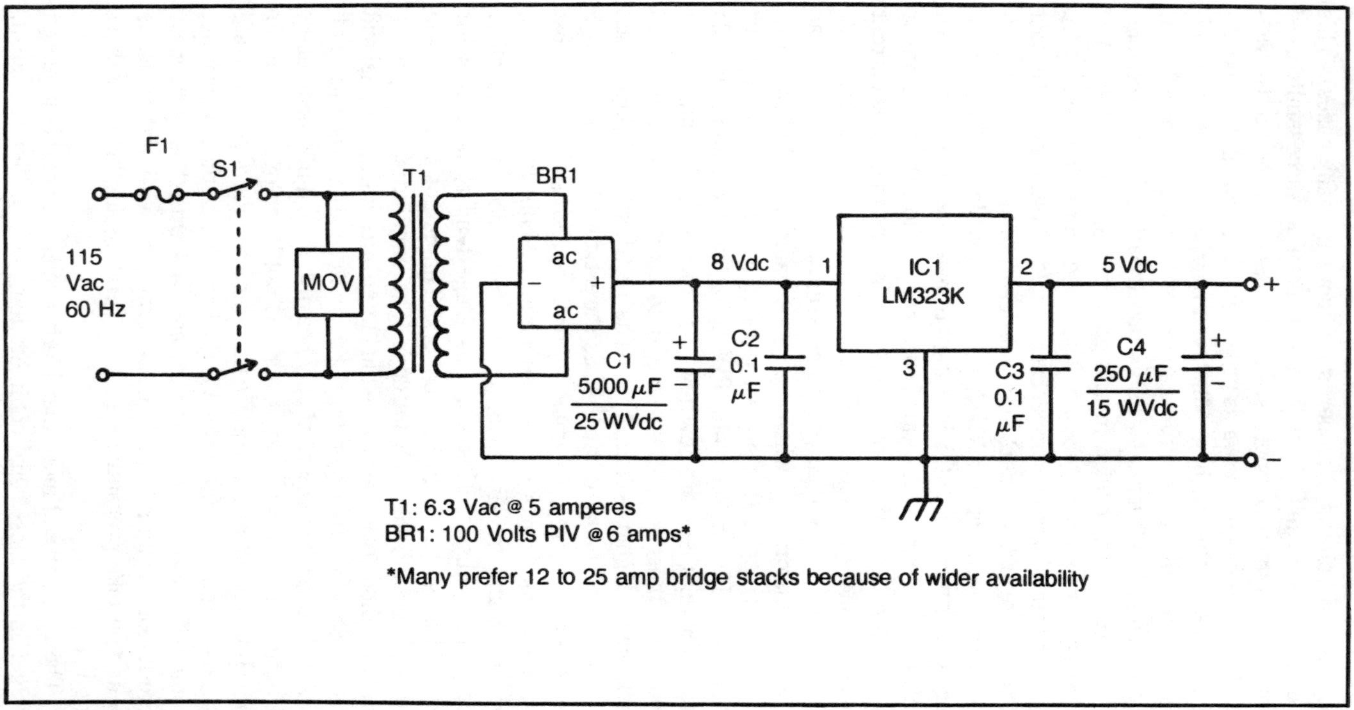

Fig. 3-9. +5 volts dc at 3 amperes.

square package with a mounting hole in the center of the flat edges. There might also be a small mounting guidepin along the bottom flat edge. Tighten the mounting screw enough to firmly hold the rectifier in place, but not so tight as to break the epoxy top surface. Also, it might be wise to use a little silicone heat-transfer grease underneath the rectifier stack. It might be wise to use a liberal heatsink if the rectifier is of the 6-ampere size.

The main filter capacitor is C1, which is rated according to a 1000 μF/ampere rule. Since 3000 μF is a little hard to find, I have selected for this one a commonly available 5000 μF model. The voltage rating of the capacitor can be as low as 15 WVdc, but 25 WVdc is recommended. It costs little or nothing extra at this capacitance to obtain the higher rating, so you might as well receive the greater reliability of the higher WVdc rating. You might also find 35 WVdc and 50 WVdc devices, but they will not confer any advantage unless one can obtain them at little increase in cost over the 25 WVdc type.

Note that it is not universally possible to obtain greater advantage from using higher-voltage filter capacitors. Besides there is a point of diminishing returns, there is also the fact that some electrolytics will not behave nicely when operated as a potential that is far below the rated potential. If you select a 450 WVdc capacitor, for example, there might be some surprises in store for you.

Capacitors C2 and C3 are used to improve the immunity of the voltage regulator to noise transients born by the ac power line, and originating in the load circuit. These capacitors can have any value between 0.1 μF and 0.5 μF, and must be mounted as physically close to the body of the regulator as possible. In fact, it is best to mount them *on* the body of the regulator! The regulator pins (1 and 2) can be used for the "hot" ends of the capacitors, while a solder lug to one of the mounting screws holding the TO-3 case to the heatsink will suffice for the ground ends. The voltage rating of C2 and C3 can be as low as 15 WVdc, but I recommend at least 25 WVdc. These capacitors are small in that size, so there is no reason not to use them. In fact, only a few types are found in the lower rating.

The output capacitor, C4, is used to improve the transient response of the voltage regulator during large changes of output current demand. The output voltage will tend to droop a little bit when the demand suddenly increases, causing a glitch. The capacitor serves as a local reservoir of current to supply the needs of the circuit for the few microseconds required for the voltage regulator to catch up with the demand. All electronic circuits have a certain response time, and this capacitor will help the circuit ignore that of the regulator.

The voltage regulator used here is the standard LM-323K device. It is rated at +5 volts dc output, at a maximum current of 3 amperes. This device is one of the "standards" sold by most hobbyist stores and mail-order outlets, so should be easily obtained.

The LM-323K comes in a TO-3 "K-package," which should be mounted to a heatsink. Use a little heatsink silicone grease between the bottom surface of the regulator package and the heatsink surface. At 3 amperes, we are getting into the range where chassis heatsinking becomes less desirable, so plan to use a generous sized finned heatsink with this project. The input/output voltage differential is approximately 3 volts, so the power dissipation will be $(V_d)(I_o)$ = (3 V)(3 amperes) = 9 watts. The same size heatsink that would be required for a 9-10 watt audio amplifier would be sufficient for this project.

This project will supply power for as many as 100 small TTL devices, so will suffice for many of the microcomputers currently on the market. If you need more current, or, want a certain margin for expansion and/or error, then use the next project instead . . . it will supply to 5 amperes at 5 volts dc.

5 VOLTS AT 5 AMPERES

The seemingly universal power supply required for all single-board microcomputers will be 3 to 5 amperes at +5 volts dc, regulated. The previous project will permit the user to draw up to 3 amperes, while this project (Fig. 3-10) allows up to 5 amperes. The two projects are essentially identical, except for a couple of components. The transformer can be the same 6.3 Vac (RMS), 5-ampere transformer used in project no. 10. I recommend, however, that you use the *Triad* 6.3/7.5 Vac (RMS), 8-ampere transformer for a little extra margin of current capacity. That transformer has a tapped primary winding that allows the user to select the tap which will yield a secondary voltage of either 6.3 Vac (RMS) or 7.5 Vac (RMS). Either may be used in this project. When rectified, the 6.3 Vac winding will produce a dc voltage of approximately 8 volts across the filter capacitor.

The main filter capacitor is C1, which in this case is a 10,000 μF unit. Capacitors as low as 5000 μF will meet the 1000 μF/ampere rule, and may be used. There will be lower ripple, and supposedly improved regulator operation however if a higher

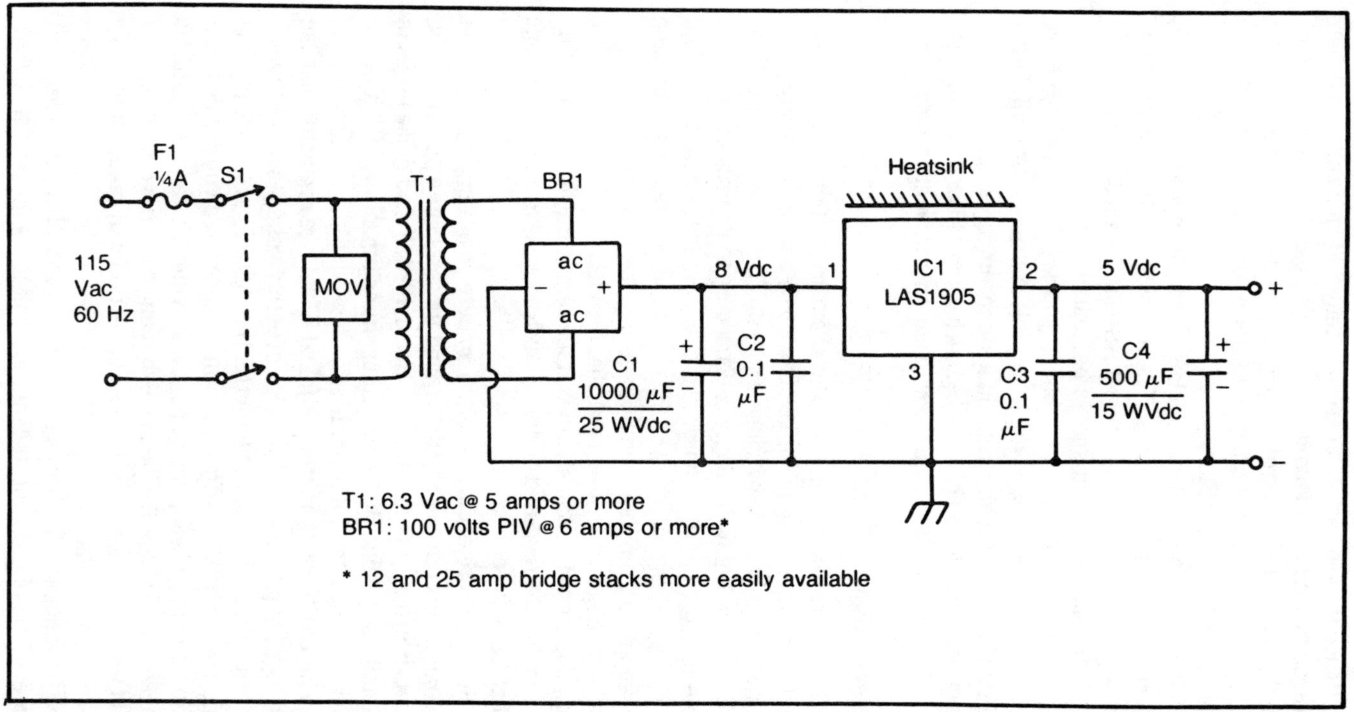

Fig. 3-10. 5 volts at 5 amperes.

capacitance is used for C1. There are many 8000 μF units available, and these might be preferred. You can also parallel connect two or more lower value electrolytic capacitors to form the required 10,000 μF. A pair of 5000 μF units, for example, yield 10,000 μF when connected in parallel.

There are only a few regulators on the market which will produce +5 volts at 5 amperes. The one selected for this project is the *Lambda Electronics, Inc.* type LAS-1905. This device comes in a TO-3 "K-package," and should be heatsinked.

All other components in this project are as in the 3-ampere version. You are advised to read that project write-up as well. I have used this circuit for a long time in a microcomputer, and found that the regulator would run cool to the touch when a liberal heatsink and small "boxer fan" (40 c.f.m.) was used for the heat transfer.

+5 VOLT ADJUSTABLE AT 5 AMPERES

There are times when a fixed regulator is not sufficient. Although the output voltage will be close to +5 volts, there are some cases where it will not be close enough. There are also situations where a +5-volt power supply is connected to its project through a couple feet (or more) of wire, and that introduces a voltage drop. The voltage at the load end of the conductor will be less than the 5 volts supplied from the fixed-voltage regulator. It is often the case that this voltage drop is so great that the circuit will not work. The solution is to use an adjustable voltage regulator, as shown in Fig. 3-11, so that the feed-end of the conductor can have a higher voltage so that the load end is exactly +5 volts.

Since this project is a 5-volt, 5-ampere power supply, all of the components except the regulator (and its associated parts) are the same as in projects 11 and 12. You are advised to read those projects for details regarding the transformer, filter capacitors, C2/C3, the output capacitor, and the rectifier.

The LM-338K is a three-terminal IC voltage regulator that can be adjusted to produce a wide range of output voltages by varying an external voltage divider network (R1 through R3). *Note*: the pinouts of this regulator IC are *different* from those of fixed voltage devices! The input voltage is applied to pin no. 2, while the "adjust" circuit is connected to pin no. 1. The regulated output voltage is the *case* of the TO-3 "K-package." Incorrect connection will result in blowing the LM-338 regulator.

The values of R1 through R3 are selected to provide a fair degree of resolution for the output voltage adjustment. Poten-

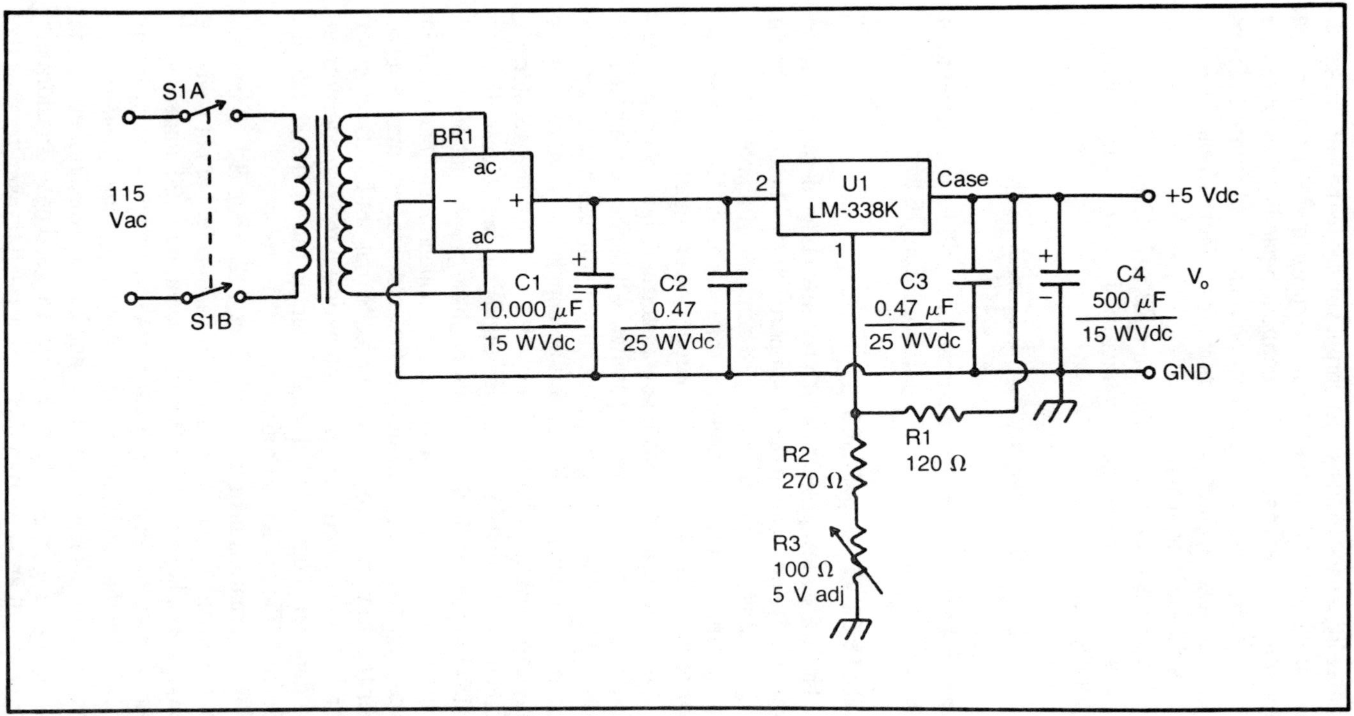

Fig. 3-11. +5-volt adjustable at 5 amperes.

tiometer R3 is used to adjust the output voltage to the value desired. This potentiometer should be a ten-turn (or more) trimmer type unless you require front-panel adjustment capability. In that case, a five-turn panel-mounted multi-turn pot can be substituted. There are ten- and up turn potentiometers available for the front panel, but they tend to be very costly.

You can calculate the values for R2 and R3 for other output voltages (provided the input voltage will allow it!) by using the formula below:

$$V_o = (1.25\text{ V}) \times \frac{R2 + R3}{120} + 1$$

(R2 and R3 are in ohms). In most cases, it will be best if the total value of R3 is a fraction of the total, e.g., from 10 to 40 percent.

+5 VOLTS DC AT 20 AMPERES

There are a few projects which require a large 5-volt regulated power supply, i.e., one in the 20-ampere range. The circuit shown in Fig. 3-12 is capable of passing 20 amperes at a potential of +5 volts, and uses a combination of a three-terminal IC voltage regulator (adjustable) and a pair of output power transistors.

The transformer is not one transformer, but two connected such that the primaries are in parallel, while the secondaries are connected in series-aiding (i.e., the output voltages *add* rather than bucking each other). If the transformer appears to produce zero (or nearly zero) ac output voltage, then it is likely that the transformer secondaries are connected in series-opposing. In that case, the two equal voltages are 180 degrees out of phase, so the net output will be zero. The solution is to reverse the connections to one winding on one (and only one) transformer. For example, you may reverse either the primary or the secondary leads (but not both!) on either T1 or T2, but not both. The junction between the two transformer secondary windings is used as the center tap of the composite secondary. This point is grounded.

An alternate transformer would be a 12.6 Vac (RMS) center-tapped filament transformer. These are available, but not nearly as commonly available as the 6.3 Vac, at least in the 20-ampere size. *Triad* makes a 20-ampere, 6.3 Vac transformer that is nearly ideal.

Still another alternative is the use of surplus transformers. There are large numbers of 6.3 Vac and 7.5 Vac transformers on the surplus market in the 15- to 25-ampere range. These transformers were (and still are!) used in high power radio transmitters for the

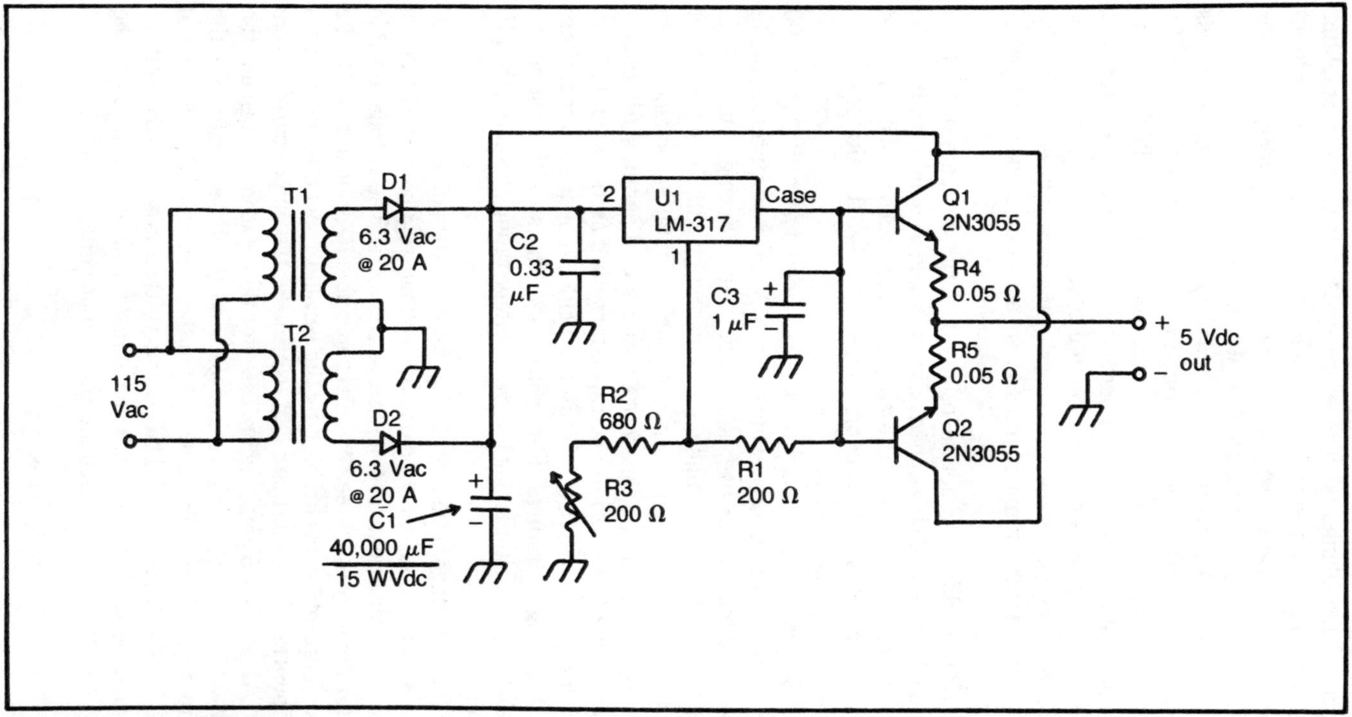

Fig. 3-12. +5 volts dc at 20 amperes.

vacuum tube filaments. There are quite a few power amplifier vacuum tubes that require, either singly or in combination, a lot of amperage at some low voltage. These transformers are often available "for a song" compared with new transformers.

The rectifiers are stud-mounted silicon diodes rated at not less than 50 volts PIV (or more, which is better), at a current of 25 amperes or more. These diodes should be of the sort which uses the mounting stud as the cathode connection. If these are used, then the two rectifier diodes can be mounted together on a common heat sink. Keep in mind, however, that the heatsink will require insulation from ground (unless you *like* to burn out diodes and transformers!).

Power diodes generate relatively large amounts of heat, so should be mounted on a finned heatsink made of aluminum. It is also advisable, if a fan is used to cool the regulator transistors (good idea!) to position circuit components such that the airstream also blows across the diodes.

The main filter capacitor, C1, is selected according to a 2000 μF/ampere rule; hence a value of 40,000 μF is required. It might also be advisable to use somewhat more filtering, but such is optional. If you cannot find a single capacitor with 40,000 to about 80,000 μF in one can, then make up the total capacitance from several smaller filter capacitors connected in parallel. Although I usually like to see 25 WVdc capacitors used as the main filter in +5-volt power supplies, it is often difficult to obtain such capacitors in these higher ratings. For that reason, the specification on the main filter capacitor is that it have a voltage rating of not less than 15 WVdc.

The voltage regulator in Fig. 3-12 consists of several components: a three-terminal adjustable voltage regulator (LM-317), a pair of power transistors, and some resistors. The voltage level is determined by the three-terminal adjustable IC regulator, while the main current load is handled by the parallel-connected series-pass output transistors (Q1 and Q2). Normally, the output current from an LM-317 is 1.5 amperes or less. For this reason, we can use the LM-317 to set the voltage and provide the base current for the output transistors. The output from the regulator IC is the *case* of the TO-3 "K-package." The output voltage is set by the voltage divider network which feeds the *adjust* terminal (pin no. 1) on the LM-317 IC. We can calculate the output voltage from:

$$V_b = (1.25\ \text{V}) \times \left[\frac{200}{R2 + R3} + 1\right]$$

(Where both R2 and R3 are in ohms.)

The output voltage from the supply will be approximately 0.6 to 0.7 volts lower than base voltage V_b because of the base-emitter voltage drops of Q1 and Q2. Potentiometer R3 is selected to provide a relatively narrow swing about the design voltage of V_b = 5.7 volts, thus, the output voltage V_o can be set precisely to 5.0 volts.

The power transistors are 2N3055 types, which are among the most popular and common low-cost power transistors on the market. Only accept those which are in TO-3 packages (same as the K-package used for voltage regulators) because they dissipate heat the best. Some of the plastic transistors which have the same type designation are not *really* very good as "2N3055" transistors, even if they are perfectly wonderful for other applications. The bases and collectors of Q1 and Q2 are connected directly in parallel, while the emitters are connected in parallel through low value fuse resistors (R4 and R5). These resistors can either be made or bought. The value of 0.05 ohms is a little unusual, but it can be made by winding resistance wire onto a higher value power resistor form. Alternatively, you can parallel-connect two 0.1-ohm, 5-watt resistors, or, four 0.22-ohm, 5-watt resistors.

Output voltage adjustment should be done initially into a "dummy load" consisting of a resistor of about 1-ohm, 50-watts rating (again, multiple resistors can be parallel-connected to form the resistance and power needed). The reason for this is to keep the higher voltages that might be expected when the potentiometer is misadjusted from damaging critical TTL circuits. Most TTL devices are not terribly tolerant of overvoltage, so must be protected. Connect the dummy load across the output, and then turn on the power supply. Adjust potentiometer R3 for an output voltage of +5 volts (±0.010 volts). Turn off the power supply, and allow the dummy load to cool off before disconnecting it! It will get quite hot, so must be treated with respect.

Capacitors C2 and C3 are used to improve the noise immunity of the voltage regulator. These capacitors should be mounted as close as possible to the body of the regulator.

0- TO +35-VOLT UNREGULATED DC POWER SOURCE

Figure 3-13 shows the circuit for a little number we used to call a *battery eliminator* in the auto radio and mobile two-way radio trade. The purpose of this type of power supply was to operate the high-current transmitter on the bench, away from the vehicle bat-

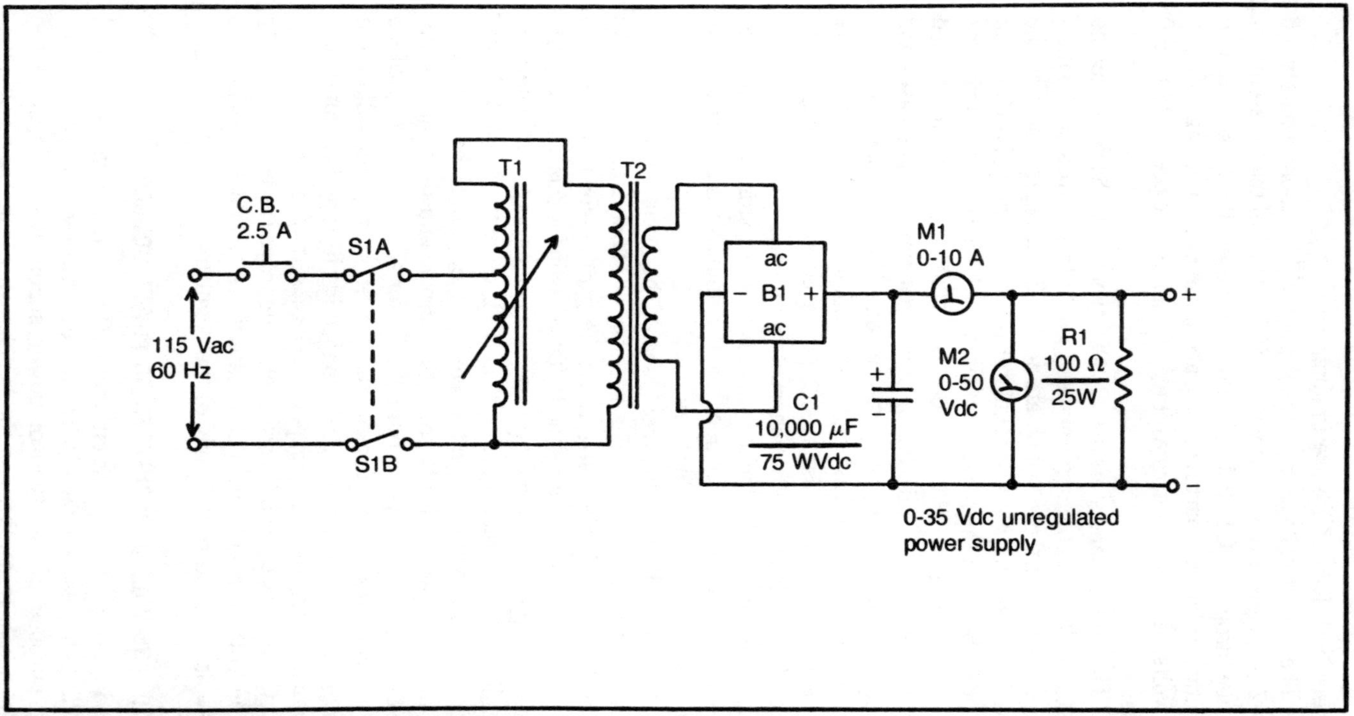

Fig. 3-13. 0-to +35-volt unregulated dc power source.

tery which normally serves to power the device. This device will provide up to 10 amperes as presently configured, and can be easily modified for much higher currents.

Transformer T1 is a device called a *variac*. T1 is an autotransformer which accepts a 115-volt ac line at the tap, and then, depending upon the setting, will deliver from 0 to 135 Vac at the output. If the small 250 VA model is selected, then this supply will deliver only 7 amperes at the full-rated output voltage (i.e., 35 volts). For this reason, the 450 and 500 VA models are recommended. A large knob will be required on the variac to permit easy adjustment of the output voltage. Also, when selecting a variac, make sure you obtain one that allows panel mounting.

The other transformer, T2, is any of several types including: signal transformer, "rectifier" transformer, and so forth. The important aspect, other than the name, is that the ratings be at least 25 Vac (RMS) output and a current rating of 10 amperes. The rectifier transformer form might be rated at the dc output, which will be 35 volts (or something close) when rectified and filtered.

The output of autotransformer T1 is connected to the primary of transformer T2. The variable primary voltage produced on T2 by T1 provides the needed adjustment over the final dc output voltage.

The rectifier is a bridge stack rated at 25 amperes or more, with a PIV rating of 100 volts or more. The PIV rating must be not less than 2.82 times the RMS ac voltage applied to the rectifier, which would be in this case about 70 volts. But 100 volts is the next higher PIV rating available in standard rectifiers.

The filter capacitor, C1, is rated according to the 1000 μF/ampere rule, and should have a working voltage of not less than 50 WVdc with 75 WVdc being preferred for reliability reasons. If it proves difficult to obtain 10,000 μF in 75 WVdc size, then make one with the more easily available 2000 to 5000 μF, 75 WVdc capacitors. Of course, a higher WVdc rating is allowable.

There are two dc meters used on the output of the power supply. Meter M1 is a 0- to 10-ampere current meter, while M2 is a 0 to 50 voltmeter. On this type of power supply it is not advisable to use high-quality meter movements unless one really needs the greater precision. The cheapie "moving coil-moving vane" type of meter is a lot more rugged than the more accurate D'Arsonval and taut-band types.

The output load resistor is used to keep the output voltage under control when there is not any load attached to the power supply. On high current power supplies the output voltage tends to rise dramatically when the output is open. In many cases, especially where 15 WVdc filter capacitors are used, this voltage is excessive and damage may result. The 100-ohm, 25-watt resistor will serve to draw sufficient current to keep the rectifiers biased properly, and the voltage in order.

Unless there is some compelling reason to ground one side of the power supply, leave the output floating above ground as shown in Fig. 3-13. The output terminals can be five-way binding posts which are insulated from the front panel. Alternatively, a heavy-duty two-terminal barrier setup can be used.

Protection of the primary circuit of the transformers is provided by a 2.5-ampere circuit breaker (CB). The use of fuses in this position is not a good idea for most users because it is virtually impossible to keep the output leads from shorting together when this supply is used on a service bench. It is one of Murphy's law's most common corrollaries that unprotected or fuse protected power supply leads will always short out at precisely the last convenient time. If a fuse is used, then the short will occur at such a time and place as to render it impossible to obtain a replacement. The answer to this problem is to select a panel or chassis mounted circuit breaker, and use it for primary protection.

+1.2- TO +22-VOLT REGULATED (ADJUSTABLE) SUPPLY AT 5 AMPERES

Another form of battery eliminator is shown in Fig. 3-14. This regulated dc power supply will work nicely on C.B. rigs, and on certain forms of amateur radio rigs (i.e., VHF-FM transceivers) and VHF-FM marine radios which have less than 70 watts of power (i.e., that will draw 5 amperes or less at 13.6 volts dc). In this case, the adjustment is provided not by a crude device such as the variac of the previous example, but by a simple front-panel potentiometer (R2).

The transformer in this project is rated at 22 volts ac (RMS) or more (25.6 Vac may be more easily obtained) with a current rating of not less than 5 amperes. No center tap is needed because full-wave bridge rectification is used. When you construct this power supply, keep in mind that the transformer, as well as the rectifier and regulator, IC, will generate a lot of heat at full output current, so mounting must be done with airflow in mind. If a fan is used to cool the components of this power supply (not needed but a darn good idea!), then try to mount the components such that the

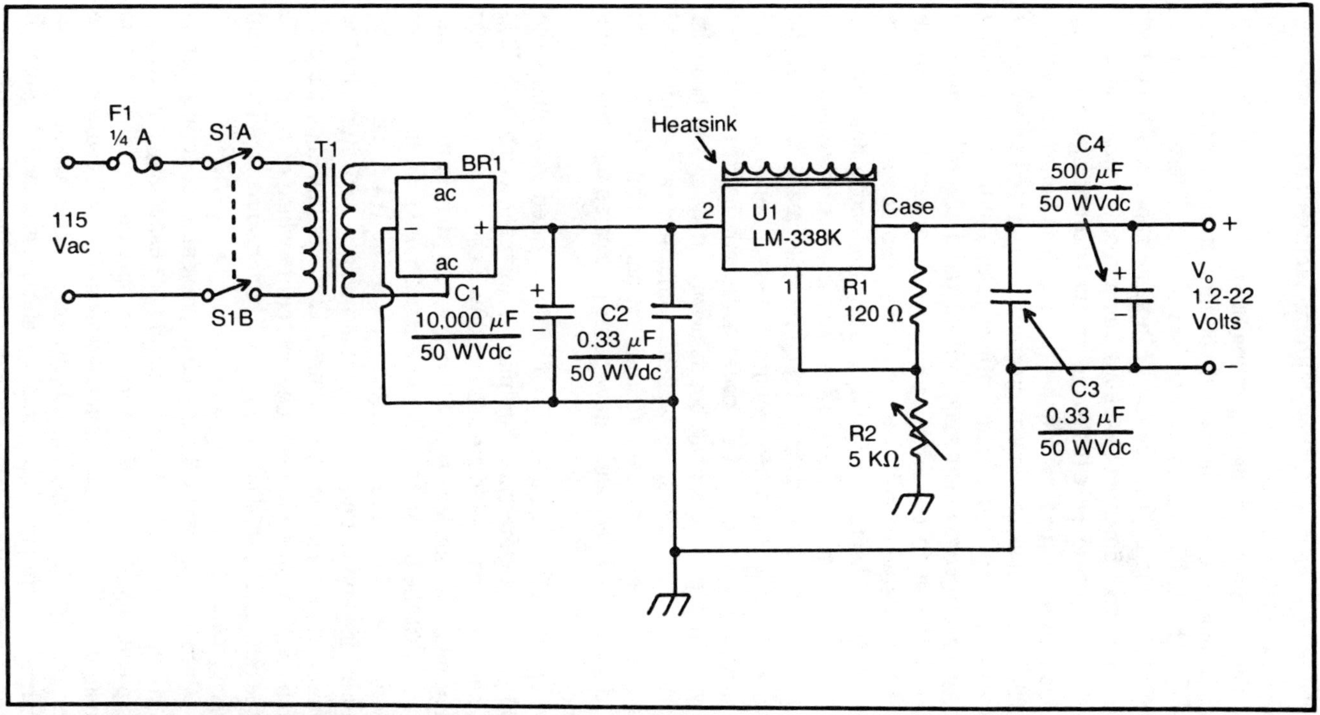

Fig. 3-14. +1.2-to +22-volt regulated (adjustable) supply at 5-amperes.

regulator gets the main airflow, with the rectifier and transformer getting at least some air.

The rectifier must have a rating of 100 volts PIV, or else it might be damaged on negative peaks of the ac from the transformer. Recall that the PIV rating of rectifiers used in filtered dc power supplies must be 2.82 times the RMS value applied. The current rating of the rectifier must be at least 5 amperes. I recommend using a bridge with a 10 ampere to 25 ampere (the latter being most easily available to most readers) for this project. It is better to "waste" capacity rather than to allow a component to run too hot because it operates at close to maximum ratings. This dictum is especially true when the component may well be rated according to some protocol similar to the ICAS rating of old vacuum tubes. If the rectifier is a stud or bolt-mounted type, then see to it that it is torqued to the heatsink properly, and that silicone heat-transfer grease is used between the rectifier and the heatsink surface. If you elect to use an axial lead form of rectifier, then make sure that you leave a lot of the lead when mounting the rectifier to the terminal strip. Also, leave plenty of airspace underneath the rectifier, especially if on a printed wiring board. Many amateur and hobbyist "ICAS" rated rectifiers, as well as some misbranded off-brand types, will safely pass the rated current only when the axial lead lengths are left intact and plenty of air circulates underneath the diode.

The main filter capacitor is C1, which is rated according to the 1000 μF per ampere rule. You may, however, elect to use 2000 μF/ampere which would mean a 20,000 μF capacitor. The voltage rating must be not less than 50 WVdc, but that is marginal and assume that the transformer is rated at not more than about 26 Vac (RMS). It would be preferred that a 75 or 100 WVdc capacitor be used in this application. If you cannot obtain a capacitor with the required capacitance in the required WVdc rating, then parallel-connect two or more lower value capacitors which *do* have the 75 WVdc rating. The total capacitance will be the sum of the individual capacitances.

Capacitors C2 and C3 are used to prevent noise from affecting the voltage regulator. These capacitors are especially important if this power supply is used to service or operate any form of radio transmitter. The rf output from the transmitter will cause possible interference with the regulator operator. For this reason, the supply should be built inside of a shielded aluminum cabinet to isolate it from the rf emanations from the transmitter. Also, there might be a good reason in that case to also include a 0.1 to 0.5 μF disc ceramic capacitor across the output terminals of the power supply (mount the capacitor directly to the output terminals of the supply). This capacitor, not shown in Fig. 3-14, will help reduce the rf that gets inside of the cabinet.

Capacitor C4 is used to prevent the output voltage from drooping when the load demand suddenly increases. The value of C4 is selected according to the 100 μF/ampere. The voltage rating should be 50 WVdc, or more. This capacitor serves as a local reservoir for current that can be drawn on when the regulator is momentarily starved during sudden increases of load. By the time C4s charge is depleted, the regulator will have caught up and be producing the required voltage and current.

The voltage regulator is a three-terminal IC adjustable type, the LM-338K device. This regulator can pass up to 5 amperes at voltages approaching 30 volts. Note that the terminals are different from nonadjustable three-terminal IC regulators. In this device, the input dc is applied to pin no. 2, while the adjustment network is applied to pin no. 2. The regulated dc output from this device is taken from the case. Because of this fact, it is essential that the case be isolated from the chassis of the power supply! If a grounded supply is used, then grounding the case will short-out the regulator output (sigh!).

Adjustment of this device is accomplished through 5000 ohm potentiometer R2. This potentiometer can be a normal panel-mounted form, but it is recommended that a 10-turn model be procured as such would provide better resolution per turn in the output voltage.

A heatsink is required for this power supply, unless you don't like reliability. A large finned heatsink is recommended, as is a fan if you can afford it. The little 3.5-inch, 40 c.f.m. "boxer" style of fan is appropriate.

So why a regulated supply when servicing modern transceivers? The difference in current demand between transmitter and receiver can cause fatal problems with nonregulated supplies. The transmitter might draw up to 5 amperes, while the receiver on squelch might need only 100 milliamperes (0.1 ampere). This 50:1 difference means that the voltage from an unregulated supply may be correct for the transmitter, but too high for the receiver (damage to the receiver can result!). Alternatively, if the voltage is correct for the receiver, then the load imposed by the transmitter will drag the potential down too low for proper operation.

The primary circuit of the power supply is protected with a ¼-ampere fuse. If you can find a ¼-ampere circuit breaker, then use it. Like the previous project, this power supply might be prone to blowing fuses because of Murphy's law. It is always the case that the output leads will short out on your workbench, causing the fuse to blow.

DUAL-POLARITY TRACKING VOLTAGE REGULATOR

An example of a *tracking* 15-volt, dual-polarity, regulator is shown in Fig. 3-15. A tracking regulator is one which uses one output voltage adjustment; the output of one regulator is fixed while the other is adjustable. In this case, the output voltage of the LM-320K-15 is adjusted by potentiometer R1. The output of the LM-340K-15 is adjusted automatically through the LM-301 operational amplifier. One input of the op amp is grounded, while the other is connected to a pair of precision 10 kohm resistors (R4 and R5) which form a voltage divider across the two outputs. When the −15 Vdc and +15 Vdc outputs are equal, then the voltage at the node of R4/R5 is zero, so the operational amplifier input is effectively zero. But if the +15-volt supply wanders off, a nonzero differential voltage is created in a magnitude and direction that will correct the error.

This design is for the regulator only, and assumes that a dual-polarity filtered power supply of at least ±18 Vdc is used ahead of the regulator. Any of the earlier designs shown in this book will suffice.

S-100 MICROCOMPUTER POWER SUPPLY

The S-100 was one of the first microcomputers on the market for amateurs and hobbyists. The S-100 bus uses distributed voltage regulation, which means that each plug-in card contains its own +5-volt regulators. In this system, there will be one to four regulators on each printed wiring card, each of which will pass 1 to 3 amperes. The high-current main power bus on the S-100 machine must, therefore, operate at more than +7.5 volts dc in order to account for the input-output voltage differential required for all regulators. The power supply shown in Fig. 3-16 is designed to provide approximately 8 volts dc at currents up to 20 amperes.

The transformer will be rated at 6.3 Vac or more. There are certain *Triad* brand transformers which will provide either 6.3 Vac or 7.5 Vac, depending upon which primary winding taps are used to connect to the 115 Vac power mains. Either voltage is recom-

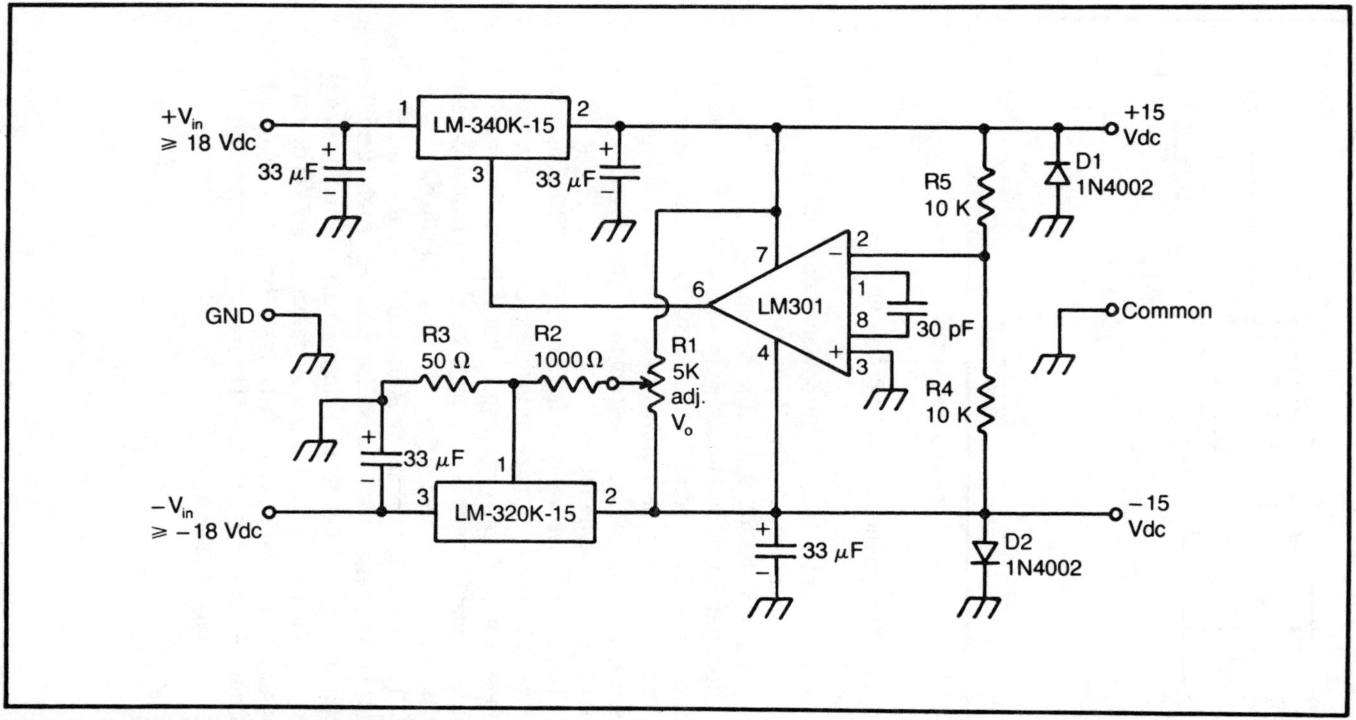

Fig. 3-15. Dual-polarity tracking voltage regulator.

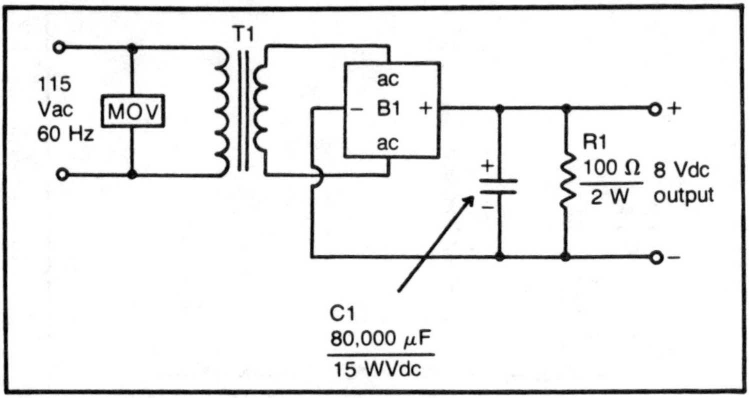

Fig. 3-16. S-100 microcomputer power supply.

mended, although the 7.5 Vac may be a little better at high current levels.

The current rating of the power supply will determine the capacity of the power supply. If you use a 25-ampere transformer, then you can make full use of the design.

Rectifier B1 is rated at 25 amperes, with a PIV rating of 50 volts or more. This rectifier wil probably be a hole-mounted form. This rectifier should be a bridge stack with terminal lugs or solder lugs, as is available. For these high current levels I personally prefer the solder-type lugs.

The filter capacitor (C1) is rated at 80,000 μF at a working voltage of 15 WVdc or more. It is difficult to obtain higher working voltage ratings at this capacitance, but it is recommended should such be available. You can connect several lower value capacitors in parallel in order to obtain the 80,000 μF specified.

The output load resistor is used to keep the output voltage under control when the load is disconnected. High current power supplies have the habit of producing excessive output voltage under zero-load conditions, and this may damage a 15 WVdc capacitor. You may mount the resistor directly to the terminals of the filter capacitor for best convenience. The mounting location is not, however, terribly critical.

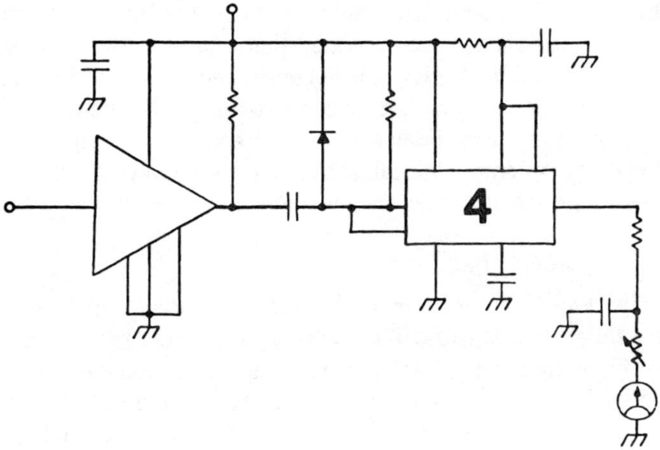

Reference Power Supplies

Thus far the power supply designs which we have considered are intended to supply power to electronic circuits to keep them going. In this chapter we are considering power supplies, but are instead intending not to supply *power* to some circuit or another, but to supply a precise voltage or current level. These power supplies, called reference voltage supplies or reference current supplies, are used to calibrate electronic instruments such as oscilloscopes and voltmeters; they are used to supply the reference potential for data converters (A/D and D/A); they are also used in other circuits to provide precise reference potentials (as in threshold level detectors).

The requirements placed on reference voltage power supplies are markedly different from the requirements for supplying power only. For one thing, rarely will a reference supply have to supply any more than a milliampere or so of current. Second, the output voltage must be a lot more precisely known than on a regular dc supply. When we use a +5-volt IC regulator to supply power to TTL circuits, for example, the actual output potential may be anywhere from +4.95 to +5.05 and still be within spec for some manufacturers. A reference supply, depending upon application, may have to be within *microvolts* and most certainly will be within ±10 millivolts (otherwise, it seems a waste to use a reference supply circuit!).

Most of the circuits in this section are based on the operational amplifier, and I assume that you know about that little gimmick. If you do not, then it is recommended that you read the section on operational amplifier basics before continuing on this discussion. I apologize for the seeming lack of concern for your index finger, but this organization for my book seemed to me most reasonable. When you finish the op-amp material, *then* come back and you will be able to understand the material on reference power supplies.

THE REFERENCE SOURCE

The reference device in most reference voltage supplies is the simple zener diode. You will recall that, in the reverse bias mode, there is a breakdown voltage that remains all but constant. We can use this as a reference source in some cases. There is, however, one tinee, little bitteee problem . . . that zener potential is *not* always stable with temperature (*sigh*!). The actual potential will wander around with changes of temperature. This phenomenon is of little consequence in dc power supply applications, but is critical in other, more precise, applications.

One solution is to bury the zener diode inside of a constant-temperature environment. This is the tactic taken by National Semiconductor in their LM-199 series of devices. The zener diode is on an IC substrate with an electronic circuit which acts as a heater element (I suspect it is a class-A amplifier with the input shorted).

Another solution to the problem is to use something like a band-gap zener. When used in an IC two-terminal "diode" device with an amplifier and other electronics the band-gap zener is capable of approaching 10 ppm/° C stability. While this figure might not thrill the designer of really classy devices, it is more than sufficient to satisfy most of us lesser lights. The Ferranti ZN454B (and related devices) is an example.

Figure 4-1 shows two reference supplies which will serve the needs of many readers. The assumption here is that you will need a 10.000-volt power supply, although adequate instructions are given below for the design of any voltage within the capability of the operational amplifier (±20 volts or so). Common voltages used for references, incidentally, are: 1.000 Vdc, 2.560 Vdc, 5.000 Vdc, 10.000 Vdc and 10.245 Vdc (care to guess why nos. 2 and 5 are popular?).

In Fig. 4-1A we see an example of a −10.000-volt reference supply, while in Fig. 4-1B is the +10.000-volt version. Both are based on Ferranti band-gap IC reference devices, or their equivalent. The circuit in Fig. 4-1A uses an operational amplifier as the

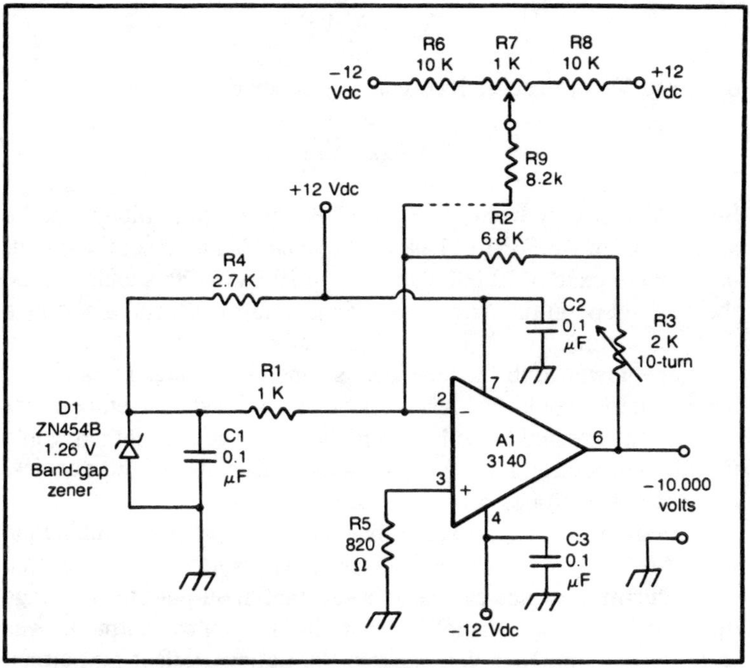

Fig. 4-1A. −10.00-volt reference power supply.

active element. This amplifier is the RCA CA3140 device. It is selected because it will produce reasonably low drift for the money, is cheap, and will not materially load down the zener diode. If greater thermal stability is required, then use one of the more costly low-drift operational amplifiers offered by such companies as Burr-Brown, Analog Devices, and so forth.

The operational amplifier, A1, is connected in the inverting follower configuration with the zener (D1) as the input voltage source. The gain of this amplifier is given by the expression:

$$A_v = \frac{-(R2 + R3)}{R1}$$

or, in this case, it will vary depending upon the setting of potentiometer R3 from $A_v = -6.8$ to $A_v = -8.8$. It is essential that the components used to set the gain of the operational amplifier have a low temperature coefficient. The usual procedure is to use precision resistors for R1 and R2, and a low tempco potentiometer for R3.

The output voltage from Fig. 4-1A will be given by:

$$V_o = A_v V_{in}$$

or, since V_{in} is fixed at 1.26 volts by zener diode D1:

$$V_o = (1.26\ V)\ (A_v)$$

Since A_v can vary from −6.8 to −8.8, the output voltage can be adjusted from −8.568 to 11.088 volts. The idea is to set the output potential to exactly 10.000 volts, or, to 10.245 volts should that be the desired potential (it may be, if the reference source serves a data converter).

The power to the zener diode is supplied through resistor R4. This resistor should also be precision, or, at least low temperature coefficient. It should also have a higher wattage rating than would usually be required for sake of reliability. Because of this requirement, I prefer to use a 2-watt resistor at R4.

Resistor R5 is approximately equal to the parallel combination of the feedback and input resistances, and is used to overcome input offset currents, which can cause a substantial output offset voltage due to the operational amplifier gain. If offset on the output proves to be a problem, then you can use the optional offset null circuit shown with dotted lines in Fig. 4-1A. This circuit will sum a small current into the inverting input of the operational amplifier, this canceling the output offset (provided that the countercurrent is of

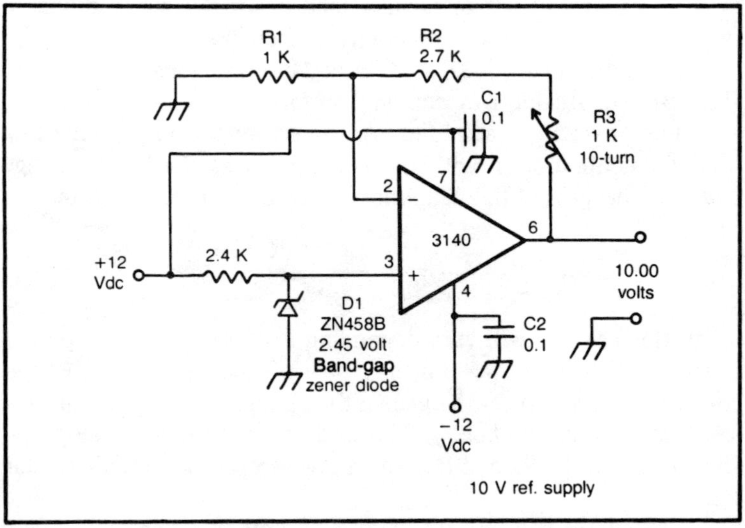

Fig. 4-1B. +10.00-volt reference power supply.

the right magnitude and polarity). Potentiometer R7 is adjusted until the output offset voltage is zero. The procedure for this adjustment is as follows:

1. Disconnect D1 and R4 from the junction R1/R4/D1.
2. Short the now-loose end of R1 (in effect setting $V_{in} = 0$).
3. Adjust R3 to approximately mid-range.
4. Adjust R9 for zero output voltage.

There is some interaction between the offset null potentiometer adjustment detailed above and the adjustment of R3. Ideally, a separate stage (with unity gain) will be used to provide offset output adjustment. Be aware of the interaction and compensate for it.

Capacitor C1 is used to reduce the noise generated by zener diode D1. The avalanche process normally used in zener diodes produces, by its very nature, tremendous amounts of white noise. In fact, this phenomenon is used by the manufacturers of "noise generators" to produce their wideband spectrum. In the reference power supply, however, that noise is a pain in the neck, so must be eradicated. This is the job of capacitor C1.

The positive voltage version of this circuit is shown in Fig. 4-1B. The basic circuit is the same, except that the operational amplifier is connected in the noninverting follower configuration. The gain of this form of circuit is given by:

$$A_v = \frac{R2 + R3}{R1} + 1$$

Given the values of the resistors in Fig. 4-1B, the gain will vary from 3.7 to 4.7. This means that the voltage will vary from 9.065 to 11.515 Vdc. Again, we adjust the potentiometer to produce exactly 10.000 volts output. All of the discussion for the previous circuit (except gain considerations) also apply to this circuit, so will not be repeated.

Adjustment of both Figs. 4-1A and 4-1B requires at least one-time access to a precision voltmeter. A model capable of 1 millivolt or better accuracy at 10 volts is a requirement. Adjust the potentiometer in either supply for exactly 10.000 volts output, ±1 mV.

Figure 4-2 shows a version of the previous circuit in which the National Semiconductor LM-399 device is used as the zener source. The LM-399 is a lower temperature range version of the LM-199 (which, incidentally, should be used in higher precision circuits!).

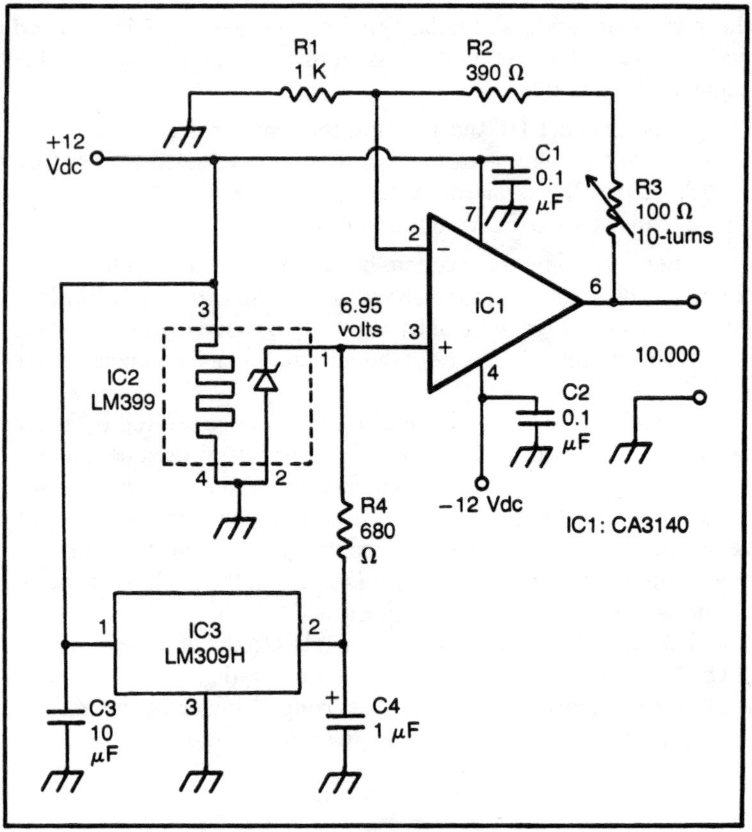

Fig. 4-2. 10.000-volt reference supply using the LM-399 device.

This circuit is very similar to the circuit of Fig. 4-1B, except that the gain is different. In this circuit, the gain is set using the same formula as above, but the results are different because of different resistor values. The gain of the circuit in Fig. 4-2 varies from 1.39 to 1.49, so the output voltage will vary from 9.966 to 10.355 volts.

The LM-399 heater terminals (3-4) are connected across a 12-volt power supply. This supply must be regulated if best results are to be achieved. The voltage for the zener diode is supplied from an LM-309H (5-volt, 100-mA) three-terminal IC voltage regulator. The LM-309H device is housed in an "H" IC package, which is identical to the TO-5 package used for transistors. Resistor R4 is for current limiting.

The operational amplifier can be the RCA CA3140 for most noncritical applications. If low drift is needed, then opt for one of the

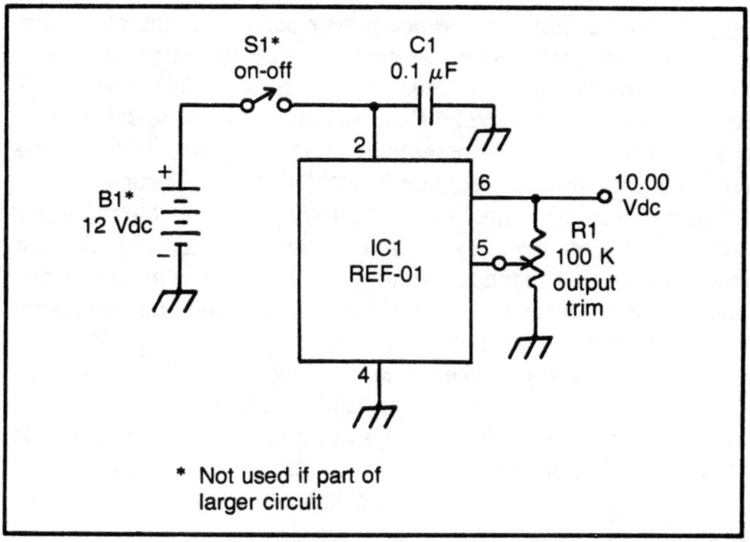

Fig. 4-3. REF-01 10.000-volt reference power supply.

more costly low-drift Burr-Brown or Analog Devices operational amplifiers (consult manufacturer's catalogs). The adjustment of this circuit is exactly like that of the previous circuit, so the instructions will not be repeated.

It is very well and good to make clever operational amplifier reference voltage circuits. But sometimes, we might want to save a few components and use a ready-made circuit. Figure 4-3 shows the use of the *Precision Monolithics, Inc.* REF-01 10.000-volt IC reference source. The REF-01 is a 10.000-volt device, while REF-02 is a 5.00-volt reference source (note: REF-02 also has a temperature output that is usable to make an electronic thermometer).

Only four terminals are needed on the eight-pin REF-01 package: V_{in} (pin no. 2), V_o (pin no. 6), *ground* (pin no. 4) and *adjust* (pin no. 5). The applied input voltage must be +12 volts, or more, but not to exceed 18 volts between pins 2 and 4. The output voltage is found on pin no. 6, and is set by the 100 K potentiometer, R1. This potentiometer should be a low temperature coefficient type, and have at least 10 turns over the resistance range. R1 is adjusted for exactly 10.000 volts output after the IC has had a few minutes to come to operating temperature.

PROGRAMMABLE REFERENCE SUPPLIES

There are times when it is desirable to digitally program an output voltage from a reference power supply. In automatic test equipment applications, for example, the computer program may call for a special output voltage other than the standard potentials listed earlier. Alternatively, the program may want to select from among several standard potentials. In this regard, we find some commercially digitally programmable "OEM" power supplies. We might also want to manually set a digitally programmable reference voltage if, for no other reason, to cite ***our*** supply as "digital" (that magical word that conjures up universal respect for otherwise bland instruments of ordinary design!). Seriously, though, manual digital control does make a reference supply look snazzy.

The idea is to produce a reference supply which will output a voltage that is proportional to an applied binary or BCD "word." There are a couple of different ways this is done, and we will discuss them both. The first (Fig. 4-4) uses a digital-to-analog converter (DAC), while the other uses switchable resistors to control an adjustable power supply.

A DAC is a circuit which will produce an output potential that is proportional to the product of a reference voltage and a fractional binary word applied to the digital inputs:

$$V_o = \frac{aV_{ref}}{2^n}$$

Where:

V_{ref} is the reference voltage (10.000 volts in this case)
n is the length of the word applied to the DAC (i.e., bit-length)
a is the decimal value of the word applied to the binary inputs of the DAC

For example, let's assume that the binary word 10001001_2 (137_{10}) is applied to the binary inputs of the DAC in Fig. 4-4. The bit-length (n) is 8-bits, so the output voltage will be:

$$V_o = \frac{(137)(10.000 \text{ V})}{2^8}$$

$$V_o = 1370/256$$

$$V_o = 5.352 \text{ volts}$$

Since the binary word will vary from 00000000_2 to 11111111, the output voltage will vary from 0 to 9.96 volts.

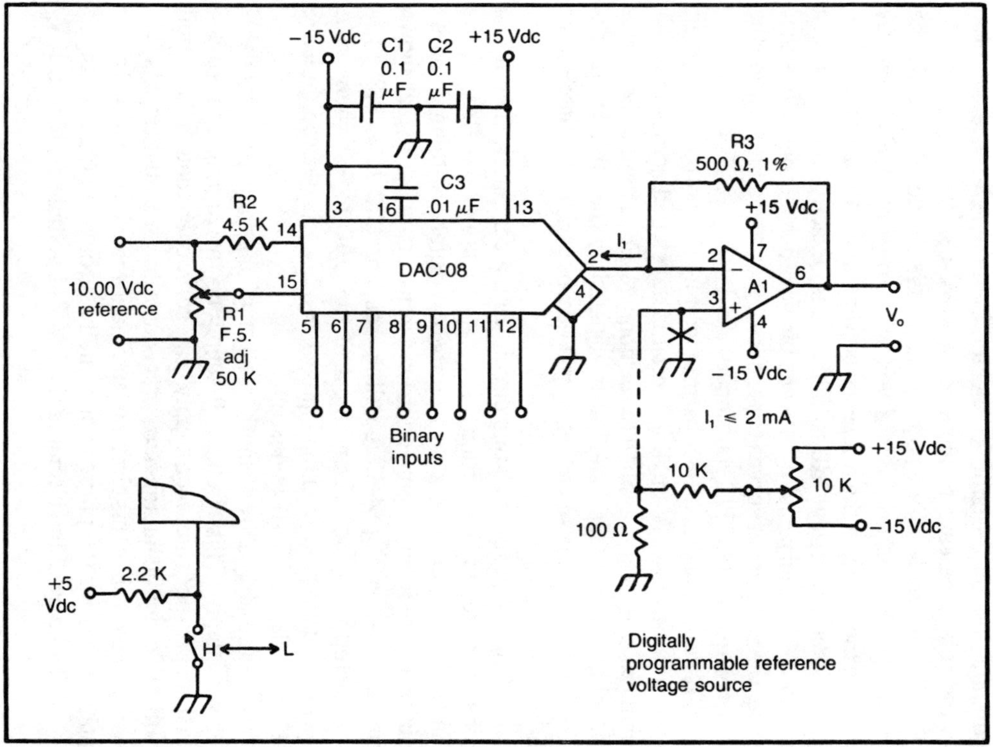

Fig. 4-4. Programmable reference power supply.

The DAC-08 was pioneered by *Precisions Monolithics, Inc.*, and was developed after the fashion of an earlier device, the *Motorola Semiconductor, Inc.* MC1408-8 device. The DAC-08 has since become the "industry standard" for low-cost 8-bit DACs, and is second-sourced by several companies. A version is even available in *Radio Shack* stores.

In the version of the DAC-08 circuit of Fig. 4-4, the binary inputs will respond to TTL input levels. This means that logical-0, or LOW, is any potential from 0 to 0.8 volts, while logical-1, or HIGH, is any potential between +2.4 and +5.0. The DAC-08 can be biased to provide other input logic levels by varying the reference potential, or, the resistors connected from the reference potential to the reference current input of the DAC-08.

If a *full-scale adjust* potentiometer is desired, then build the circuit as shown here. But if you wish to provide *F.S. Adj.* by varying the output of the reference power supply, then eliminate the potentiometer, change R2 to 5000 ohms (exactly) and provide a second 5000-ohm resistor between pin no. 15 and ground. In either case, the adjustment serves to produce +9.96 volts output when the input word is 11111111_2 (all inputs HIGH).

A zero adjustment can be provided by the dotted-line circuit shown in Fig. 4-4. A small potential is developed across the 100-ohm resistor, and is applied to the noninverting input of the operational amplifier. This potential is adjusted by the 10-kohm potentiometer, which is used to set $V_o = 0.00$ when the input binary word is 00000000_2.

If you want to drive this circuit from a computer, then it is a simple matter of connecting the binary inputs of the DAC to the lines of a latched output port on the computer. Nothing could be simpler. The manual connection uses a circuit similar to that shown in the inset to Fig. 4-4. A pull-up resistor (2.2 K to 3.3 K) is connected from *each* binary input of the DAC to +5 volts dc. A switch is used to ground the input when a LOW is desired, or leave it open when a HIGH is desired.

Figure 4-5 shows another form of digitally programmable power supply. This circuit will supply up to 1.5 amperes, so is a little out of the "reference supply" category, even though the circuit will suffice for all but the most precision reference applications. We can, however, use this circuit in automatic test equipment and other applications for the purpose of digitally setting an output voltage.

The heart of the circuit is the LM-317 voltage regulator. If V_{in}

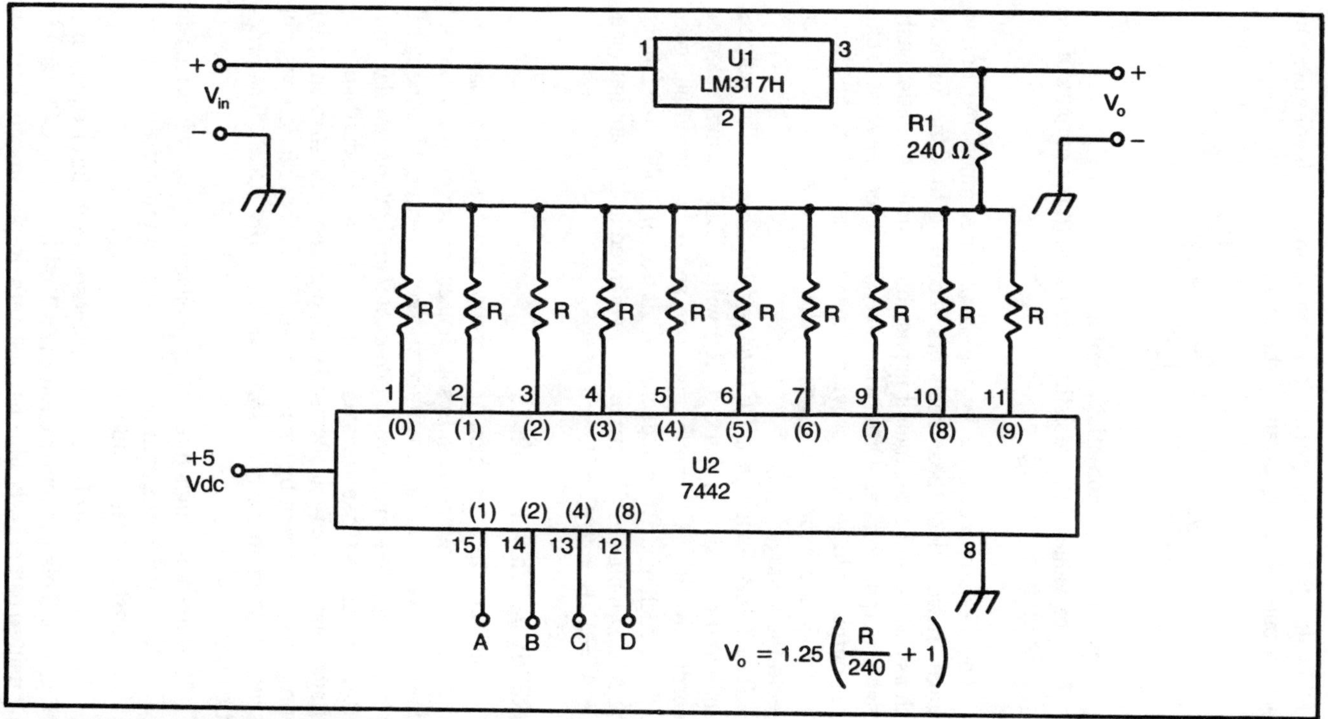

$$V_o = 1.25\left(\frac{R}{240} + 1\right)$$

Fig. 4-5. High-current programmable reference power supply.

is +28 volts, then this regulator can output up to +25 volts dc. The exact output voltage is given by the expression:

$$V_o = (1.25\ \text{V})\left[\frac{R}{240} + 1\right]$$

Where:

V_o is the output potential in volts

R is the value of the resistors connected to the outputs of the 7442 device in Fig. 4-5

The resistor selected by the 7442 determines the output voltage. If we select these resistors according to some formula, such as the 1-2-4-8 rule, then it is possible to make a continuously variable output voltage. Otherwise, we will select several "standard" output voltages and provide resistors that will make them. The voltages are tabulated against the 4-bit digital code which selects the voltage.

The 7442 device is a *BCD-to-1-of-10-decoder.* The outputs (connected to resistors marked "R") will go LOW when selected according to the BCD input code; one and only one output goes LOW for each binary coded decimal digit from 0 (0000) to 9 (1001). The ABCD inputs of this circuit are weighted in the familiar manner: A = 1, B = 2, C = 4, and D = 8.

CURRENT REFERENCE SOURCES

Thus far, our discussion of reference power supplies has centered on voltage references. In some cases, however, it is a current source that we want, not a voltage source. An example is the usual form of integrated circuit *analog-to-digital* or *digital-to-analog* converter. Most of those devices want to see a standard *current*; voltage sources are typically used with precision resistors in order to produce the current by Ohm's law (E/R). In this section, we will discuss common forms of *constant current source* (CCS). The properties of a CCS are:

1. Extremely high internal impedance so that variations in load impedance will not affect the output current
2. Constant current output

The classic form of constant current source is the circuit shown in Fig. 4-6. This particular circuit is a 1 milliampere CCS, and is built using a pair of NPN silicon transistors. Replacements for these are available in the Motorola HEP-line, by the way. In this circuit,

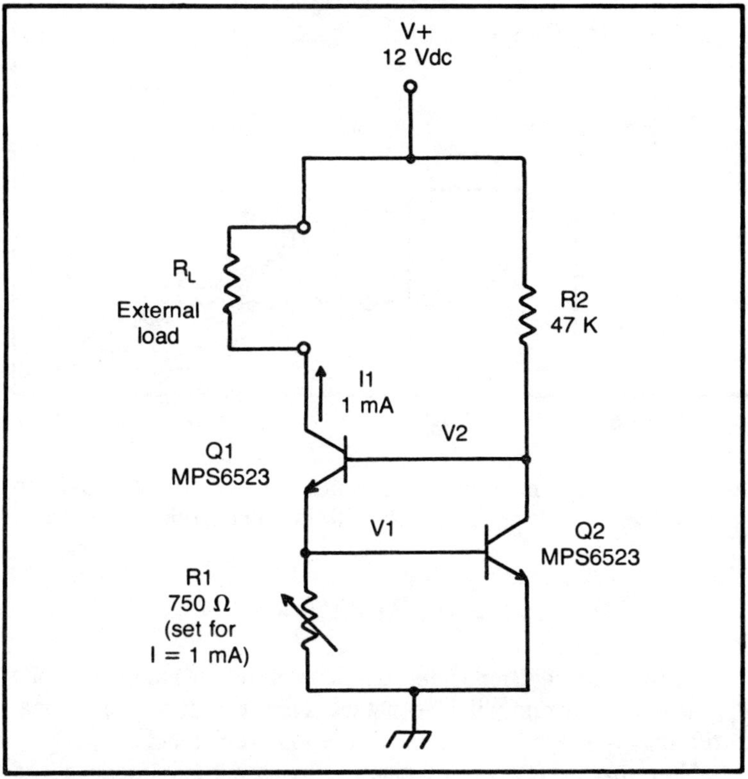

Fig. 4-6. 1-mA constant current source (CCS).

the bias on transistor Q2 is controlled by the voltage drop across resistor R1, which in turn is controlled by the collector-emitter current flowing in transistor Q1. If the load (R_L) decreases, then the current will increase. This action causes the voltage drop (V1) across R1 to rise. When V1 increases, transistor Q2 is biased on harder, so its collector potential drops. Since the collector voltage (V2) of Q2 is also the bias for Q1, we find that Q1 is now turned less hard, so the current (I1) decreases. For increases in R_L, exactly the opposite action occurs—but I will let you figure that one out for yourself.

The actual output current (I1) can be trimmed to exactly 1 milliampere with potentiometer R1.

The *Precision Monolithics, Inc.* REF-01 is normally thought of as a voltage reference IC, but it can also be used to make a constant current source. Figure 4-7 shows the usual connection for this type of service. This particular version of the circuit causes an output

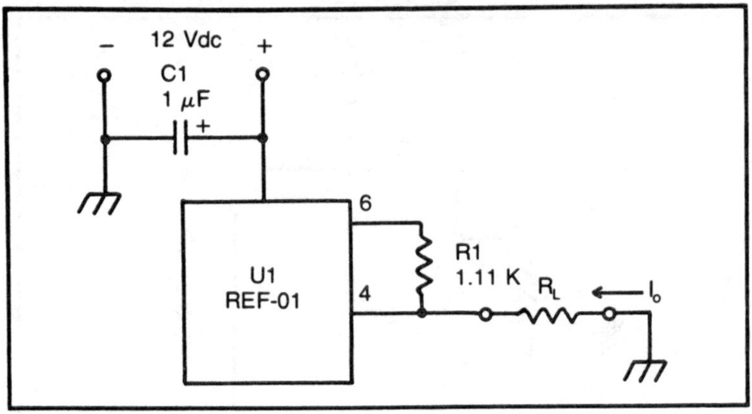

Fig. 4-7. Adjustable CCS.

current between the negative terminal of the REF-01 and the ground side of the dc power supply. The output current will be equal to:

$$I_o = (1\ \text{mA}) + (10/\text{R1})$$

As you can see, for wide variations of output load resistance R_L, the output current will be constant. This current is proportional to the bias resistor R1, but not to the load resistance.

Note that, as R1 gets very large, the output current produced by this power supply will converge on 1 mA.

TRANSDUCER EXCITATION POWER SUPPLY

Many forms of electronic measurement depends upon the use of a resistance that varies with some external parameter. The *thermistor*, for example, is a device which exhibits resistance change with changes in temperature. Under the right circumstances, the thermistor can be used for measurement of temperature. For this, a fixed reference voltage is required. Other forms of transducers use a piezoresistive strain gauge to measure factors such as displacement, force, pressure, and so forth. Vibration transducers are often strain gauges. In most cases, the strain gauges are made elements of a Wheatstone-bridge circuit (of which, more in a later section of this book). Piezoresistivity, incidentally, is the change of resistance with mechanical deformation. Piezoresistive strain gauge transducers require a fixed reference potential for proper operation.

Most Wheatstone-bridge transducers have a sensitivity specification of the form: N *μV/V/U*, where N is a number, *μV* is output microvolts, *V* is the applied excitation voltage and *U* is the applied stimulus. For example, a popular human blood pressure transducer used in hospital monitoring of arterial pressure has a specification of 50 μV/V/mm Hg of pressure. Thus, with the usual 7.5 volts applied as the excitation potential, we will see a voltage as follows when the pressure is 120 mm Hg:

$$V_o = \frac{50\ \mu V}{V^*mm\ Hg} \times 7.5\ V \times 120\ mm\ Hg$$

$$V_o = (50\ \mu V)(7.5)(120)$$

$$V_o = 45{,}000\ \mu V = mV = 0.045\ V$$

In the circuit of Fig. 4-8 we see a power supply which can be used for excitation of any resistive Wheatstone-bridge transducer up a potential of 7.5 volts. In this circuit, there are two trimmable, switch-selectable potentials of 5.0 and 7.5 volts. Since this transducer can provide up to 500 mA when the LM-317H device is used, the circuit will power any transducer with a looking-back resistance of 10 ohms or more (@ 5 volts), or 15 ohms or more (@7.5 volts). If you use a transducer that is normally excited with direct current

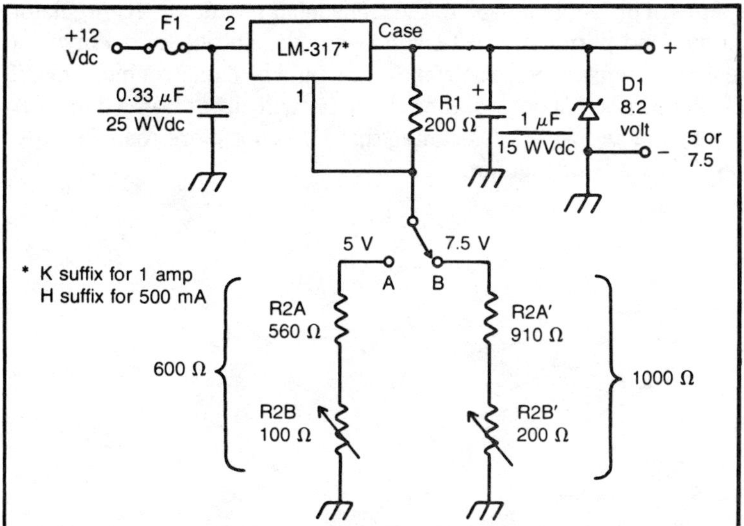

Fig. 4-8. Transducer excitation power supply.

(dc), then this supply is a candidate to relieve the headache of messing with batteries all of the time. A remarkable number of scientific and engineering personnel who use certain forms of transducers put up with the problem of battery care too long for the lack of a simple power supply such as this. Originally, this supply was used with a *Grass* FT-3 force-displacement transducer (another project on that transducer is shown later in this book).

The heart of this circuit is the LM-317 adjustable IC voltage regulator. This same IC was used in some of the power supply projects earlier in this book. The device will produce an output voltage equal to:

$$V_o = (1.25\ V)\left[\frac{R1}{R2A + R2B} + 1\right]$$

(*note*: R2A′ is used in place of R2A, and R2B′ is used in place of R2B when the switch is in the "B" position).

In this project, we use a switch, S1, to select from two different resistor banks. When the switch is in position "A," then the output voltage will be 5.0 volts dc, while position "B" yields 7.5 volts dc. You may also elect to replace one or both R2 resistor networks with a 1-kohm potentiometer that will allow you to produce any set voltage between 1.25 volts and 7.5 volts.

The zener diode across the output of this power supply is used mainly to prevent damage to the transducer should the IC regulator go bad and permit the +12 volts onto the output line. For most resistive transducers, such an event would be catastrophic. This, if a fault such as that occurs, then the zener will either clamp it to +8.2 volts or blow fuse *F*1 (assuming that F1 is not more than 750 mA, and D1 is a 5-watt zener.

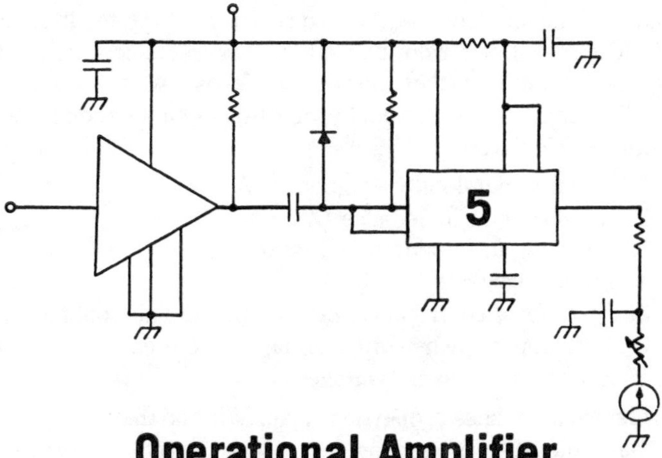

5

Operational Amplifier Circuits: The Basics

The design of amplifiers for analog signals was once the province of highly skilled electronic engineers—none other (save a few highly motivated amateur hobbyists who missed their calling) possessed the requisite skills for making an amplifier work properly. I recall a physics professor who had a B.S.E.E. (electrical engineering) in addition to his M.Sc. and Ph.D. in physics. One day I walked into his laboratory and he was tinkering with a hugish breadboard loaded with transistors and resistors. He was trying to make a gain-of-100 amplifier to process the 400 Hz signal that would be produced by some experiment his freshman students were required to perform. He had little trouble making the amplifier work, but was having a dickens of a time getting precisely ×100 gain with uniform results. With that, I introduced him to a little IC device that makes such design chores little more than trivial: the *operational amplifier.*

The operational amplifier (also known as the "op amp") is truly ubiquitous. It has found a niche in modern analog electronic design like no other component. It is so universal, that it is found in products throughout all of industry. Originally, the operational amplifier was intended to perform mathematical operations on analog signals in, of all things, analog computers. Today, the op amp is used in virtually everything except analog computers (which are no longer popular)!

One reason why the operational amplifier is so darn popular, and so very easy to "tame," is that it allows gains to be set merely through the manipulation of the two resistors of the feedback network. The actual gain from dc to the upper frequency limit of the operational amplifier is merely the ratio of those two resistors! This ability is a function of the ideal properties of an ideal operational amplifier, which are:

1. Infinite open-loop voltage gain ($A_{vol} = \infty$)
2. Infinite input impedance ($Z_{in} = \infty$)
3. Zero output impedance ($Z_o = 0$)
4. Infinite bandwidth
5. Zero noise contribution by the operational amplifier
6. Both inputs (when differential) stick together, i.e., are always at the same voltage

Just what do these properties mean? What do they imply? What are the achievable properties in real-life, practical operational amplifiers (in other words, how do real op amps compare with the ideal op amp of textbook discussions?). Although the real operational amplifier will not approach either infinite or zero on any of these properties, the approximations are good enough to be exciting, and permit performance of the operational amplifier very nearly that predicted for the ideal case. For example, infinite open-loop gain, which is the gain when the feedback resistors (or other components in the feedback network) are disconnected, can approach 1,000,000 in relatively low-cost devices and will be at least 50,000 in even garbage-grade devices. Zero output impedance is approximated by real output impedances of 50 to 200 ohms, with most running around 75 ohms. Of course, for an op-amp output terminal to approach being a perfect voltage source, it is necessary to have an output impedance close to zero relative to the load impedances it drives.

Let's consider property no. 2: infinite input impedance. What does this mean? The input impedance of any amplifier is merely $Z = V_{in}/I_{in}$. For this value to approach what we pretend to call infinity, the value of I_{in} must approach zero. This requirement means that the input can *neither sink nor source any current* (very difficult when there are transistors inside to bias!). This assumption is very important to our analysis, and will in large part determine how any given operational amplifier departs from ideal behavior. Almost all op amps can offer an input impedance of 1,000,000 ohms or more, while those with JFET or MOSFET transistors (BiFET and BiMOS, respectively) for the input circuit will offer the best compromise, with 10^{12} ohms of input impedance. These devices come very near the ideal, and can be used for such exotic high impedance applications as biophysical amplifiers, electrotet amplifiers, electrometers, and so forth.

Zero noise contribution is one of those ideal properties which real op amps rarely even approximate. As long as we are dealing with relatively high-level signals, i.e., 500 μV or more, this is not usually a problem. But, as we deal with smaller and smaller signals the noise contribution becomes extremely important. If you have a need for either very low noise, or, very low drift, then there are special premium-grade devices on the market which will serve the function. The Burr-Brown Corporation makes some of these, and their catalog will be worthwhile reading for most readers who need that type of product.

Infinite bandwidth is also not met by the real operational amplifier. In at least one respect, we might not want such wide bandwidth because infinite bandwidth also means infinite amplification of noise! Also, wide bandwidth means the operational amplifier stands a chance of oscillating at any frequency where the internal phase shifts occur and the gain is more than unity. Most operational amplifiers have a gain-bandwidth product considerably less than infinite—the high audio range being common. Operational amplifiers which are advertised as being unconditionally stable (e.g., the 741 device) get that way by beating the dickens out of the frequency response. The 741, in return for its unusual stability, will not amplify at frequencies more than a few kilohertz. The 741 device can be used for communications-grade speech amplification (3000 Hz upper limit on frequency), but not for high-fidelity amplification (15,000 Hz upper limit on frequency).

Property no. 6 is the one which will be used most in our analysis of how operational amplifier circuits work. The typical operational amplifier has *differential inputs.* Such inputs produce equal but opposite effects on the output terminals. Thus, if 1 volt is applied to both the inverting input and to the noninverting input, then the two output contributions cancel each other out and the output voltage is zero. This phenomenon will prove very useful for eliminating certain forms of interference (e.g., 60 Hz) or other forms of unwanted signal. An implication of property no. 6 is that the voltages on these two inputs are always the same. If you apply +1 volt to, say, the inverting input, then you must treat the noninverting input as if 1 volt were also applied to it. This statement is not merely some theoretical device to make our equations come out nicely, but is an observable fact! If you take a voltmeter and measure the voltage on one input when a potential is applied to the other, then you will measure the same voltage on *both* input terminals! We shall use that bit of information later.

Figure 5-1 shows a simplified schematic of a typical bipolar transistor input stage for an operational amplifier. This circuit uses a differential input scheme, and is very nearly what is found inside many op-amp ICs. Of course, there will be additional stages in cascade with this type of stage in order to achieve the gain and stability required.

Current *I*3 is from a constant current source not unlike the circuit shown earlier in this section. The current I3 will not change despite changes of applied voltages (V− and V+) or changes in the load impedance presented by transistors Q1 and Q2. Since current I3 produces the emitter-collector currents of Q1 and Q2, we can conclude:

$$I3 = I1 + I2$$

When the voltages applied to the bases of Q1 and Q1 (which

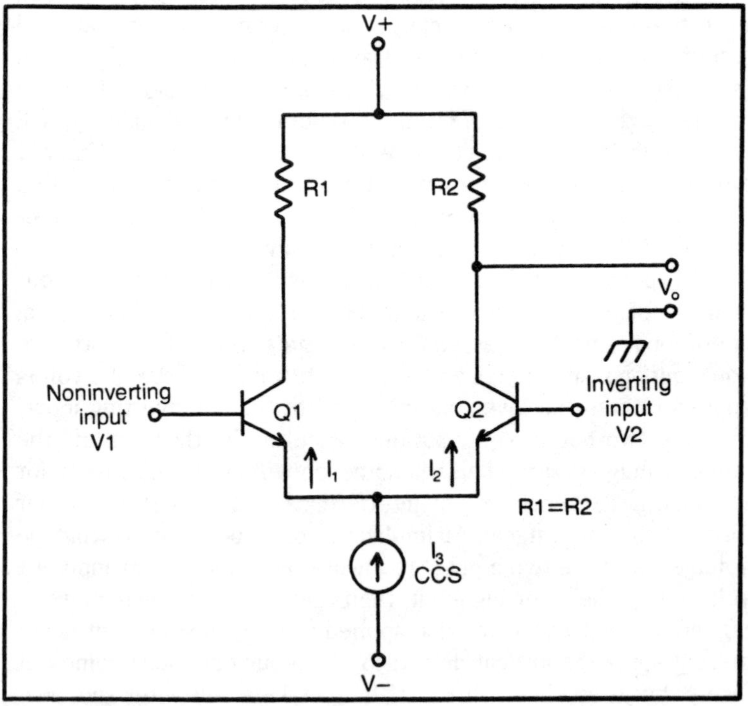

Fig. 5-1. Differential amplifier input stage.

form the noninverting and inverting inputs of the op amp, respectively), are equal, then

$$I1 = I2$$

Under that circumstance, the voltage drops across resistors R1 and R2 are equal (assuming R1 = R2). Output voltage V_o is the difference between the applied input voltage ((V+) − (V−)) and the drop across R2 (I_2R_2). Thus, when V1 = V2, we find that V_o is zero. But, when there is a differential voltage applied to these inputs (i.e., V1 = V2), then V_o will be nonzero. Let's consider two other scenarios: V1 > V2, and V1 < V2.

When V1 is greater than V2, we are in essence applying a positive voltage to the noninverting input. Let's see what happens. Applying a positive voltage to the base of Q1 will cause current I1 to increase. But since I3 = I1 + I2, and I3 is constant, this can only result in reduction of I2. Since I2 is now reduced, the voltage drop I_2R_2 goes down thereby permitting V_o to rise. Since a positive-going input voltage on the base of Q1 permits a positive-going output voltage, the base of Q1 is a noninverting input (i.e., there is no phase reversal). Similarly, applying a positive voltage to the base of Q2 is the same as saying V1 is less than V2. In this case, I2 is increased, so the voltage drop I_2R_2 is also increased. This makes the voltage V_o go negative. Thus, the base of Q2 is an inverting input.

Figure 5-2 shows the two forms of circuit symbol normally used for operational amplifiers. The straight-backed version of Fig. 5-2A is the one normally used by most users, and that which will be used in this book. But some users, notably the American Radio Relay League (a ham radio group) and Burr-Brown Corporation, use the version shown in Fig. 5-2B. The inverting input is marked "−" in both versions in order to signify that a 180-degree phase reversal takes place between this input and the output terminal. The noninverting input is marked "+" for exactly the opposite reason: there is no phase reversal between this terminal and the output terminal.

Note that some ambiguity may be caused by using the symbol of Fig. 5-2A, even though its use is all but universal. All other forms of IC amplifiers (except the Norton CDA amplifier) use exactly this same symbol, and may even have the "−" and "+" symbols if their inputs are differential.

The typical bipolar dc power supply used for operational amplifiers is shown in Fig. 5-3. Although batteries are shown here (and indeed are used for many operational amplifier instruments), elec-

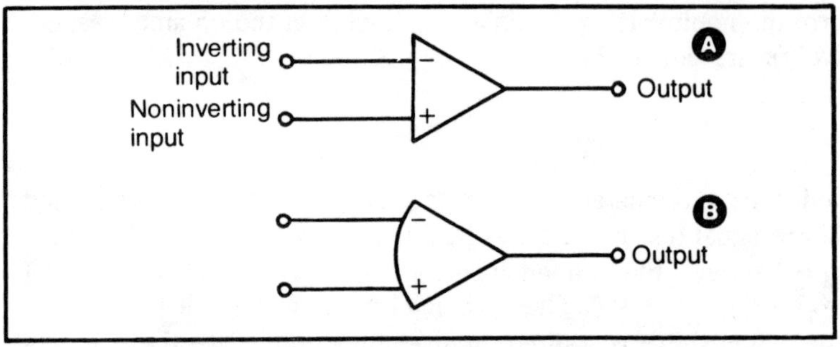

Fig. 5-2. Operational amplifier circuit symbol A) usual symbol, B) alternate symbol.

tronic power supplies can be substituted for those who want to operate from the 110-volt ac power mains. The *V+* terminal is positive with respect to ground, while the *V–* terminal is negative with respect to ground. These are two separate power supplies, and are *not* two different (positive and negative) ends of the same power supply. This is shown by the fact that the common/ground terminal is at the junction of batteries (or power supplies) B1 and B2.

Most operational amplifiers have a large degree of immunity from changes in the power supply potentials. Regardless, however, it is recommended that operational amplifier projects be operated from regulated dc power supplies. Most operational amplifiers will operate over the range ±4.5 volts to ±18 volts, with but one limitation. The limitation is that the differential power supply voltage, (V+) - (V–) must not exceed some magical number. For the 741 device, this power supply differential voltage is 30 volts. Thus, if we use ±15-volt power supplies, we just meet the specification. If we make one power supply greater than 15 volts, how-

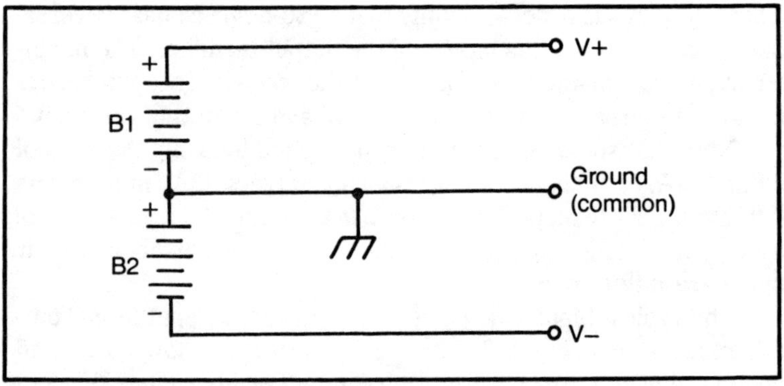

Fig. 5-3. Operational amplifier power supply arrangement.

ever, we must reduce the other power supply voltage proportionally. For example, if V+ is 18 volts dc, then V− must be no larger than (−30)+(18 V), or −12 volts. Three-terminal IC voltage regulators can be used for this purpose.

The power supplies should be bypassed to ground for decoupling purposes, especially if ac or varying dc signals are being used. There are usually 4.7 to 10 μF capacitors and 0.1 μF capacitors at each operational amplifier terminal, and up to 100 μF at each power supply terminal. In many cases, however, we find that the 0.1 μF capacitors are sufficient.

INVERTING FOLLOWERS

There are two basic configurations for operational amplifiers: inverting follower and noninverting follower. These designations refer to which of the two differential inputs are used, as well as the relationship of the output signal phase to the input signal phase. In the inverting follower configuration, signal is applied to the inverting input and the output signal will be 180 degrees out of phase with the input signal. Figure 5-4 shows an example of a typical inverting follower circuit.

The inverting follower shown in Fig. 5-4 applies a feedback resistor from the output to the inverting input, and applies the signal to the "free" end of the input resistor. The noninverting input is grounded. According to ideal operational amplifier property no. 6, the inverting input is also at ground potential—a fact that sometimes leads to a bit of confusing terminology, i.e., the concept of a "virtual ground." Such a ground is a point which is at ground potential, and must be treated both mathematically and physically

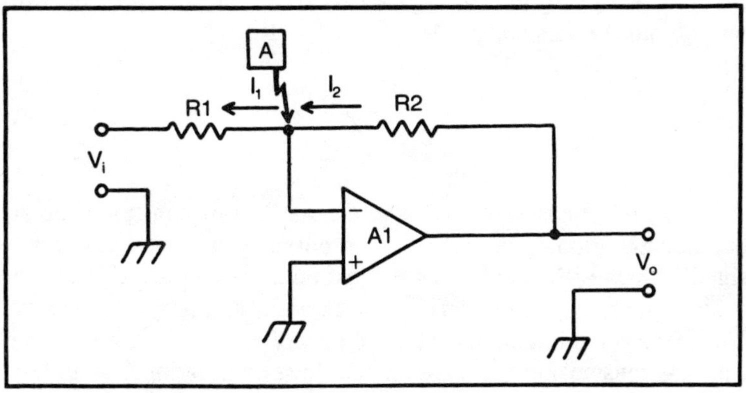

Fig. 5-4. Inverting follower amplifier.

as if it were at ground potential, yet is not connected directly to ground. Since ground potential is zero volts, we find that point "A" in Fig. 5-4 is at a potential of zero volts. In the analysis which follows, our arithmetic will revolve around considering point "A" as if it were actually (instead of "virtually") grounded.

Since one property of the operational amplifier is that the inputs neither sink nor source current, the only currents in the circuit of Fig. 5-4 are those shown, I1 and I2. Current I1 is created when the input voltage V_{in} is applied to the input resistor, I1. Current I2 is generated from the output voltage being applied to the feedback resistor.

From Kirchhoff's current law (KCL) we know that the currents into and out of a junction must sum to zero. Thus, we can deduce that currents I1 and I2, being the only currents at summing junction "A," must be equal:

$$I1 = -I2 \qquad \textbf{(a)}$$

From our discussion above, we know that the input current, I1, is created by the action of input voltage V_{in} on resistor R1. In other words:

$$I1 = V_i/R1 \qquad \textbf{(b)}$$

Also, from above we know that current I2 is created by the action of the output voltage on feedback resistor R2:

$$I2 = V_o/R2 \qquad \textbf{(c)}$$

If we substitute Equations (b) and (c) into Equation (a), we will obtain the following:

$$\frac{V_i}{R1} = \frac{-V_o}{R2} \qquad \textbf{(d)}$$

Now let's find the ***transfer equation*** for the inverting follower. A transfer equation is the ratio of the output signal over the input signal, which in the case of a voltage amplifier is V_o/V_i. This figure is also called the ***voltage gain*** of the system, and is symbolized by "A." If we solve Equation (d) for the ratio V_o/V_i, then we obtain the gain expression for the inverting follower operational amplifier configuration:

$$\frac{V_o}{V_i} = \frac{-R2}{R1} \quad \textbf{(e)}$$

Since the ratio V_o/V_i is also described by the symbol A_v, we sometimes see Equation (e) written in the form:

$$A_v = -R2/R1 \quad \textbf{(f)}$$

Another form of Equation (e) is the expression which relates output voltage to both the input voltage and the gain; in other words:

$$V_o = -V_i R2/R1 \quad \textbf{(g)}$$

or,

$$V_o = -V_i A_v \quad \textbf{(h)}$$

An implication of the above discussion which makes it so easy for nonengineers to easily design simple amplifier circuits is that the voltage gain over the range dc to the upper frequency limit of the amplifier is controlled merely by the ratio of two resistors: R2/R1 (note, the "–" sign in the equations merely indicates the output voltage is 180 degrees out of phase with the input voltage). In designing circuits around Fig. 5-4 we would take into consideration both the required gain and the minimum required input impedance. In order to not load the preceding circuit, the input resistance must be not less than 10 times the source impedance of the driving circuit. Thus, if a preceding circuit has an output impedance of 100 ohms, then the minimum permitted input impedance for Fig. 5-4 would be (10)(100 ohms) = 1000 ohms; even greater impedance is desirable. In Fig. 5-4, with "A" at virtual ground, we find that the input impedance of the circuit is the value of R1.

Example:

We require an inverting amplifier with an input impedance of 1000 ohms or more, and a gain of ×100. Find the correct values for R1 and R2.

1. Since the minimum acceptable input impedance is 1000 ohms, we are permitted to set R1 = 1000 ohms.
2. Plugging this value into Equation (f), we obtain the following:

$$A_v = \frac{-R2}{R1} \tag{i}$$

$$-100 = \frac{R2}{1000} \tag{j}$$

Rearranging algebraically:

$$R2 = (100)(1000 \text{ ohms}) \tag{k}$$
$$R2 = 100{,}000 \text{ ohms} = 100 \text{ kohms} \tag{l}$$

We will obtain a gain of 100 and a minimum input impedance of 1000 ohms with the following values of components: R1 = kohm, R2 = 100 kohms.

The limitation on the use of the inverting follower is often a matter of input impedance. When gains get high, then it is often difficult to obtain the input impedances needed with the values of resistors commonly available. There are also problems with real operational amplifiers which often make the use of high value resistors not advisable (more of that in a few moments). The common 741 series operational amplifiers, for example, often misbehave a little bit when resistors over 1 to 3 megohms are used in the feedback network. Using a BiMOS device, such as the RCA CA3140 device, solves that problem, but that in itself is not always advisable.

A solution to the input impedance problem is the noninverting follower circuits of Figs. 5-5 and 5-6.

NONINVERTING FOLLOWERS

The noninverting follower circuit uses the noninverting input

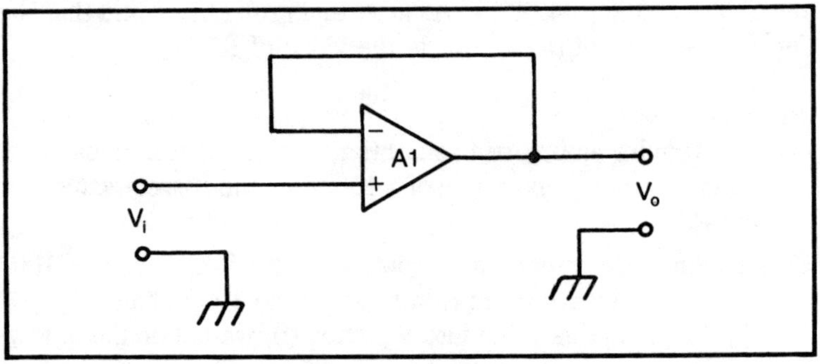

Fig. 5-5. Unity gain noninverting follower amplifier.

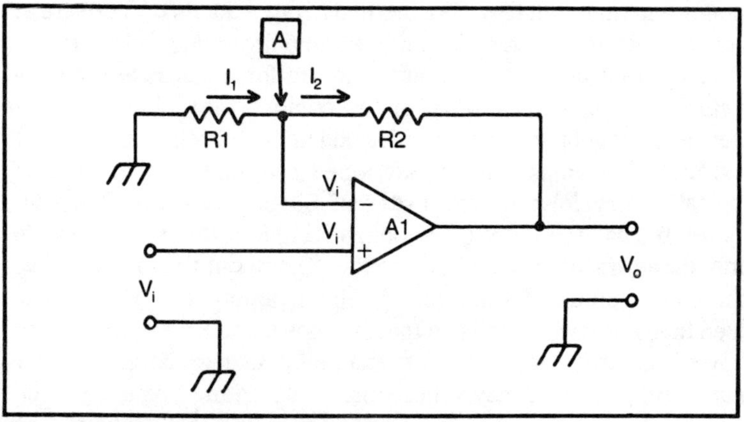

Fig. 5-6. Noninverting follower with gain circuit.

of the operational amplifier, so the output voltage will be in-phase with the input signal. There are two configurations for dc noninverting follower amplifiers: *unity gain* and *gain*. The unity gain follower is shown schematically in Fig. 5-5. In this circuit, the input signal is applied directly to the noninverting input, while the output signal is fed back directly to the inverting input without an attenuation network (i.e., as in R1 and R2 in the previous example). The result of this is a gain that is very nearly unity (1). Let's assume that the open-loop gain (A_{vol}) of the operational amplifier is 1,000,000. Using the normal equation for feedback amplifiers of any type we can see the effect of this circuit. If A_{vol} is 1,000,000 and B is 1, the equation solves as follows:

$$A_v = \frac{A_{vol}}{1 + (A_{vol}B)} \tag{m}$$

$$A_v = \frac{10^6}{1 + (10^6 \times 1)} \tag{n}$$

$$A_v = 1{,}000{,}000/1{,}000{,}001 \tag{o}$$

$$A_v = 0.999999 \tag{p}$$

I suspect that most people would consider 0.999999 close enough to unity for all practical purposes!

So what's the use of a gain-of-one amplifier that doesn't even invert the signal. On first blush that might seem to be a waste of money and time, but there is a method to our madness. There are at *least* two good uses for the unity gain noninverting follower: impedance transformation and buffering. As for impedance transformation, the input impedance of the noninverting follower is immense; 1 megohm on the cheapies and up to 10^{12} ohms for BiMOS and BiFET op amps. The output impedance, on the other hand, is typically very low; on the order of 50 to 200 ohms. We can, therefore, use the unity-gain-noninverting follower as a stepdown impedance transformer (high-Z to low-Z) without the loss of either phase or amplitude information. A "buffer" amplifier is one which is used to isolate two circuits so that changes in the load circuit cannot effect the driver circuit. For example, an oscillator or other waveform generator may suffer shifts in operating frequency if the load impedance changes. In order to overcome that problem, the designer will often place a unity-gain-noninverting-follower between the output of the oscillator and its load. The input impedance of the operational amplifier will only lightly load the oscillator and will not change with changes of output load impedance.

The other configuration is the noninverting-follower-with-gain circuit shown in Fig. 5-6. In this circuit, we have replaced the direct feedback path of Fig. 5-5 with an attenuator network similar to that used in the inverting follower of Fig. 5-4. In this case, however, the "free" end of the input resistor is grounded. The noninverting input is at a potential V_i, so we find that V_i also exists at the inverting input (i.e., point "A"). Using an analysis similar to that used for the inverting follower, with appropriate changes due to the noninverting configuration will yield the transfer function for the noninverting operational amplifier follower circuit.

From KCL, we know that the two currents in Fig. 5-6 must be equal and in the same direction:

$$I1 = I2 \qquad \textbf{(q)}$$

The "input" current, I1, is created by the potential at summing junction "A," which is V_i, on resistor R1:

$$I1 = V_i/R1 \qquad \textbf{(r)}$$

The feedback current I2 is created by the action of the potential difference between the output and the inverting input on feedback resistor R2:

$$I2 = (V_o - V_i)/R2 \qquad \textbf{(s)}$$

Substituting (s) and (r) into (q) yields:

$$\frac{V_i}{R1} = \frac{(V_o - V_i)}{R2} \qquad \textbf{(t)}$$

Solving Equation (t) for the transfer equation, V_o/V_i, produces:

multiply both sides by R2

$$\frac{V_i R2}{R1} = \frac{(V_o - V_i)}{R2} \times R2 \qquad \textbf{(u)}$$

$$\frac{V_i R2}{R1} = V_o - V_i \qquad \textbf{(v)}$$

Add V_i to both sides

$$\frac{V_i R2}{R1} = V_o - V_i \qquad \textbf{(v)}$$

$$\frac{V_i R2}{R1} + V_i = V_o \qquad \textbf{(w)}$$

factor out V_i

$$V_o = V_i \times \left[\frac{R2}{R1} + 1\right] \qquad \textbf{(x)}$$

Equation (x), therefore, is the expression which relates the output voltage to the input voltage and the gain of the amplifier. This is, of course, a form of the transfer function. The gain of this circuit, A_v, is the portion of the expression:

$$A_v = \frac{R2}{R1} + 1 \qquad \textbf{(y)}$$

Let's take a look at Equation (y) and see some of the implications of that expression. For one thing, like noninverting follower circuit, the voltage gain of the amplifier is set by the ratio of two resistors, the feedback resistor (R2) and the input resistor (R1). Furthermore, there is a limit on how low the gain can become. If we make the feedback resistance equal to zero, then the gain expression resolves to $A_v = (0 + 1) = 1$. In this case, the voltage gain is unity, or the same as the noninverting unity gain follower circuit of Fig. 5-5.

We are allowed somewhat more freedom in selecting resistors for the noninverting gain follower because the input impedance of the amplifier is not related to the values of resistors used in the feedback network (i.e., R1). The input impedance will be quite large no matter what resistor values are used. Thus, we can use 100 ohms for R1, if that is what we want, thereby gaining an order of magnitude or more in the possible ranges with commonly available resistor values and within the resistance value limitations placed on the amplifier.

OPERATIONAL AMPLIFIER PROBLEMS

The practical kind-you-can-buy operational amplifier departs somewhat from the ideal operational amplifier discussed earlier in this section. We have certain problems on real operational amplifiers, and these sometimes insinuate themselves into our design and must be solved. One of the most commonly encountered is the matter of output offset voltages, i.e., voltages that are nonzero at a time when the output voltage should be zero. If we ground the signal input of the operational amplifier circuit, then the output voltage should go to zero. If there is a problem, which is frequently the case, then that voltage will not go to zero but to some specific offset voltage level (which may be either positive or negative).

One source of output offset voltage is input offset current. The input terminals of an ideal operational amplifier will neither sink nor source current, but the terminals on real op amps will provide a current because of the necessity of biasing the input transistors (a problem that tends to go away on BiMOS and BiFET operational amplifiers). If this is a problem, and it will be on 741 and related devices, then some means of nulling the resultant output offset voltage must be provided. This requirement becomes a lot more urgent in the case of a high gain circuit because the output voltage produced will be proportional to circuit gain. Thus, at a gain of ×10 the output offset may be so small as to be ignorable, but the same operational amplifier at a gain of ×100 or more will produce an output offset voltage so high that it cannot be ignored.

In the case of an output offset produced by the bias current of the input transistors we can use a single fixed-value resistor to null the output voltage. Those bias currents tend to be the same in both inputs, so will produce equal but opposite potentials at the output.

These potentials are caused by the bias current flowing in the input and feedback resistors. If these resistors have high values, then the resultant voltage drop is larger and the output offset voltage is also larger (this is the reason why some bipolar input transistor op amps cannot work well with high value resistors in the feedback network).

A solution to the problem is the use of a compensation resistor between the noninverting input and ground. Normally, the noninverting input (in an inverting follower) is grounded. When the resistor is placed in the path between the input and ground, a voltage (see Fig. 5-7) will be developed equal to $I_{bias}R3$. If we conspire to make resistor R3 equal to the parallel combination of R1 and R2, then this voltage will be equal to that applied to the inverting input from the action of $-I_{bias}$ on the inverting input. Since the two inputs will produce equal but opposite results on the output, the output offset voltage will be nulled by this process.

Some operational amplifiers have a pair of special terminals that are intended to be used to null the output offset voltage, regardless of the source of the voltage. Figure 5-8 shows the use of a typical pair of *null* terminals on an operational amplifier. The potentiometer is connected such that its ends are attached to the null terminals, and the wiper is connected to the V− power supply voltage. This potentiometer is adjusted with the input(s) grounded (i.e., $V_i = 0$) to produce a zero output voltage. Not all operational amplifiers have null terminals, or, the null terminals may be used for some other purpose. In those cases, we will use another tactic.

The other tactic is shown in Fig. 5-9. Here we see a third current applied to the summing junction (point "A") that has a

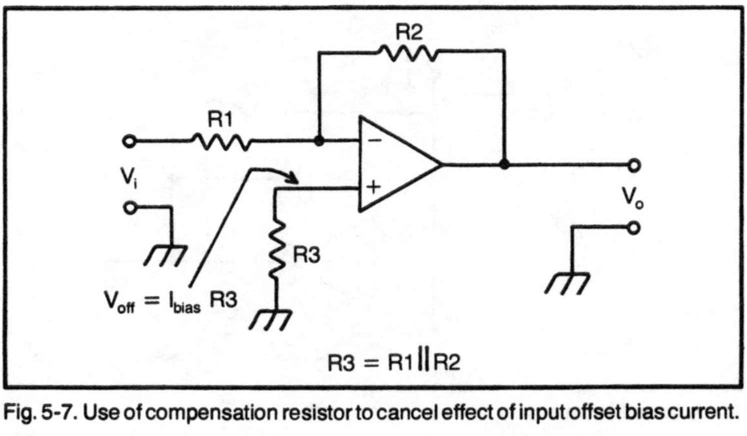

Fig. 5-7. Use of compensation resistor to cancel effect of input offset bias current.

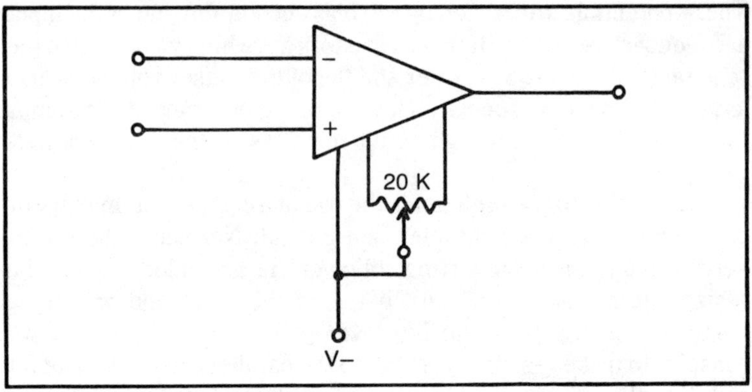

Fig. 5-8. Use of the offset null terminals.

magnitude and polarity to cancel the output offset voltage. Current I3 is created by the voltage at point "B," which is created by potentiometer R4. This potentiometer is adjusted to produce a zero output potential when the signal inputs are grounded.

The capacitor in Fig. 5-9 is used to shape the frequency response of the circuit. This capacitor reduces the feedback at frequencies above the −3 dB point, as defined by:

$$F_{Hz} = \frac{1}{2 \pi R_2 C_1} \qquad \textbf{(z)}$$

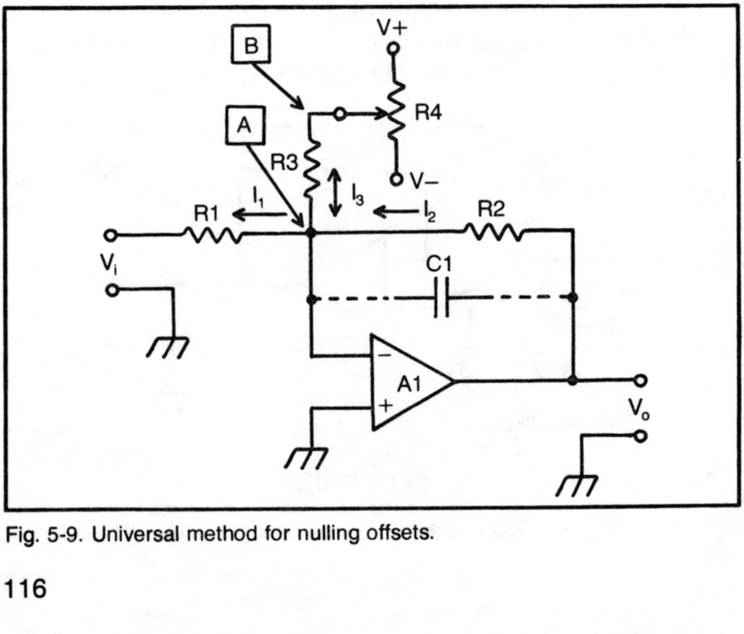

Fig. 5-9. Universal method for nulling offsets.

Where:

F_{Hz} is the frequency in hertz
R_2 is the resistance of R2 in ohms
C_1 is the capacitance of C1 in farads (note: $1\ F = 10^6\ \mu F$)

The frequency response of the amplifier circuit will fall off at a rate of about −6 dB/octave above the frequency given in Equation (z).

There are also other means of reducing the frequency response of operational amplifier circuits. Some operational amplifiers have *lag* or *lead* terminals which can be used for this purpose. But these methods are unique to the type of op amp in use.

If an operational amplifier is described as "unconditionally stable," then it has a greatly reduced frequency response. The factor which causes the response to fall off is that internal phase shifts (due to input capacitance and resistances) will add to the 180-degree phase shift inherent in the inverting configuration to produce a total phase shift of 360 degrees at a frequency where the loop gain of the circuit is unity or greater. These are the main criteria for any oscillator: the op amp will oscillate madly. In those cases, we will use one or the other technique to reduce the gain to below unity at that frequency.

DC DIFFERENTIAL AMPLIFIERS

A *differential amplifier* is one in which the output is proportional to the gain of the amplifier and the *difference* between two input voltages, *V*1 and *V*2. The operational amplifier (usually) has differential inputs, so can easily be used to form various types of differential amplifier circuits. Since these devices are capable of operating from dc to their upper frequency limit, the term for these circuits is dc differential amplifiers."

Figure 5-10 shows a basic dc differential amplifier based on the use of a single operational amplifier device. Note that both inputs are used. There are two input resistors (R1 and R2) and two "feedback" resistors (R3 and R4), although one might not be recognized as such because it is grounded (R4). If the circuit is balanced, i.e., R1 = R2 and R3 = R4, then the gain of the circuit will be given by:

$$A_v = R3/R1 \qquad \textbf{(aa)}$$

or,

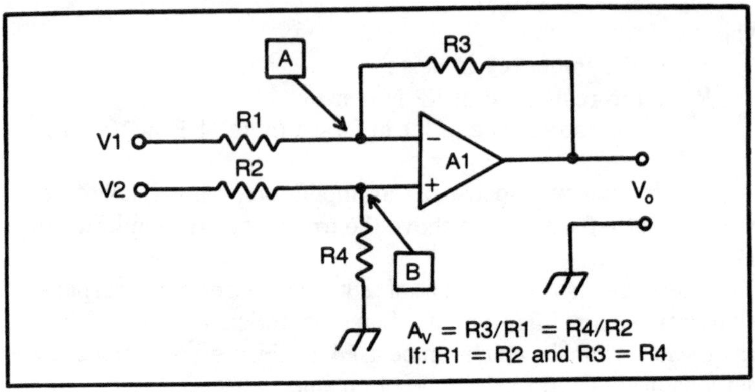

Fig. 5-10. Dc differential amplifier.

$$A_v = R4/R2 \tag{bb}$$

The input to this circuit is differential, while the output is single-ended. Thus, the circuit can be used to interface inherently differential sources (such as a Wheatstone-bridge transducer) to single-ended devices.

The input resistance of this circuit is limited by the values of R1 and R2. If this impedance is too low for some applications (it may well be), then it may be preferred to use some other form of differential amplifier circuit, e.g., one in which a noninverting follower is used as the input stages. One candidate is the instrumentation amplifier of Fig. 5-11.

The classic instrumentation amplifier (Fig. 5-11) uses three operational amplifiers. Two of these operational amplifiers, A1 and A2, should be part of a dual operational amplifier package, such as the RCA CA3240. The idea in that specification is to keep the two input amplifiers in the same thermal environment, thus reducing drift.

There are two stages to this amplifier, and they each have their own gain expression. When combined, the two amplifier stages have a gain equal to the product of the individual gains:

$$A_v = \left[\frac{2R_2}{R_1} + 1\right]\left[\frac{R_6}{R_4}\right] \tag{cc}$$

Where:

A_v is the differential gain of the amplifier
R_1 is the resistance of R1

R_2 is the resistance of R2
R_4 is the resistance of R4
R_6 is the resistance of R6

Note: All resistances must be in the same unit, i.e., ohms, kohms, or megohms.

Provided that: R2 = R3, R4 = R5 and R6 = R7. It is interesting to note that resistor R1 can be used as a gain control, provided that its value does not get too low. For this application, many designers will use a potentiometer in series with a fixed resistor for R1. Thus, there will be a minimum value for R1 that limits gain to some practical value.

Common Mode Rejection. A *common mode* signal is one which affects both inputs equally. When V1 = V2, then the net output should be zero. This is due to the fact that each input provides the same amount of gain, yet are 180 degrees out of phase with each other *viz.* the output terminal. Unfortunately, the suppression of common mode signals at the output is not total and complete: there is always some output potential caused by common mode signals. The ratio of differential gain to common mode gain is a measure of the differential amplifier's ability to reject common mode signals. This phenomenon is called the *common mode rejection ratio,* or CMRR.

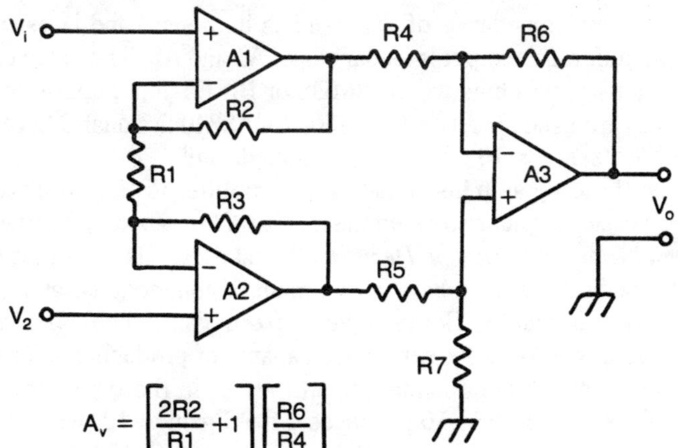

$$A_v = \left[\frac{2R2}{R1} + 1\right]\left[\frac{R6}{R4}\right]$$

If:

R1=R2
R4=R5
R5=R7

Fig. 5-11. Instrumentation amplifier circuit.

Common mode rejection ratios of all operational amplifiers are at least 60 dB, while most good quality devices offer 100 to 120 dB of CMRR. This figure can be degraded, however, by external circuitry. If the balance of the resistors is not maintained, for example, there will be a degradation. Oddly, in the instrumentation amplifier of Fig. 5-11 we find that the CMRR is not seriously degraded by imbalances between R2 and R3 (although a gain error will occur in that case!). The CMRR of Fig. 5-11 will be degraded, however, by imbalances in the second stage (R4 = R5 and R6 = R7). In many circuits, the designer will make R7 variable so that the CMRR can be adjusted to optimum value. In many such circuits, resistor R7 will be a series combination of a fixed resistor and a potentiometer; the value of the potentiometer is usually about 10 percent of the total combination resistance, R7. Adjustment of this CMRR control is simple:

1. Short together the two input terminals, and then connect them to the output of a signal generator. Set the generator output to about 1 volt.
2. Observe the output signal on an oscilloscope.
3. Adjust the CMRR control for minimum output signal. Keep increasing the sensitivity of the oscilloscope as the output gets lower, and continue to do so until no further improvement is possible.

The input impedance of Fig. 5-11 is immense, and is essentially the differential input impedances of A1 and A2. The value can be as much as 10^{12} ohms when BiMOS or BiFET input operational amplifiers are used for A1/A2. The RCA CA3240 is a dual CA3140, which boasts 1.5×10^{12} ohms input impedance!

The three-op-amp instrumentation amplifier is often replaced with IC versions that are now on the market. *Burr-Brown, National Semiconductor* and *Analog Devices* all make IC instrumentation amplifiers. Such devices have a minimum of connections, and are easily programmable. Some have a fixed gain. The National Semiconductor LM-363-x series are capable of producing gains of ×10, ×100, or ×500 depending upon the "x" in the type number suffix. Thus, an LM-363-10 produces a fixed gain of 10, while the LM-363-100 produces a fixed differential gain of 100. Another device in the LM-373 family is the LM-363-AD device, which produces fixed gains of ×10, ×100, or ×1000, depending upon which of three external connections are shorted together. Thus, a three-position, single pole switch can select three different gains (10, 100, and 1000). The Burr-Brown INA-101 is a dc differential instrumentation amplifier producing a gain of ((40/R)+1), where *R* is an external programming resistor (and corresponds to R1 in Fig. 5-11, where R2 = R3 = 20 kohms).

DIFFERENTIAL OUTPUT AMPLIFIERS

There are occasional uses for an amplifier which has a differential, or "balanced," output. Most audio or public address applications can benefit from this type of circuit. We can use a transformer with a 10-kohm primary and a 200- to 600-ohm secondary winding in order to provide the needed balanced output. Unfortunately, transformers are bulky and (if you want wide audio bandwidth with minimum phase shift) expensive. An operational amplifier circuit, without a transformer, is shown in Fig. 5-12. This circuit uses a pair of operational amplifiers, which should, incidentally, be part of a dual, triple, or quad op-amp package.

The circuit operates on the theory that the phase shift time difference between A1 and A2 is very small, compared with the period of the audio signals being processed. Thus, we can use inverting amplifier A1 to reamplify the output of noninverting amplifier A2. If R3 = R4, then the gain of A1 is −1, while the gain of A2 is +1, thus we have a differential output with a gain of unity. For most such amplifiers, both R3 and R4 are 10 kohms.

The output impedance of the circuit is approximately 600 ohms when the output impedances of A1 and A2 are added to the resistances of R1 and R2. This value is consistent with the 600 ohm line

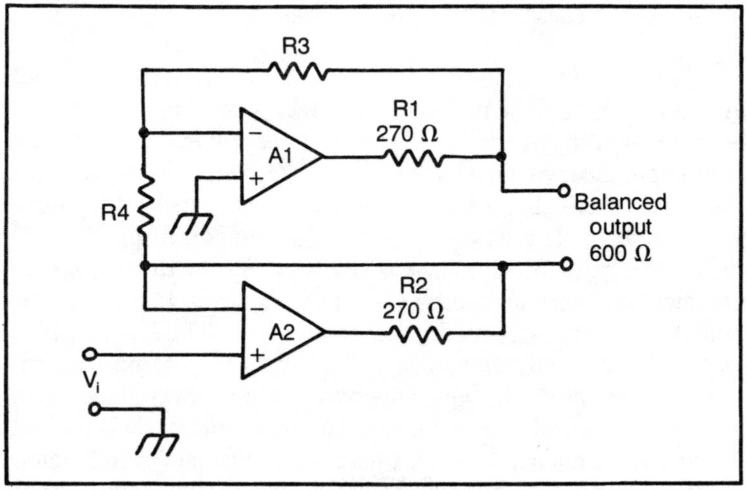

Fig. 5-12. One form of balanced output amplifier.

typically used in audio applications of public address and broadcasting work.

AC AMPLIFIERS

Operational amplifiers work from dc to some upper frequency limit. We can make the response limited to some lower frequency above dc by using any of the tactics of Fig. 5-13. The version shown in Fig. 5-13A uses capacitor coupling of both the input and output circuits. Capacitor C1 blocks dc from the noninverting input. The purpose of R1 is to keep the input bias current (which exists on nonideal practical op amps) from charging the capacitor. If this resistor is not provided, the charge on C1 will build up and eventually (often in a very short period of time) cause the amplifier to saturate. This potential is seen by the operational amplifier as a valid input signal! The resistor continually drains this capacitor.

The output circuit contains a similar circuit, consisting of C2 and R2. This circuit is not often used, but is a definite possibility.

The frequency response of the circuit is usually set by the input network, R1/C1. The lower frequency response is governed by:

$$F_{Hz(1)} = \frac{1}{2\pi R_1 C_1} \qquad \textbf{(dd)}$$

Where:

$F_{Hz(1)}$ is the lower −3 dB frequency, in hertz
R_1 is the resistance of R1 in ohms
C_1 is the capacitance of C1, in farads

The input impedance of the amplifier is set by resistor R1, so it will be considerably less than is otherwise possible with a noninverting follower circuit. In most cases, the minimum acceptable input impedance will govern the value of R1, while the capacitance of C1 will be calculated from this value and the needed frequency response. In some cases, such as biological/medical applications, the low end frequency response is nearly dc. An electrocardiograph amplifier, for example, needs an input impedance of 10 megohm, or more, yet requires a frequency response down to 0.05 Hz in order to not lose critical information in the patient's waveform. The response cannot be dc, however, because there is always a dc offset potential created whenever a conductive metal electrode is applied to electrolytic human skin. This potential is typical 0.5 to 2 volts, while the amplifier gain is ×1000 or more (the human ECG

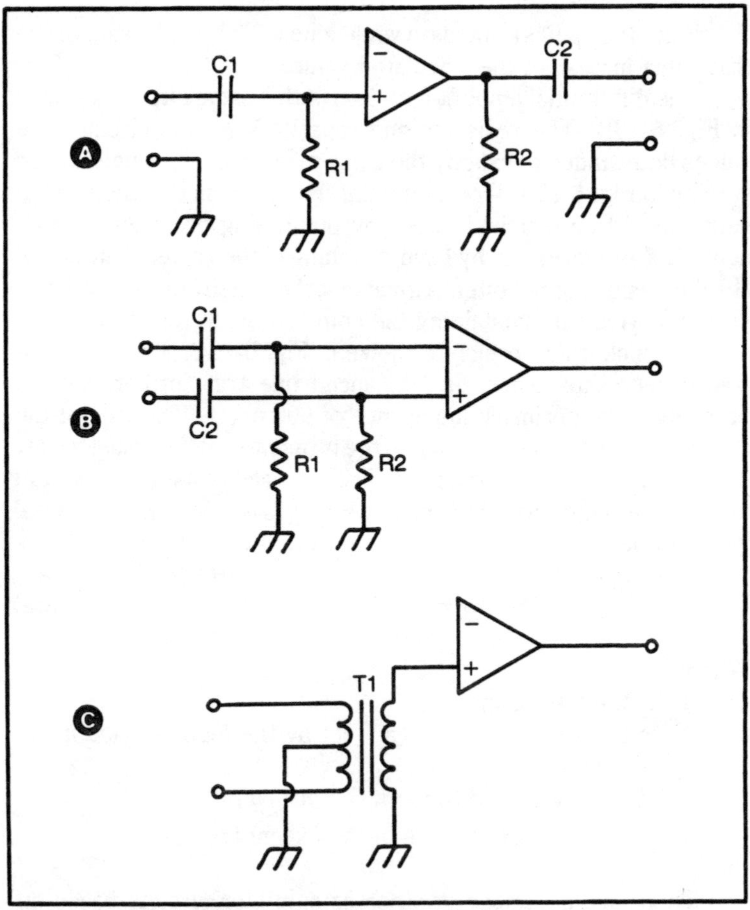

Fig. 5-13. Ac amplifier circuits A) capacitor, single-ended, B) differential capacitor, C) transformer coupled.

waveform is typically 1 mV, peak). Thus, we need a lower end frequency response of 0.05 Hz in order to capture needed data while blocking a dc error voltage. For a 10 megohm input impedance and a frequency response down to 0.05 Hz, we would solve Equation (dd) for C1:

$$C_1 = \frac{1}{(6.28)(10^7 \text{ ohms})(0.05 \text{ Hz})} \qquad \textbf{(ee)}$$

$$C_1 = 1/3,140,000 = 3.19 \times 10^{-7} \text{ farads}$$

$$= 0.32\ \mu\text{F}$$

Since 0.33 μF is a standard value, one will probably want to use that value instead of the calculated value.

The differential amplifier version of this same circuit is shown in Fig. 5-13B. The resistors and capacitors in this circuit have values determined in exactly the same manner as the single-ended version in Fig. 5-13A. One constraint, however, is that the components should be matched either by purchasing precision components (1% or more), or, by hand-matching of the values. The actual absolute value are not often critical, but the match between value is very important to maintaining the common mode rejection ratio.

A transformer version is shown in Fig. 5-13C. The particular transformer shown here is a balanced line transformer, so will probably have a primary impedance of 600 ohms. The gain of the circuit goes up by a factor equal to the turns ratio of the transformer. If N_p is the number of turns in the primary, and N_s is the number of turns in the secondary, then the turns ratio is N_p/N_s. Thus, the total gain will be:

$$A_{vt} = (A_v)(N_p/N_s) \qquad \textbf{(ee)}$$

Where:

A_{vt} is the total gain
A_v is the normal voltage gain set by the feedback and input resistors
N_p is the number of turns in the primary
N_s is the number of turns in the secondary

There are a number of other ac amplifier circuits based on operational and other linear IC amplifiers. Other TAB books discuss these circuits in greater detail, including several by this same author.

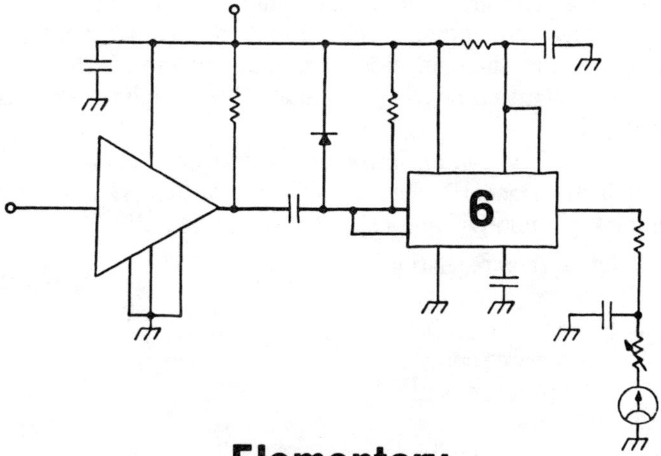

Elementary Operational Amplifier Projects

In this chapter we will provide some practical information that will help you put the theory into practice. The circuits shown in this section are derived directly from the theoretical, and are easily designed or modified to suit almost any application which you have in mind. Each discussion will be given in terms of the theory so that you can relate how we arrived at the component values used.

In some cases, there will be more than one project using the same circuit diagram. This was done to limit the clutter in the book. Where only a value change or two will make the project work to some new specification, then I have usually elected to simply tell you which components to change, and what change to make. You can fill in the values on your own diagram when you actually go to build the project.

Only a few of the circuit drawings show the power supply connections for the operational amplifier. Make no mistake about it, however, that those power supply connections (V+ and V−) are to be made by you. Standard drawing practice calls for elimination of the power supply connections, except where special treatment or bypassing is required, in order to make the drawing less "busy," thus easier to read.

Unless otherwise specified, the operational amplifier used can be the "universal" (well—almost universal!) type 741. These devices are low in cost, and are available almost everywhere. Unfortunately, they are not the best operational amplifier on the market.

Where a premium-grade device is required for the application, I have specified some other type. If you do not see a specification, then assume that the operational amplifier selected is the 741, or one of its derivative types (i.e., the dual 741, also known as the 1458 device).

Another little caution involves the matter of pinouts shown. The "industry standard" pinouts are those of the 741 in the 8-pin metal or 8-pin miniDIP package. These are as follows:

1. Offset/compensation*
2. Inverting input
3. Noninverting input
4. V− power supply
5. Offset/compensation*
6. Output
7. V+ power supply
8. Offset compensation*

*Note: Not on all devices.

It is your responsibility to ascertain whether or not any operational amplifier which you select in substitution fits the same pinouts. Although most operational amplifiers in either 8-pin metal or 8-pin miniDIP packages use this pinout structure, there are sufficient numbers which do not give one cause for concern. If in doubt, look in the data manual or specifications sheet for that particular operational amplifier!

Power Supply Note. The standard rule of thumb for operational amplifier power supplies is that the supply must be higher than the maximum allowable output voltage. If you expect an output voltage swing of ±10 volts, then the V+ and V− power supplies must both be higher than 10 volts dc. Just how much higher depends upon the type of operational amplifier used. In normal bipolar transistor-based operational amplifiers, it is necessary to count the number of PN junctions between the output terminal and each power supply terminal in order to determine the output-vs-power-supply-rail differential. In some 741 devices this value can be as much as 3.5 volts, assuming 0.7 volts per PN junction. In the RCA BiMOS operational amplifiers, some models can get to within 0.5 volts of the supply rail. Thus, a 10 volt output on a 741 might require a power supply of 13.5 volts or more, while on the BiMOS operational amplifier a power supply voltage of 10.5 volts or more will suffice. Keep this limitation in mind when specifying the op amp for any given use. If, for example, you opt for a 9-volt transistor radio battery as the power supplies, then the maximum 741 swing on the output might be as little as (9-3.5) Vdc, or, 5.5 volts!

Also be aware that reversed polarity can kill most operational amplifiers in a quick hurry! Unless you really want to make an expensive silicon-to-carbon converter, then make sure that reverse power supply polarity does not occur. There are some schemes for limiting this type of damage (e.g., rectifier diodes in series with each power supply terminal), but those methods have enough other problems that they are not always desirable.

Now let's look at some projects!

0 TO ×1 INVERTING AMPLIFIER

The circuit shown in Fig. 6-1 is an operational amplifier stage that will provide a variable gain from zero to minus one (0 to −1). This type of stage is often used as the output stage in larger projects, or, where some adjustment is needed in order to make the output some specified value. One possible application for this type of stage is to make an "output span" control for an output display device such as an oscilloscope, digital panel meter or strip chart, and/or X-Y paper recorder. Recall from our discussion of operational amplifier theory that the gain of an operational amplifier is given by the simple expression:

$$A_v = -R2/R1$$

Where:

R1 is the "input" resistor
R2 is the "feedback" resistor

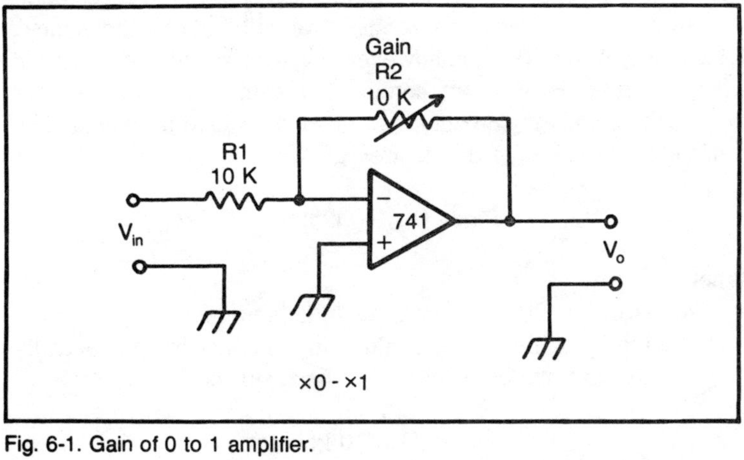

Fig. 6-1. Gain of 0 to 1 amplifier.

(assuming that R1 and R2 are in the same units, i.e., ohms, kohms, or megohms)

Thus, when we plug the values into the standard formula, we find the following:

at R2 = 10 K

$A_v = -R2/R1$
$A_v = -(10\text{ K})/(10\text{ K})$
$A_v = -1$

at R2 = 0

$A_v = -R2/R1$
$A_v = -(0)/(10\text{ K})$
$A_v = 0$

The minimum value of R2 produces a gain of zero, while the maximum value of R2 produces a gain of −1. At values in between zero and maximum, therefore, we will have a value of less than one but greater than zero.

0 TO ×10 AMPLIFIER

The gain of the amplifier as configured in Fig. 6-1 is zero to unity. For some applications, however, this is an insufficient range. We can make the gain anything we wish, within reason, by changing the values of the two resistors. It is probably not wise to change the value of R1, since that directly affects the input impedance of the circuit (which is equal to R1). Input impedances less than 10 kohms are not usually advisable, so I suspect we will find that the value of R1 is a sacred cow. We can, however, change the value of R2 almost with impunity. If we want some other gain, therefore, we can modify the standard gain equation, given above, to provide a solution for R2 in terms of the desired gain:

$$R2 = (A_v)(10\text{ kohms})$$

Where:

R2 is the feedback resistance, in kohms

If we plug in the gain of 10 (keeping in mind that it is actually −10 that will be achieved!), the result is:

$$R2 = (10)(10\text{ kohms})$$

$$R2 = 100 \text{ kohms}$$

Thus, if we provide a 100 kohm resistor for R2 in Fig. 6-1, the gain will go from 0-1 to 0-10. A 1 megohm potentiometer will make the gain 0-100.

The potentiometer used in any Fig. 6-1 project can be either a trimmer type mounted on a printed wiring board, or, a panel mounted type. The particular type which you select will depend upon the application. Obviously, if the device is to be a "set and forget," then a trimmer potentiometer should be selected. If the adjustment is intended to be either frequent or made by the end user, then it should be a panel-mounted type.

FIXED GAIN LESS THAN UNITY INVERTING AMPLIFIER

The circuit shown in Fig. 6-2 is a variation on the circuit of Fig. 6-1, in which the gain is less than unity, but is fixed. The standard equation for the gain of an inverting amplifier based on the op amp is:

$$A_v = -R2/R1$$

Thus, if resistor R2, the feedback resistor, is less than R1, the input resistor, then the gain will be less than unity. Of course, for R1 = R2 the gain will be exactly unity. This was the theory on which variable gain amplifier was based. In the example shown, the gain will be exactly 0.1, so V_o will be $V_i/10$. Resistances for other gains less than unity are shown below:

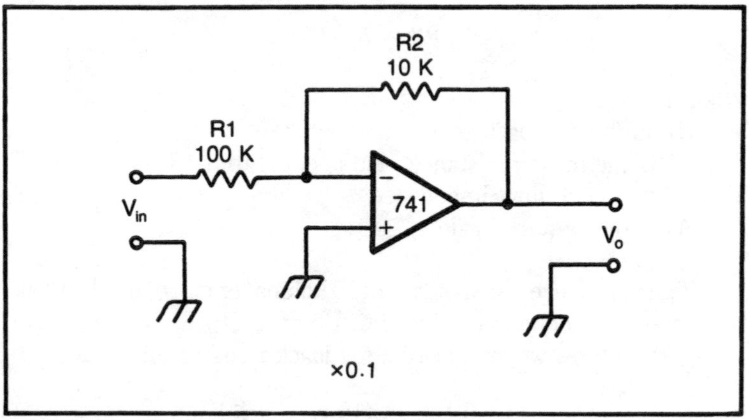

Fig. 6-2. Gain of 0.1.

GAIN	*R2*
−0.05	5 K
−0.1	10 K
−0.2	20 K
−0.33	33 K
−0.5	50 K
−0.66	66 K*
−0.75	75 K
−1	100 K

*It may be necessary to use two 33 K in series

GAIN OF ×100 INVERTING AMPLIFIER

The gain of – ×100 amplifier is sometimes used as a fixed gain block in analog circuits, or, as an interface to display instruments and so forth. We can, for example, sometimes use the 10 mV output of some Wheatstone-bridge transducers; e.g., those which connect directly to a recorder to drive a 1-volt input standard display device (strip-chart recorder, some X-Y recorders or oscilloscopes). The circuit in Fig. 6-3 is one such circuit.

The gain of any inverting amplifier based on the op amp is given by the expression:

$$A_v = -R2/R1$$

so, when we want to specify a standard gain and an input impedance, we can use the derived version:

$$R2 = A_v R1$$

Where:

*R*2 is the feedback resistance*

*R*1 is the input resistance, and is equal to not less than the required input impedance

A_v is the required gain

Thus, if we use our standard of 10 kohms, or more, for the input resistance, the equation reduced to $R2 = A_v$ (10 K). For a gain of −100, therefore, we will need a feedback resistor of:

$$R2 = A_v(10\ K)$$
$$R2 = (100)(10\ K)$$

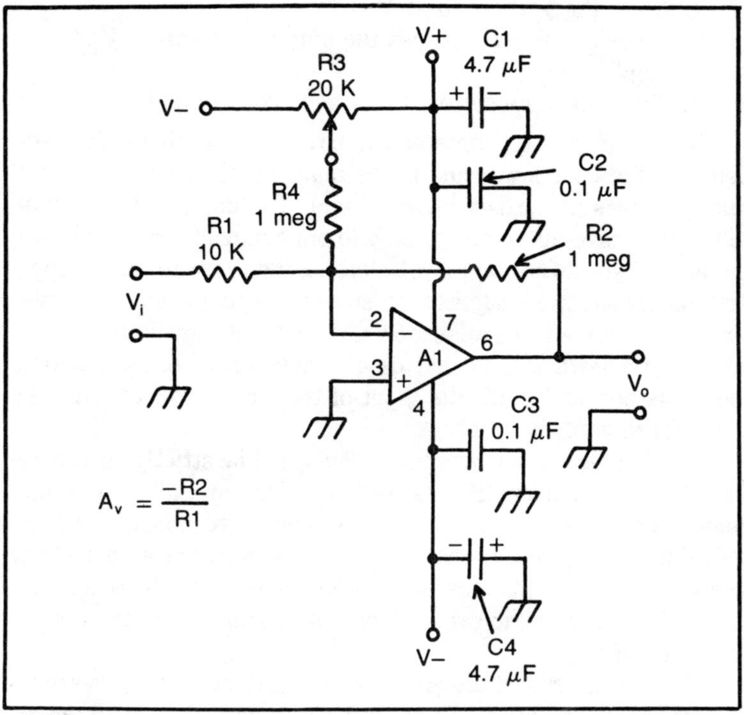

Fig. 6-3. Gain of −10.

$$R2 = 1{,}000{,}000 \text{ ohms, or } 1 \text{ megohm}$$

If we use a 1 megohm resistor in the feedback loop, and a 10 kohm resistor in the input circuit, we will achieve the desired input impedance of 10 kohms and a gain of −100.

Unfortunately, the gain may give us another little problem: offset output voltage! The output voltage will not be zero when the inverting input is shorted to ground (like the noninverting input). The higher gain of this circuit makes this problem more acute than with previous amplifiers.

We can, however, eliminate the output offset by using the potentiometer labeled R3. This potentiometer is configured in the circuit to cause a counter offset voltage. The gain of the circuit will be approximately −1, so we can adjust the voltage at the wiper of the potentiometer to be approximately equal to the output offset voltage in magnitude, but opposite in polarity. In actual practice, we would use the following adjustment procedure:

1. Short the V_i input to ground.
2. Place a voltmeter across the output to measure V_o.
3. Apply power.
4. Adjust potentiometer R3 for zero volts output.

The range of the voltmeter will have to be continuously readjusted to allow for lower and lower ranges until the output offset voltage shows no further improvement. Of course, some digital voltmeters have autoranging, so will automatically make this adjustment. If you do not have a voltmeter, but do have an oscilloscope (an unlikely situation, admittedly, since most people buy a voltmeter before buying an oscilloscope!), then use dc coupling to make this same adjustment. In that case, make sure that the position control is adjusted (with the input of the 'scope vertical channel shorted!) to zero the trace.

The bypassing shown in Fig. 6-3 may not be strictly necessary if a 741 operational amplifier is used. The 741, after all, is unconditionally stable, which means that its frequency response is shot to hell. But, if you prefer to use an operational amplifier with a wider frequency response than the 741 provides, then it may be necessary to use the bypassing in order to prevent oscillation of the operational amplifier.

The 0.1 μF capacitors are used for high frequency bypassing, and are the most critical. These capacitors should be mounted as close as possible to the body of the operational amplifier. Do not use paper, mica, or Mylar capacitors for this application. Disc ceramic is fine, as are those mylar types which are dipped. There are also a large number of newer types on the market, and those are especially designed for digital and analog bypassing use.

The 4.7 μF capacitors are tantalum types, and are intended for low frequency bypassing. These capacitors are not as necessary as are the 0.1 μF units, and are frequently deleted except in the most critical applications.

GAIN OF 500 INVERTING AMPLIFIER

Once the gain of an operational amplifier circuit increases above 100 or so, the builder must take more care in matters such as layout, power supply bypassing and so forth. Output offset problems also become more critical, so must be addressed where previously they had been ignored. At gains of less than 100, we can often totally ignore the output offset problem, but at gains of more than 100 the problems insinuate themselves more forcefully onto our little cloud. Figure 6-4 shows a circuit which will provide a gain of

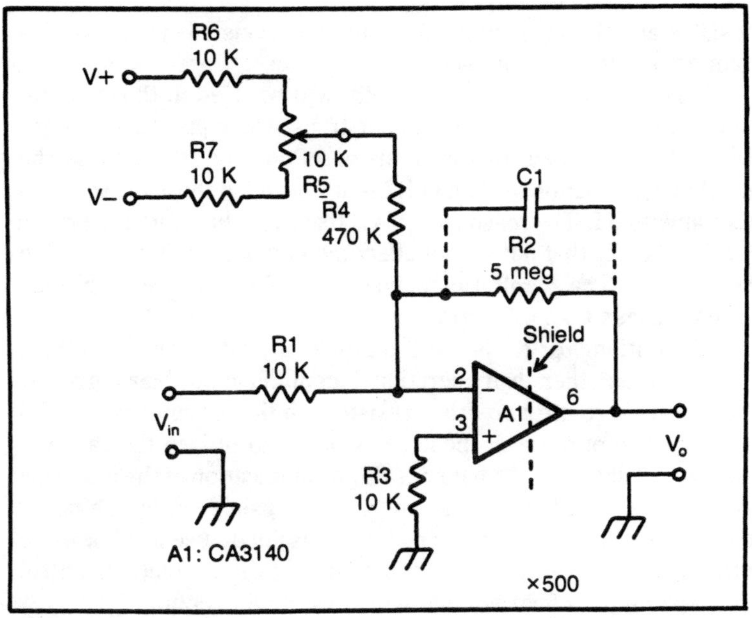

Fig. 6-4. Gain of −500.

−500, and takes into account some of the problems mentioned above.

For gains in the range of 500, I recommend that a 741 not be used. It is possible to use more premium devices, such as the LM-301, but I prefer the use of devices such as the RCA BiMOS operational amplifier, type CA3140 or its equivalent. In that case, we can probably delete resistor R3.

The gain of the circuit is fixed by the standard equation for inverting amplifiers:

$$A_v = -R2/R1$$

For a gain of 500, therefore, we will want a R2/R1 ratio of 500. If we want to make the input impedance 10 kohms (a good idea unless more impedance is needed), then we will solve the standard equation as follows:

$$R2 = A_v(10\text{ K})$$
$$R2 = (500)(10\text{ K})$$
$$R2 = 5{,}000{,}000 \text{ ohms, or, } 5 \text{ megohms}$$

The feedback resistor will, therefore, be 5 megohms. Both this resistor and the input resistor should be precision types unless a gain error can be accommodated.

The compensation resistor, R3, will be used in those operational amplifiers which use bipolar transistor input stages (e.g., LM-301). The value of this resistor is usually specified as the parallel combination of R1 and R2, but we find in this case that it is the same as R1. The reason for this is that the value of R2 is so much higher than R1 that it does not affect the parallel combination value very much. The parallel combination would be more than 9980 ohms, or less than 0.2 error.

Potentiometer R5 is used to cancel any offset output voltage from sources other than operational amplifier input bias currents. Note that there are two other resistors in this circuit, namely R6 and R7. The purpose of these resistors is to reduce the range of voltage permitted at the wiper of R5; an implication of the immense gain of this circuit. Keep in mind that the gain seen by the offset potentiometer is more than ten. It is advisable to use a 10- or more turn potentiometer for R5 in order to gain the resolution control over the output offset voltage which we might require.

Some builders, especially where layout space is tight, may want to use a shield between the inputs and the output of the operational amplifier. Because of the large gain of this circuit, it is likely that some layout problems can cause feedback that will lead to oscillation. The shield can be a little piece of copper foil, or, a piece of printed circuit board "adopted" for the purpose. In some cases, the shield will be designed into the foil pattern on the printed circuit board. In most cases, it will surround the input leads.

The frequency response can be tailored to some special range, or, it may be reduced for purposes of improving the stability of the operational amplifier (no small problem at these gains!). Although some operational amplifiers have provision for reducing the frequency response, the capacitor shown in Fig. 6-4 (as optional) will work for all cases. The −3 dB point in the frequency response is given by:

$$F_{Hz} = \frac{1}{6.28\ R_2\ C_1}$$

Where:

F_{Hz} is the −3 dB frequency, in hertz (Hz)
R_2 is the value of resistor R2
C_1 is the capacitance of C1

Since we know the value of R2 (5×10^6 ohms), and we will specify the value of F_{Hz}, we should rearrange this equation to find C1:

$$C_1 = \frac{1}{(6.28)(5 \times 10^6)(F_{Hz})}$$

or,

$$C_1 = \frac{1}{(3.14 \times 10^7)\ (F_{Hz})}$$

when the constants are lumped into one value. Note that this version is usable *only* when R2 is 5 megohms.

Example

We want to design an operational amplifier with a gain of −500, an input impedance of 10 kohms and an upper −3 dB frequency response point of 100 Hz. Calculate the value of C1 in Fig. 6-4.

$$C_1 = 1/(3.14 \times 10^7)(F_{Hz})$$
$$C_1 = 1/(3.14 \times 10^7)(100)$$
$$C_1 = 1/3.14 \times 10^9)$$
$$C_1 = 3.19 \times 10^{-10} \text{ farads}$$

Of course, we don't usually specify capacitors in farads, so must convert to either microfarads or picofarads. Since $1\ \mu F = 10^{-6}$ farads, and $1\ pF = 10^{-12}$ farads, we will probably find *this* value most conveniently specified in picofarads:

$$C_1 = 3.19 \times 10^{-10}\ F \times \frac{10^{12}{}_{PF}}{F}$$

$$C_1 = 319_{PF}$$

Of course, it is difficult to obtain a 319 pF capacitor, so we will use the nearest standard value, which is 330 pF, and live with the error in frequency response. This value results in an approximately 3.5 percent difference in actual frequency response.

GAIN OF 1000 INVERTING AMPLIFIER WITH HIGH-IMPEDANCE INPUT

All of the circuits presented thus far have suffered from one minor, well maybe major, problem: low input impedance. In the inverting follower amplifier, the input impedance is limited to the value of the feedback resistor. If we want to provide a high input impedance, then we must use the noninverting follower configuration. But there are times when noninverting follower is not what we need: we might want an inverting amplifier. Figure 6-5 shows a circuit which will provide a gain of approximately −1000, is inverting, and provides an extremely high input impedance.

The input impedance in Fig. 6-5 is set by the input amplifier, an RCA BiMOS CA3140 device. The input impedance of the BiMOS amplifiers is on the order of 1.5×10^{12} ohms (that's 1.5 terraohms!), which is as high as most applications require. The overall gain of this amplifier is set by the equation:

$$A_v = \left[\frac{R2}{R1} + 1\right]\left[\frac{R4}{R3}\right]$$

Where:

$R1$ is the resistance of resistor R1
$R2$ is the resistance of resistor R2
$R3$ is the resistance of resistor R3
$R4$ is the resistance of resistor R4

Given the values of R1 through R4 in Fig. 6-5, we can calculate the actual gain from the following:

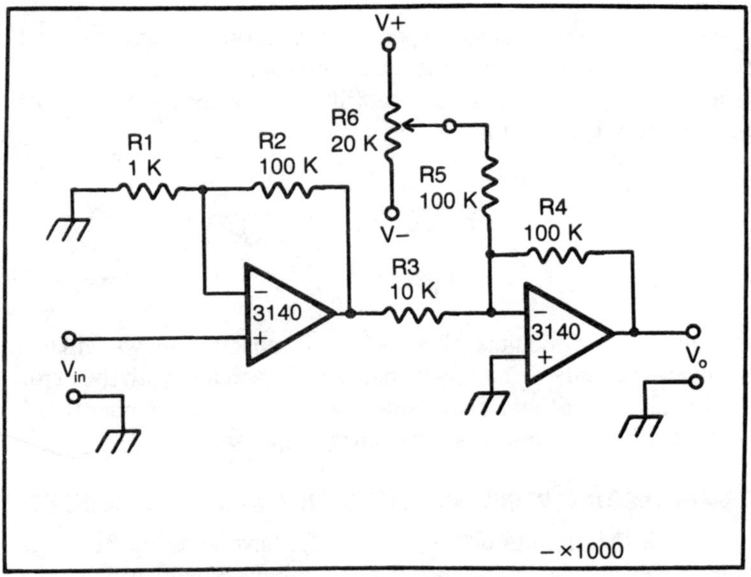

Fig. 6-5. Gain of −1000.

$$A_v = \left[\frac{100\text{ K}}{1\text{ K}} + 1\right]\left[\frac{100\text{ K}}{10\text{ K}}\right]$$

$$A_v = (100 + 1)(10)$$

$$A_v = (110)(10)$$

$$A_v = 1100$$

This figure represents a gain error of 10 percent, which can easily be trimmed to exactly −1000 by making either R1, R2, and R4 a potentiometer. In most cases, the designer will probably want to make the resistor chosen for gain variation a combination of a fixed resistor and a potentiometer which, when connected in series, equal the value that will provide the required gain. The potentiometer should have a value of approximately 20 percent of the total required resistance.

The output offset voltages that *will* exist in almost all ideal versions of this circuit can be nulled out using the potentiometer, R6. The adjustment of this circuit is as follows:

1. Adjust R6 to approximately mid-range.
2. Short the V_{in} input to ground, setting V_{in} effectively to 0 volts.
3. Connect a voltmeter to V_o to monitor the output voltage.
4. Apply power
5. Adjust R6 to produce $V_o = 0$.

The range of the voltmeter should be reduced as the value gets close to zero, and the adjustment repeated. This will allow you to home in on the correct value.

UNITY GAIN BUFFER AMPLIFIER

Perhaps the easiest of all amplifier projects to be formed with the IC operational amplifier is the unity gain noninverting follower. This circuit (Fig. 6-6) is formed merely by connecting the output of the operational amplifier to the inverting input, thereby producing 100 percent negative feedback. It can be shown arithmetically, that the gain of this circuit is so close to unity that only a small gain error exists. If, for example, the open-loop (i.e., no feedback) gain of the operational amplifier is 500,000 the closed-loop gain is 0.999998—which is close enough to unity for all but the most severe applications.

The uses of the unity gain noninverting follower are primarily impedance transformation and buffering. In the former case, we take advantage of the fact that the operational amplifier in the

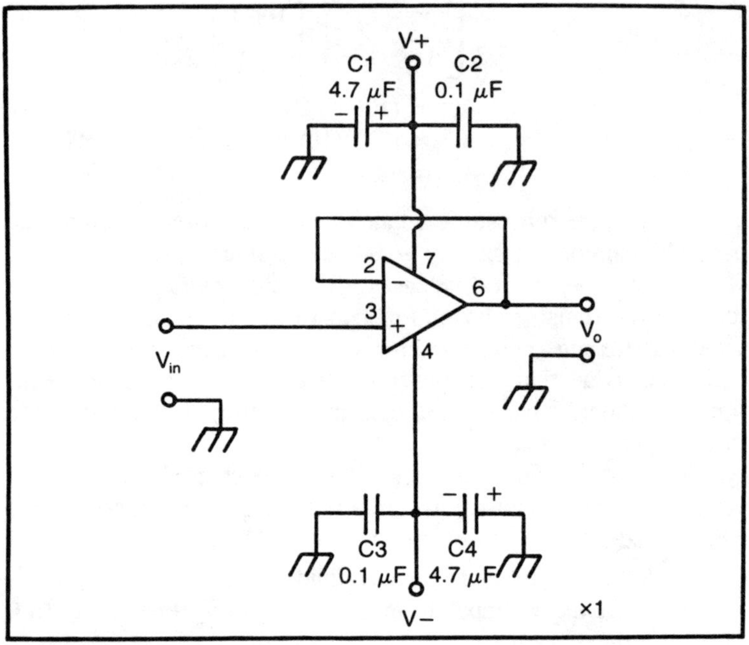

Fig. 6-6. Gain of 1.

noninverting follower configuration has a tremendous input impedance (1 megohm to 1.5 terraohm, depending upon model), while the output impedance is typically less than 100 ohms. Thus, the signal will see an extremely high impedance, so the source will not be loaded. Yet, at the same time, the output impedance is low enough for the operational amplifier to look like a near-perfect voltage source to circuits which follow.

Buffering, which is the second major use of the noninverting unity gain amplifier, uses the operational amplifier in order to isolate a signal source from the load. In an oscillator circuit, changing load impedances may be within the drive capacity of the oscillator, but the oscillator will, nevertheless, experience a frequency shift in the process. In other cases, the drive power or drive current capacity of the circuit may not be sufficient. There are also cases where we want to prevent a short or other anomaly in the "outside world" from affecting the signal source. In some medical instruments, for example, there are multiple displays (e.g., oscilloscope, strip-chart recorder and computer input) connected to the same signal source (e.g., an electrocardiograph amplifier). If we develop a short in the input circuit of any of these display devices, then it could conceivably shut down the entire instrument. If that instrument is critical to the patient, then the failure can be of major consequence. By using a buffer stage between the output of the ECG amplifier (or other signal source) and the instrument output which drives each different peripheral, then we will avoid that problem. A short in the cable from the bedside monitor to the computer, for example, will not affect the oscilloscope or strip-chart recorder.

For most instrumentation purposes, the ordinary 741 operational amplifier is sufficient for this purpose. The 741 device will work well to a frequency of several kilohertz, so will work nicely in low frequency circuits. As the signal component frequency increases, however, we must become more and more aware of the rather low frequency response of the 741-family of devices. In those cases, we will want to use an operational amplifier which will pass the frequency involved. Note, incidentally, that the operational amplifiers ideal properties assume a very high open-loop gain. Just because some op amp provides *some* amplification out to, say, 15 MHz, doesn't mean that the device behaves like an operational amplifier at those frequencies. One device, for example, boasts a gain-bandwidth product of 25 MHz. But when we set some almost arbitrary gain below which operational amplifier-like behavior is degraded, then we find that the "real" frequency response is a lot less. If that gain is taken to be 100,000, then the "real" available gain is a lot less:

$$F = \frac{25 \text{ MHz}}{100{,}000}$$

$$F = 0.00025 \text{ MHz} = 250 \text{ Hz}$$

The capacitors in Fig. 6-6 are needed when the frequency response is large, and are used to decouple the power supply. These capacitors are also well recommended if there are more than two stages in cascade. The 0.1 μF capacitors are used to decouple high frequencies, while the 4.7 μF capacitors are for low frequencies. This double capacitor scheme is needed because the 4.7 μF tantalum electrolytics do not work well at high frequencies. They tend to act like capacitors at low frequencies and at dc, but as the frequency increases above the audio range they begin to act less like a capacitor and more like a block of wood.

The V+ and V− power supplies should be regulated in most cases, and must have a minimum value which reflects the maximum desired output voltage. If the differential between the power supply voltage and the maximum output voltage is 3.5 volts (for example) then the power supply potentials must be 3.5 volts higher than the maximum required output voltage. We must also keep in mind the maximum permissible voltage applied to any one power supply lead, and the maximum voltage between the power supply leads. Rarely will an operational amplifier provide a lead-to-lead voltage that is twice the maximum applicable to any one lead. If, for example, we wish to use +18 volts for V+, and the maximum lead-to-lead voltage is 30 volts, then the maximum value of V– is (30-18), or –12 volts dc.

The voltage gain of Fig. 6-6 is unity, i.e., 1. But the power gain is considerably higher. Since the signal loses no voltage, but the impedance is a lot lower, then the power available at the output must be higher; consider the $P = E^2/R$ relationship for the reason why this is true. Thus, we can claim that the unity gain (unity *voltage*, that is) is actually a power amplifier. Most operational amplifiers will provide from 25 milliwatts to 200 milliwatts of output power, with some units offering considerably more. Some devices, intended as servo or audio power amplifiers, will produce several watts even while working in the manner of an operational amplifier.

GAIN OF FIVE NONINVERTING AMPLIFIER

The noninverting follower configuration of the operational amplifier can produce amplifiers with gains from unity up to some limit which varies with each different type. Figure 6-7 shows the basic circuit for the noninverting follower with the values of R1 and R2 selected to produce a gain of +5. Recall from our earlier theory discussion that the gain of the noninverting follower is given by:

$$A_v = \frac{R2}{R1} + 1$$

In this case, therefore, the gain of the amplifier is set by:

$$A_v = \frac{1100 \text{ ohms}}{270 \text{ ohms}} + 1$$

$$A_v = 4.07 + 1$$

$$A_v = 5.07$$

The small gain error of this circuit is expressed by:

$$A_v = \frac{(5.07 - 5)(100\%)}{(5.00)}$$

$$A_v = (0.07)(100\%)/5 = 1.4\%$$

The procedure for finding the resistor values in an amplifier like Fig. 6-7 is somewhat different from the case of the inverting follower because we don't have to worry about a minimum value for R1 (which sets the input impedance in the inverting amplifier circuit). In the noninverting circuit, therefore, we have a little more flexibility in setting these values. We can solve the standard gain equation to a form which gives us the resistor ratio, and then plug in standard values from the table of resistor values to find the values required. The rearranged equation is

$$R2/R1 = A_v - 1$$

Where:

R1 and R2 are the resistances (expressed in exactly the same units!), while A_v is the voltage gain required

The circuit shown in Fig. 6-7 does not include the power supply circuitry. For the sake of simplicity, most operational amplifier schematics delete those connections, but they are there nonetheless. Look at Fig. 6-6 for an example of the power supply connections. The capacitors used in Fig. 6-6 will apply to the Fig. 6-7 circuit, and for exactly the same reasons.

The maximum achievable gain of this circuit is dependent in part upon the nature of the operational amplifier used. If we use a

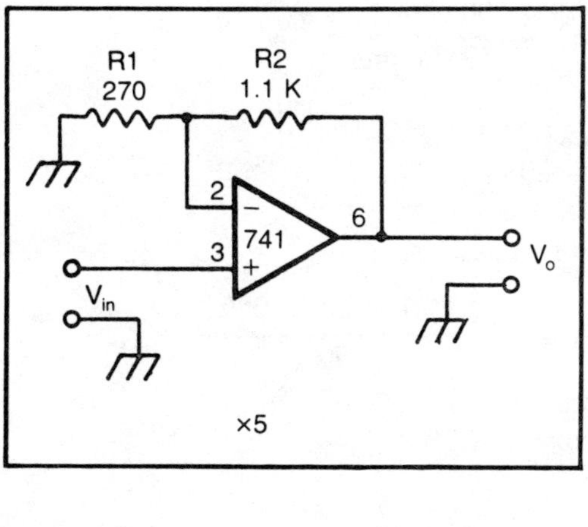

Fig. 6-7. Gain of 5.

741 device, then we will find limitations on gain from both the frequency response perspective and from the dc output offset problem. The greater the gain, the more profound the effects of the offset problem. In the case of higher gain, therefore, we might want to use a different form of operational amplifier. A circuit follows in the next section.

GAIN OF 500 NONINVERTING FOLLOWER

Just where the breakpoint comes between high and low gain amplifiers may be open to some debate, and is surely affected by the extraordinary steps the designer must use to overcome problems such as drift and offset potentials.

A couple of decades ago, the designer would look upon ×500 as a high gain, while today some consider this a medium gain. The reason for this change is, perhaps, that operational amplifiers are available at low cost that will work nicely at closed-loop gains of ×500 without the need for any form of compensation or other profound modifications. The RCA CA3140 family of devices (there are several in the family) are such a family. These devices will even operate with very low drift if they are operated at suppply potentials of ±5 volts, especially when Wakefield "top-hat" heatsinks are used on the metal can of the operational amplifier. Figure 6-8 shows a gain of 500 amplifier based on the RCA BiMOS CA3140 device (CA3160 usable as well, as is the CA3240 if a dual operational amplifier is needed).

The gain of this circuit is, according to the usual equation:

$$A_v = \frac{R2}{R1} + 1$$

$$A_v = \frac{1,000,000 \text{ ohms}}{2000 \text{ ohms}} + 1$$

$$A_v = 500 + 1$$

$$A_v = 501$$

The gain error in this case is:

$$\text{Error} = \frac{(501 - 500)(100\%)}{500}$$

$$\text{Error} = (1)(100\%)/(500)$$

$$\text{Error} = 0.2\%$$

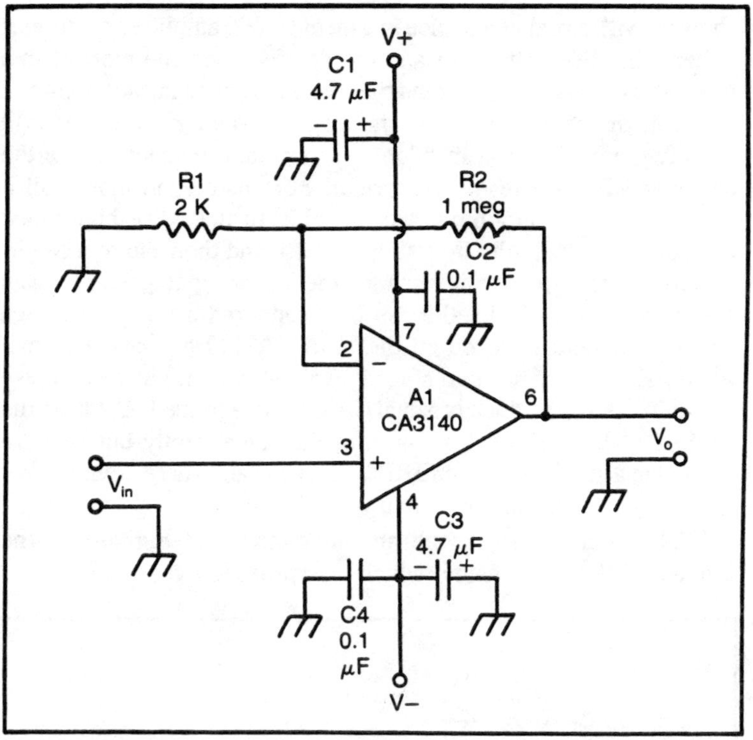

Fig. 6-8. Gain of 500.

Notice the effect of the resistor ratio error on total error in the noninverting follower circuit as gain goes up. If the resistor ratio were lower, as in lower gain circuit, then the error caused by the "+1" factor is a lot greater. In this circuit, however, it is a lot less because R2/R1 is much larger than 1.

As the gain increases in a circuit, then so does the need for power supply bypassing, especially when two or more stages are connected in cascade. The capacitors are needed in this circuit for the reasons given earlier. Because of the high gain, however, the effects are a lot more acute, making the capacitors even more necessary. Mount the 0.1 μF capacitors as close as possible to the body of the operational amplifier, and the 4.7 μF units close by; this is probably the most often overlooked problem in layout in operational amplifier circuits.

GAIN OF 1000 NONINVERTING AMPLIFIER

The problems which affect operational amplifiers begin to

show up with a real vengeance in gain of ×1000 amplifier circuits. It is here that the drift factor and the dc offset become more than a mere annoyance. Keep in mind that the dc operational amplifier will see drift or offset signals on the input as valid signals, and will amplify them with wild abandon . . . much to the annoyance of the designer who was less than careful! Fortunately, in most multi-stage high gain amplifiers, we are able to use very high-grade components in only the first stage or two, and then can reduce our requirements to lesser devices for the following stages. The gain ×1000 amplifier of Fig. 6-9 can be considered a medium-to-high gain circuit, and is based on the RCA CA3140 devices. Alternatively, since there are two stages involved, you might want to use the RCA CA3240, which is a dual CA3140 (it is to the CA3140 as the 1458 is to the 741 device—both of the same family but two are housed in a single 8-pin miniDIP IC package and share common V− and V+ power supply terminals).

The gain of the overall amplifier circuit of Fig. 6-9 is the product of the stage gains, which is expressed by:

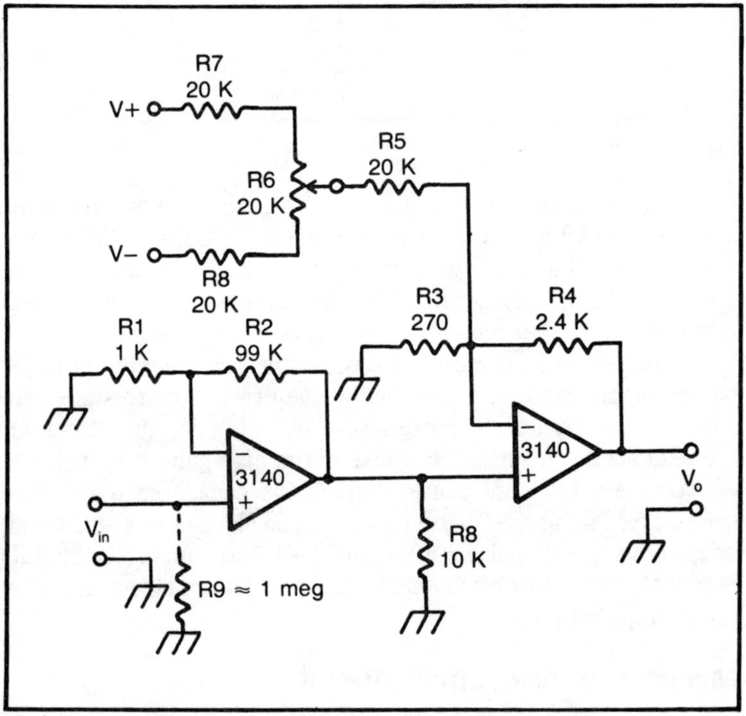

Fig. 6-9. Gain of 1000.

$$A_v = \left[\frac{R2}{R1} + 1\right]\left[\frac{R4}{R3} + 1\right]$$

With the values of the resistors shown, therefore, the overall amplifier gain is:

$$A_v = \left[\frac{99\ K}{1\ K} + 1\right]\left[\frac{2400\ ohms}{270\ ohms}\right]$$

$$A_v = (99 + 1)(8.89 + 1)$$

$$A_v = (100)(9.89) = 989$$

The gain error, then, is

$$Error = \frac{(1000 - 989)(100\%)}{(1000)}$$

$$Error = (11)(100\%)/(1000)$$

$$Error = 1100\%/1000 = 1.1 \text{ percent}$$

If you need the exact gain, then change the input resistor of the second stage (i.e., R3) to a potentiometer, or, a potentiometer and fixed resistor combination with a total value of 250 ohms or slightly less (see the equation above). Alternatively, do the same thing with the feedback resistor in either stage. Make sure, when the maximum and minimum values of the potentiometer or pot/resistor combination are plugged into the gain equation, they bracket the desired gain. If that is done, then you will know that the gain is correctly settable because the potentiometer will vary the gain above and below the correct gain.

At a gain of ×1000, the offset problem becomes more acute than at smaller gains, even in a low offset device such as the BiMOS operational amplifier. This offset voltage can be canceled by the potentiometer resistor network R6-R8. This network will inject a counter-offset into the output stage which will cancel the main offset potential. The adjustment procedure is as follows:

1. Set R8 to approximately mid-scale.
2. Short the input, thereby setting V_{in} to zero volts.
3. Connect a dc voltmeter across the output to monitor V_o.
4. Apply power.
5. Adjust potentiometer R6 until V_o is zero.
6. Keep readjusting R6, with the voltmeter set to progres-

sively lower scales, until no further improvement results.

Resistor R9, across the input of the amplifier, is optional. It will reduce the input impedance, but at the same time it will allow us to use capacitor coupling of the input stage. It will also keep the output voltage at zero when the input is disconnected.

GAIN OF 0 TO 1 NONINVERTING AMPLIFIER

The gain of ×0 to ×1 amplifier shown in Fig. 6-10 is usable as a universal "rear end" (no pun intended) for operational amplifier projects. This project is equipped with a gain control and a "DC BAL" control, and can be equipped with a "position" control if this circuit is to be used with a strip-chart recorder or an oscilloscope. The input impedance of this circuit is 10 kohms, which is more than enough when the amplifier is used as a "rear end" for some other

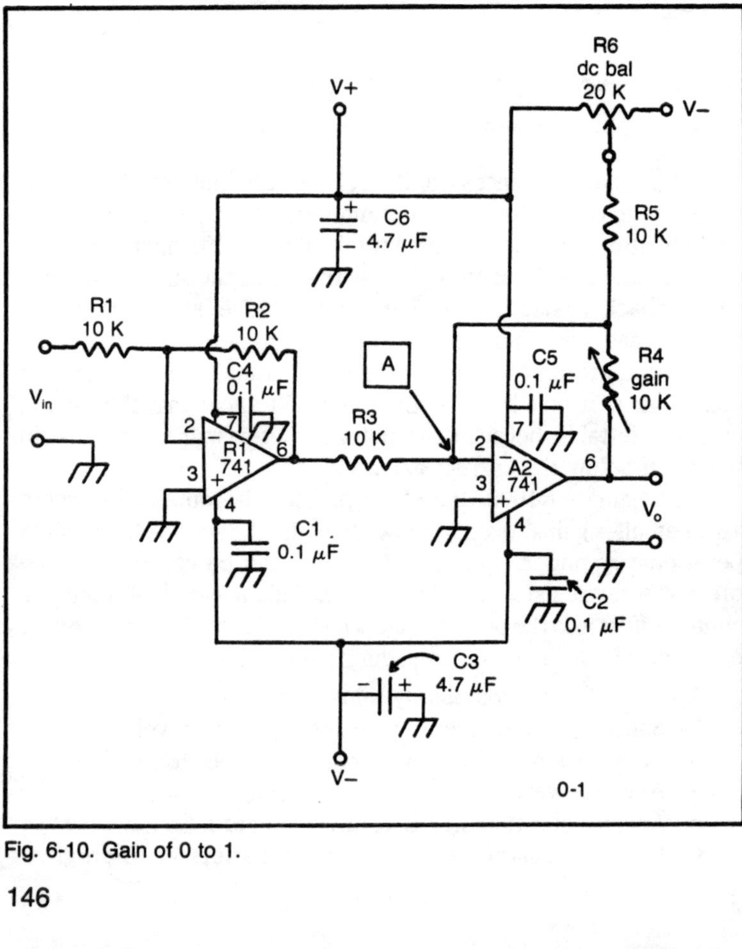

Fig. 6-10. Gain of 0 to 1.

op-amp circuit. For most cases, the signal source will be a voltage amplifier output, so the source impedance will be on the order of 100 ohms. Thus, a 10 kohm input impedance for this circuit presents no problem. The gain of the circuit is the product of the gains of A1 and A2:

$$A_v = (A_{V1}) \times (A_{V2})$$
$$A_v = (-R2/R1)(-R4/R3)$$
$$A_v = (-10\text{ K}/10\text{ K})(-R4/10\text{ K})$$
$$A_v = (-1)(-R4/10\text{ K})$$

From the above equation, we can see that the gain will be zero when the potentiometer R4 is at zero ohms, and the gain is 1 when the potentiometer is at maximum; for the maximum case:

$$A_v = (-1)(-10\text{ K}/10\text{ K})$$
$$A_v = (-1)\,(-1)$$
$$A_v = +1$$

Potentiometer R4 is, then, the gain control. This potentiometer might be a trimmer type mounted on the wiring board, but if this is used as a rear end for some instrument, the potentiometer should be a panel-mounting type so that the operator can control it. In any event, I recommend a multi-turn model so that the gain can be adjusted with a fair degree of resolution.

The DC BAL control is "DC balance." This control is used to null any offset components in the input signal which may tend to create a shift of the base line on output display instruments. For example, if you connect the output to an oscilloscope, then adjust the sensitivity/gain control through its range from zero to one, you will notice the output waveform may tend to shift up or down as the output offset affects the base line. The base line is, of course, the output voltage at zero input, and the term is derived from the appearance of the trace on an oscilloscope of strip-chart recorder. In most instruments, the DC BAL control is made inaccessible to the user, or, may be screwdriver adjustable. The only time it should be mounted on the front panel with a knob is when the rest of the instrument design causes a tremendous dc offset voltage. While this condition is sometimes found naturally, it is usually the case that it really means the design is poor.

The 0.1 μF capacitors should be mounted as close as possible to the body of the operational amplifier. These capacitors help decouple and prevent oscillation at high frequencies. If the 741-type of operational amplifier is used, then it may be true that these capacitors are not needed. The frequency response of an unconditionally stable amplifier like the 741 is low enough that high frequency oscillation rarely becomes a problem. In noncompensated devices, i.e., those which are *not* "unconditionally stable," these capacitors become necessary.

DC DIFFERENTIAL AMPLIFIER (GAIN OF UNITY)

The dc differential amplifier produces an output that is proportional to the differential voltage gain and the *difference* between two input signal voltages. One voltage is applied to the inverting input, while the other is applied to the noninverting input. If V1 and V2 are the voltages, and the differential voltage gain is A_{vd}, then the output signal will be equal to:

$$V_o = (A_{vd})(V1\text{-}V2)$$

An example of a simple dc differential amplifier is shown in Fig. 6-11. This circuit is based on a single operational amplifier device. Note that the usual operational amplifier is well-suited to this service because it already has a differential input circuit, i.e., the inverting and noninverting inputs form a differential pair. Assuming that the circuit is balanced, i.e., R1 = R3 and R2 = R4, the gain is set by:

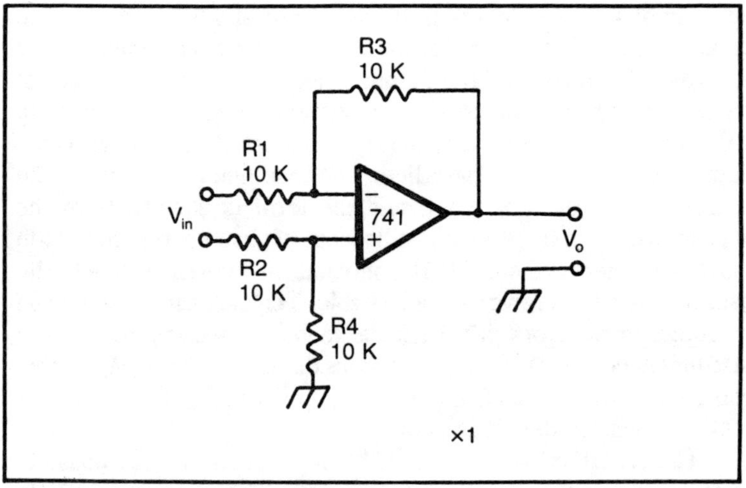

Fig. 6-11. Gain of 1.

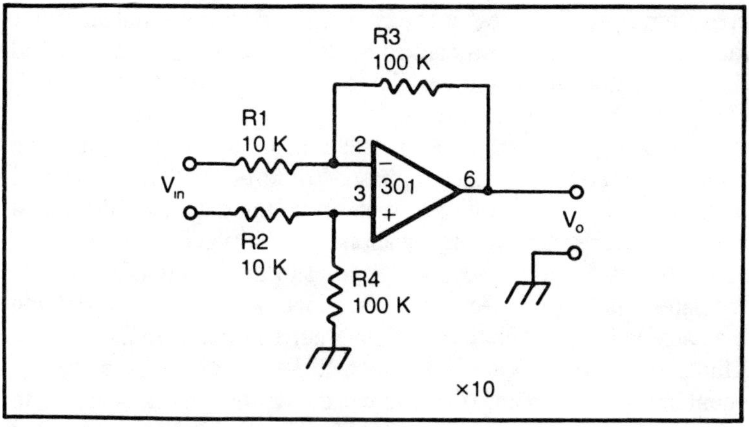

Fig. 6-12. Gain of 10.

$$A_{vd} = R3/R1$$

or,

$$A_{vd} = R4/R2$$

or whichever combination of a feedback resistor over an input resistor turns you on! In the case shown here, all four resistors are equal values, so the gain is unity:

$$A_{vd} = (10\ K/10\ K)$$
$$A_{vd} = 1$$

This stage can be used to buffer a differential output while at the same time it produces a single-ended output from a differential drive signal.

The simple dc differential amplifier can be built with a higher gain by increasing the value of the feedback resistors, or, by decreasing the value of the input resistors. Keep in mind, however, that the input impedance is a function of those input resistors, so there will be a practical minimum value for their value. That value would be set in part by the output impedance of the signal source, and should be at least ten times the drive source impedance.

An example of a gain-of-10 dc differential amplifier is shown in Fig. 6-12. This is the same circuit as we saw in Fig. 6-11, except that the gain is increased to ×10 by increasing the value of the feedback resistors (R3 and R4) to 100 kohms. Like any other operational amplifier stage, the gain of the dc differential amplifier may have a practical limit caused by the bias currents, offsets, and the drift factors for any given device. Gains up to 200 or so with cheap components are feasible.

Before going further, let's talk a little about common mode rejection of the differential amplifier. A differential signal is formed by applying different voltages to the two inputs of the differential amplifier. If the same voltage is applied to the two inputs, or, if two separate but equal voltages are applied to the inputs (functionally the same thing as connecting the two inputs together and applying the same voltage), then there will be a zero output from the op amp. This phenomenon is caused by the fact that the two inputs provide equal but opposite gains (in other words, A_{V1} has the same absolute value as A_{V2}, but opposite polarity). In Fig. 6-13 we see a model of this situation. Voltages V1 and V2 represent the individual input voltages, and in this case the differential voltage would be (V1-V2). The *common mode voltage* is the one which affects both inputs (V_{cm}). If the operational amplifier were ideal (it's not!), then the output voltage due to this common voltage would be zero. But in real, practical kind-you-can-actually-buy operational amplifiers the common mode gain is not zero, but will have some small value. Thus, there will be a small voltage at the output due to V_{cm}.

The common mode rejection ratio of the operational amplifier gives us some indication of the extent of this problem. Figures for CMRR vary from 40 dB to 120 dB, depending upon the quality of the device; the larger the number the better the suppression of common mode signals.

But we also find that external components will affect the CMRR of the amplifier. If the condition of equality is not met (i.e., R1 = R2 and R3 = R4), then the CMRR will be deteriorated. We can change the basic circuit of Figs. 6-11 and 6-12, however, and insert a CMRR ADJ control. Resistor R4 in both cases is connected between the noninverting input and ground. Replace this resistor with either a potentiometer, or, a series combination of a poten-

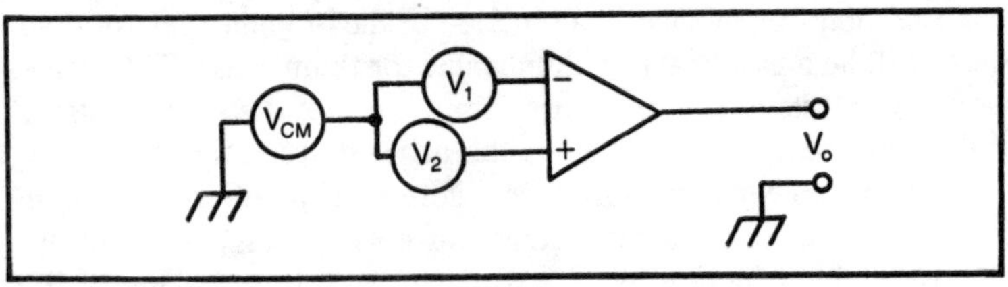

Fig. 6-13. Model of the differential amplifier.

tiometer and a fixed resistor (latter method preferred). Make the potentiometer a value that is approximately 20 percent of the total value of R4, and the value of the fixed resistor about 90 percent of the total. This arrangement will allow the actual resistance to be varied from 0.9R3 to 1.1R3, thereby allowing the fine tuning of the CMRR. You will see this scheme in several projects to follow. Adjustment of the CMRR ADJ control is as follows:

1. Set the potentiometer about mid-range.
2. Short the two inputs together (not at the op amp, but at the "input end" of the input resistors, i.e., R1 and R2 in Fig. 6-12).
3. Apply a signal to the shorted-together inputs, usually 1-volt P-P is sufficient; frequency should be 50 to 1000 Hz.
4. Use either a voltmeter or oscilloscope to monitor the output ('scope preferred).
5. Apply power.
6. Adjust the oscilloscope input sensitivity to show a readable signal on the output of the op amp.
7. Adjust the potentiometer in the "R4" slot for minimum signal output. Note that the signal will reduce as you adjust the pot, but will increase again after you pass the zero point.
8. Repeat step no. 7 several times, using successively lower input deflection factors on the 'scope, until no further improvement is possible.
9. Disconnect the signal source and remove the short between the two inputs.

In most cases, the CMRR ADJ potentiometer will be a multiturn (10 or more) trimmer potentiometer, and will be mounted on the printed circuit board. This adjustment rarely requires readjustment over the life of the equipment, so will not require any panel mounted controls. In fact, most companies secure the screw on the potentiometer by using a little *RTV* or *Glyptol* to prevent it from shaking loose.

GAIN OF 500 DC DIFFERENTIAL AMPLIFIER

The problems of high gain amplifiers with differential inputs are pretty much the same as for single-ended amplifiers, with the possible exception (as some textbooks proclaim) that the drift characteristics with some operational amplifers may be worse. Thus, we will want to keep in mind the same sort of factors as in the previous cases.

Figure 6-14 shows the circuit for a dc differential amplifier with a gain of ×500. The input resistors are 20 kohms, but these can be reduced to 10 kohms (with corresponding change in the feedback resistors, or, change in gain) without unduly affecting the driver stage. Like any differential operational amplifier, the circuit in Fig. 6-14 has a gain of:

$$A_{vd} = R3/R1$$

or,

$$A_{vd} = (R4 + R5)/R2$$

Also like the other operational amplifier circuits of this type,

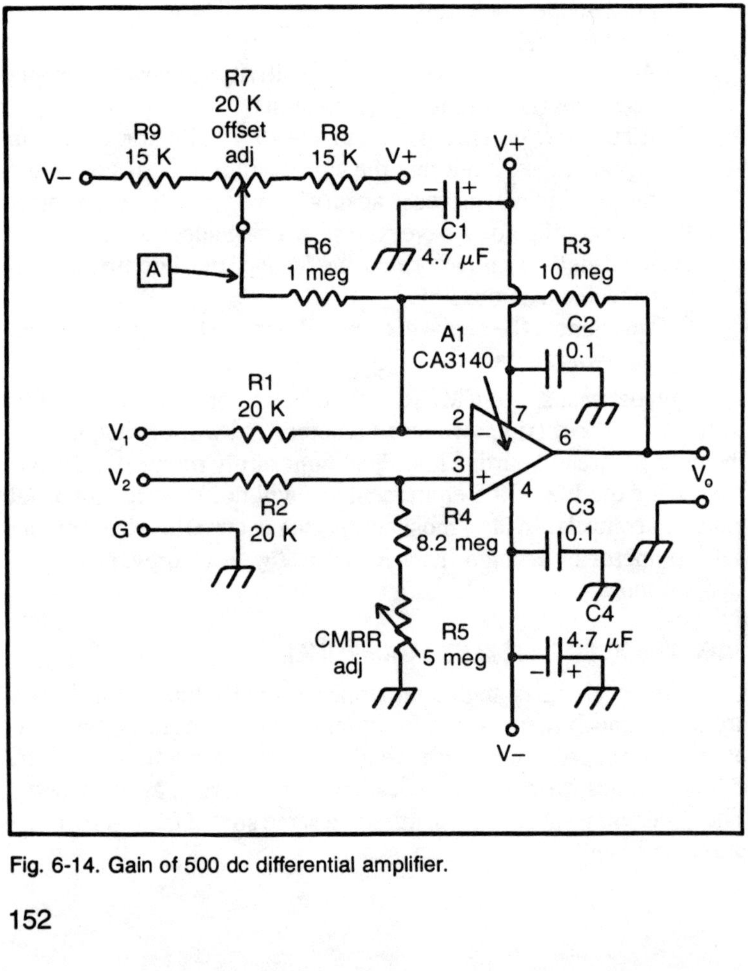

Fig. 6-14. Gain of 500 dc differential amplifier.

there must be a balance such that R1 = R2, and R3 = (R4 + R5); resistor R5 is adjusted to permit the combination R4 + R5 to be equal to R3. Potentiometer is the CMRR ADJ control for this amplifier.

Potentiometer R7 in Fig. 6-14 is used to null the output offset voltage, and works by injecting a counter-offset into the system. The combination of the two offsets produces a net result of zero volts output. Adjustment of the circuit is as follows:

1. Set potentiometer R5 to approximately mid-range.
2. Set potentiometer R7 to approximately mid-range.
3. Short V1 to V2.
4. Connect a dc-coupled oscilloscope or voltmeter to V_o; if a 'scope is used, center the zero-volts base line at mid-screen.
5. Apply a signal (see instructions given earlier) to V1/V2.
6. Adust R5 for minimum ac signal on the output; if the effect of R7 is great enough to place the signal off-screen, then readjust R7 to bring the ac signal on-screen.
7. Repeat step no. 6 using successively more sensitive settings of the output indicator ('scope or VM), until no further improvement is possible.
8. If the steps above are not possible, then the problem might be too great a voltage at point A. This voltage is produced by the potentiometer R7. Adjust the pot for zero volts at point A and repeat steps 5-7.
9. Disconnect the signal source and connect V1 and V2 to ground.
10. Adjust R7 for zero volts output.
11. R5 and R7 may tend to interact a little bit, so you might want to repeat the adjustments several times, or until you are satisfied that no further improvement is possible.

The capacitors used in Fig. 6-14 are used for decoupling, and tend to prevent oscillation of the amplifier. These capacitors are especially needed if the operational amplifier used has a high frequency response, or, is uncompensated. Such amplifiers tend to oscillate under some circumstances. The 0.1 μF capacitors are especially critical in this regard, and must be mounted as close to the operational amplifier as possible. The 4.7 μF units are for lower frequencies, and can be mounted close to the operational amplifier, but are not as critical in this regard as the 0.1 μF units.

The V+ and V– power supplies must have voltages higher than the maximum required output voltage. In the CA3140 device used here, however, the difference between the power supply voltages and the maximum output voltage are a lot less than in, say, 741 devices. The positive output voltage can come within 0.5 volts of the positive supply rail (V+), while the negative output voltage may be within about 1 volt of the V− supply rail.

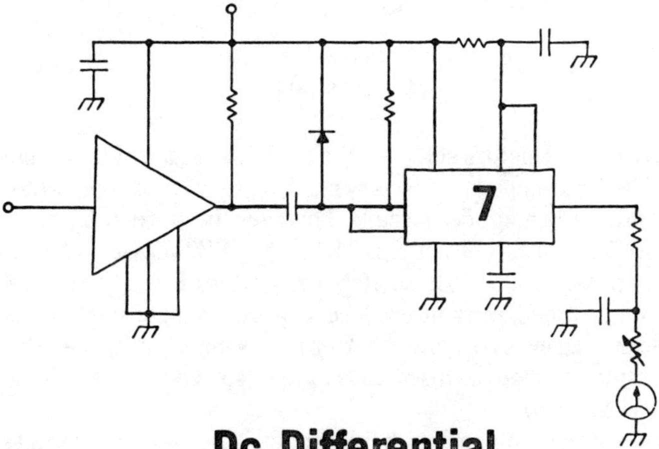

Dc Differential Instrumentation Amplifiers

The dc differential amplifier is the basis of most operational amplifier instrumentation circuits. These amplifiers are characterized by extremely high input impedance and gains that range from low to very high. There are also certain special purpose operational amplifier circuits which fall into this category, but are sometimes classified as "computer programmable amplifiers," "signals acquisition amplifiers" and so forth. These different types of amplifiers are considerably more advanced than some of the simple circuits considered earlier, so are part of this separate chapter. The simple dc differential amplifier and the classical three-op-amp instrumentation amplifier were introduced at the end of Chapter 6. Some of that material is reviewed here.

HIGH-IMPEDANCE DIFFERENTIAL AMPLIFIERS

The simple dc differential amplifiers shown thus far in this section suffer from the same problem, although not as acutely, as the inverting follower: low input impedance. The differential input impedance is set by the two input resistors, so can be quite low, especially at high gains. We can, however, produce a very high input impedance version by using the scheme of Fig. 7-1. Here, we have two unity gain noninverting follower stages, A1 and A2, driving the inputs of the simple dc differential amplifier. Since the gains of A1 and A2 are both unity, the gain of this stage is set by the usual equation:

$$A_{vd} = R3/R1$$

or,

$$A_{vd} = R4/R2$$

With the resistors selected in Fig. 7-1, the gain of this stage is ×10. The input impedance is very much higher than our previous dc differential ×10 amplifier circuits, however, because of A1 and A2. If these devices are either a BiMOS or BiFET device, then this input impedance might reach 10^{12} ohms or so. Even the cheapest operational amplifiers will produce 1 megohm input impedance in this circuit. Drift is controlled better in this circuit if A1 and A2 are in the same package, as in the case of the 1458 and CA3240 devices (latter preferred).

One of the problems with this circuit, however, regards waste of components. It would be better if we could use the two input amplifiers to provide some of the gain. If we did that, we would have the classic three-op-amp instrumentation amplifier (IA) of Fig. 7-2.

CLASSIC INSTRUMENTATION AMPLIFIERS

We can use three operational amplifiers (introduced in Chapter 6) in a circuit like Fig. 7-2 in order to obtain a high gain, flexible differential amplifier with extremely high input impedance specifications. Three operational amplifiers are used, A1–A3. The two

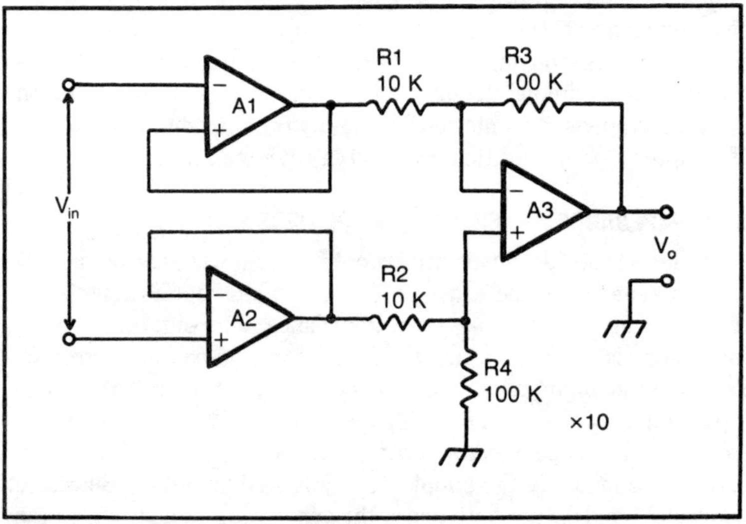

Fig. 7-1. Gain of 10 instrumentation amplifier.

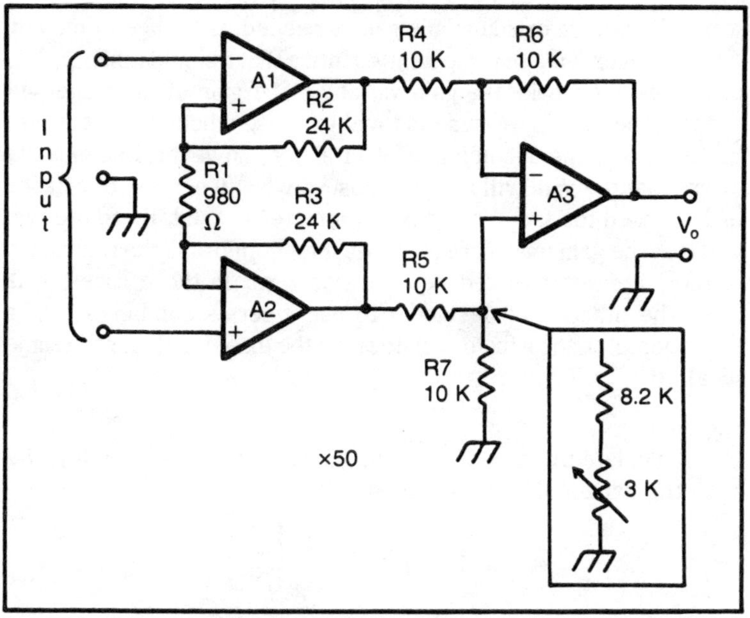

Fig. 7-2. Gain of 50 instrumentation amplifier.

input operational amplifiers, A1 and A2, should be identical with each other and are often found in the same package. An RCA CA3240 device, for example, would provide both the common thermal environment and the extremely high input impedance which is the hallmark of the ideal instrumentation amplifier. The two input operational amplifiers form one stage, while the output amplifier (A3) forms the second stage. The overall gain of the amplifier is the product of the individual gains of these two stages:

$$A_{vd} = \left[\frac{2R_2}{R_1} + 1\right]\left[\frac{R_6}{R_4}\right]$$

Where:

R_2 is the resistance of R2
R_1 is the resistance of R1
R_4 is the resistance of R4
R_6 is the resistance of R6

Note: all of the resistances must be in the same units, ohms, kohms, or megohms.

In most cases, we find that amplifier A3 will have a fixed gain, while, if variable or adjustable gain is needed, the stage formed by A1/A2 will provide that capability. Notice that a single resistor, R1, can be used to make the gain variable. The gain of the stage will reduce when R1 increases, and will increase when R1 is reduced. Since R1 is in the denominator of a fraction, however, one must be aware that the gain will increase absurdly fast when the resistance is decreased too far. When a potentiometer is used, therefore, we find that the gain may increase too fast for comfort as the resistance of the potentiometer reduces towards zero. In those cases, it is often the practice of designers to use a series combination of a potentiometer and a fixed resistor (see the inset for R7 for the same idea).

Example

Calculate the gain of the instrumentation amplifier in Fig. 7-2 with the values of resistors shown.

$$A_{vd} = \left[\frac{2R2}{R1} + 1\right]\left[\frac{R6}{R4}\right]$$

$$A_{vd} = \left[\frac{(2)(24)}{(0.98)} + 1\right]\left[\frac{10}{10}\right]$$

$$A_{vd} = \left[\frac{(48)}{(0.98)} + 1\right]\left[1\right]$$

$$A_{vd} = (48.98 + 1)$$

$$A_{vd} = 49.98$$

The gain error of this circuit is expressed by:

$$\text{Error} = \frac{(50 - 49.98)(100\%)}{(50)}$$

$$\text{Error} = \frac{(0.02)(100\%)}{(50)}$$

$$\text{Error} = 2/50$$

$$\text{Error} = 0.04 \text{ percent}$$

The equation given above for the gain of the classic instrumentation amplifier depends upon the following conditions:

$$R2 = R3$$
$$R4 = R5$$
$$R6 = R7$$

Interestingly enough, the match R2 = R3 does not create a high CMRR, but is necessary to a low gain error. The other matches, R4 = R5 and R6 = R7, are important to the maintenance of a high common mode rejection ratio (CMRR).

Like any differential amplifier, we are interested in the instrumentation amplifier in maintaining the CMRR as high as possible. We can adjust out certain forms of errors by replacing R7 with a series combination of two resistors, as shown in the inset. The potentiometer is adjusted for maximum CMRR using the procedures given earlier in this section for other amplifiers.

We can standardize the circuit of Fig. 7-2, and thereby create a nearly "universal" amplifier in which the gain is set by a single resistor. This is the method of the IC instrumentation amplifiers which we will discuss in a few minutes. If, for example, we keep R4, R5, R6, and R7 as they are in the figure, and make R2 and R3 some well-behaved easy to remember number—like 20 kohms for example, then we can write a version of the basic equation which will result in a value for R1 when the desired gain is plugged in (which is the way most readers will need the equation—not the textbook format given earlier!). If we make R2 = R3 = 20 kohms, then:

$$A_{vd} = \left[\frac{(2)(20\text{ K})}{R1} + 1\right]\left[\frac{(10\text{ K})}{(10\text{ K})}\right]$$

$$A_{vd} = \left[\frac{40\text{ K}}{R1} + 1\right]\left[1\right]$$

$$A_{vd} = \frac{40\text{ K}}{R1} + 1$$

$$A_{vd} - 1 = \frac{40\text{ K}}{R1}$$

$$R1 = \frac{40\text{ K}}{A_{vd} - 1}$$

Where:

$R1$ is the resistance of R1 in kohms

A_{vd} is the gain of the amplifier

GAIN OF 1000 INSTRUMENTATION AMPLIFIER

The classic instrumentation amplifier can be used to provide gains up to ×1000, or so, with little in the way of headaches. Figure 7-3 shows such a circuit. Amplifiers A1 and A2 are parts of a common CA3240 device, which means that the input impedance will be extremely high (on the order of 10^{12} ohms). The gain of the circuit is given by:

$$A_{vd} = \left[\frac{(2)(24\ K)}{(0.98)} + 1\right]\left[\frac{(200\ K)}{(10\ K)}\right]$$

$$A_{vd} = \left[\frac{(0.98)}{(48)} + 1\right]\left[20\right]$$

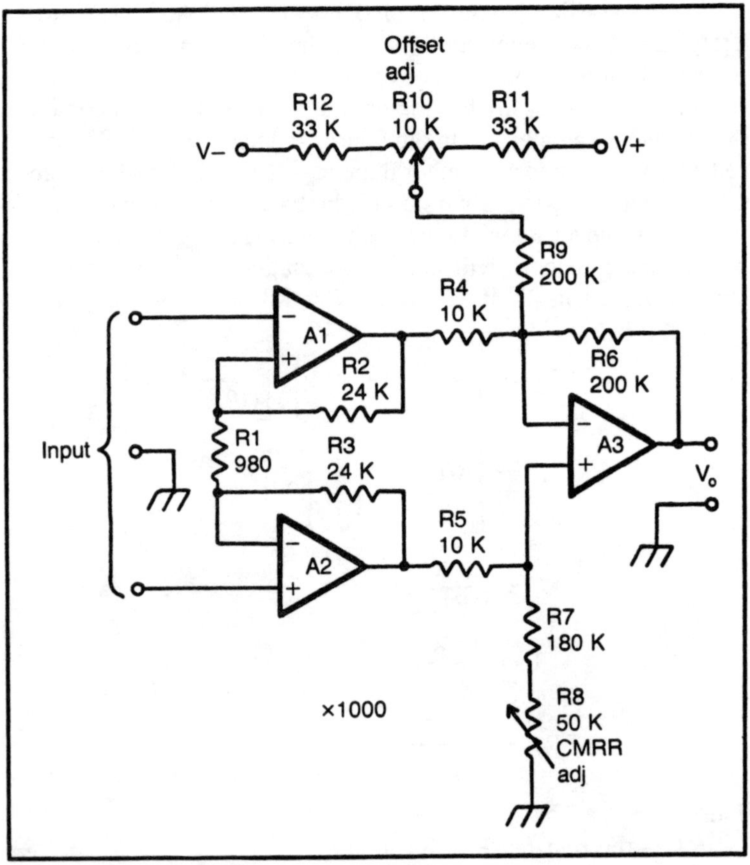

Fig. 7-3. Gain of 1000 instrumentation amplifier.

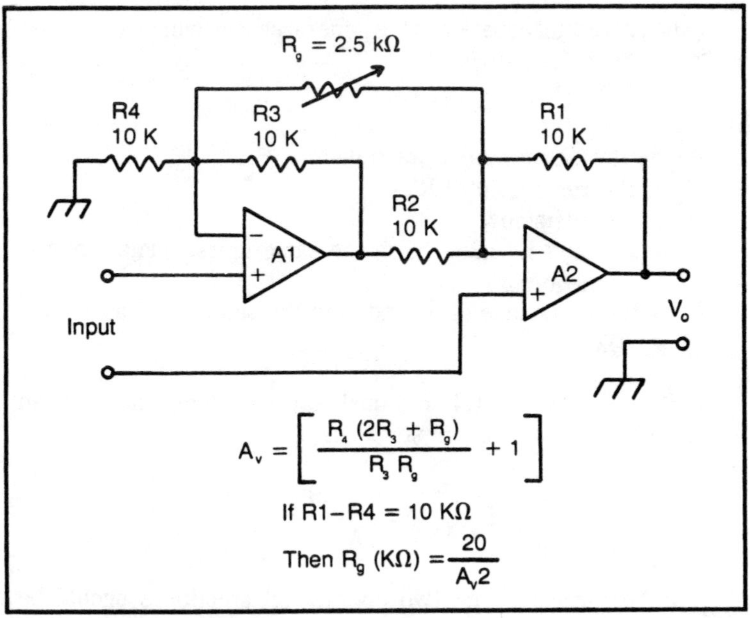

Fig. 7-4. Two-amplifier IA.

$$A_{vd} = (48.98 + 1)(20)$$
$$A_{vd} = (49.98)(20)$$
$$A_{vd} = 999.6$$

The gain error is:

$$A_{vd} = \frac{(1000 - 999.6)(100\%)}{(1000)}$$

$$A_{vd} = (0.4)(100\%)/1000$$
$$A_{vd} = 40\%/1000 = 0.04 \text{ percent}$$

TWO-OP-AMP INSTRUMENTATION AMPLIFIER

Another form of instrumentation amplifier, less used than the previous circuit, is shown in Fig. 7-4. This circuit uses just two operational amplifiers, both of which are in the noninverting follower configuration. The gain of this circuit is given by:

$$A_{vd} = \frac{(R_4)(2R_3 + R_g)}{R_3 R_g} + 1$$

(The preceding equation is the general case in which we can set the values of the resistors.)

Where:

A_{vd} is the differential voltage gain
R_3 is the resistance of R3
R_4 is the resistance of R4
(R3 and R4 must be in the same units, ohms, kohms, or megohms)
R_g is the resistance of R_g and is in the same units as R3/R4 R3/R4

If R1, R2, R3, and R4 are equal, say 10 kohms, then we can reduce the equation to a simpler form:

$$R_{g(kohms)} = \frac{20}{A_{vd} - 2}$$

For best results, the two operational amplifiers should be matched, and it is highly recommended that a dual (or more) operational amplifier be used. The 1458 (a 741-family device) and the BiMOS CA3240 devices are considered good selections in their own areas of competence.

AC INSTRUMENTATION AMPLIFIERS

The differential amplifiers which we have presented are mostly dc amplifiers. This *does not* mean that they will not amplify ac signals, for they will do that nicely. The ac signals will be amplified up to the frequency limit of the amplifier. But the circuit also amplifies dc signals, which may or may not be an advantage. In some cases, we find that the ac signal is what is of interest, yet there may be a substantial dc component on the signal which tends to obscure the real interesting data. In the medical electronics field, for example, we often see huge dc offset potentials in the presence of tiny ac signals. The electrocardiograph and electroencephalograph (ECG and EEG, respectively) use metallic electrodes applied to the patient's skin in order to pick up tiny biopotentials. The typical EEG signals are approximately 5 to 100 microvolts, with frequency components from 0.05 Hz to 30 Hz (some authorities use 250 Hz). The ECG signal is standardized at a bandwidth of 0.05 Hz to 100 Hz, with a peak amplitude of the main feature (i.e., the QRS complex) of about 1 millivolt. Unfortunately, these signals are acquired using metallic electrodes applied to essentially electrolytic skin through electrolytic electrode jelly or paste . . . the result is a tiny battery at the electrode skin interface. The electrode potential can be as much as 2 volts, but is usually on the order of 0.5 to 1 volt. Consider the implications. In EEG amplifiers, the electrode half-cell potential is a dc offset that is 10,000 to 20,000 times higher than the signal source! In the ECG case, the 1-volt half-cell potential is 1000 times higher than the signal. Obviously, an amplifier which will be suitable for EEG or ECG signals will be totally saturated by the dc half-cell potential if a dc amplifier is used. The solution seems to be an ac amplifier.

While we recognize that the amplifier for such applications as mentioned above must be accoupled, it is also true that the low-end −3 dB point on the frequency response curve must be very near ac—0.05 Hz in the case of the biopotentials amplifiers. The reason for this is the low frequency nature of the signals would cause major features of the waveform to disappear if the lowest frequencies were obliterated. In designing the low frequency amplifier, then, we must use a frequency response close to dc.

Figure 7-5A shows the basic circuit for the ac-coupled input stage for the classic instrumentation amplifiers shown earlier. There is an RC network added to each input such that:

$$R4 = R5$$

and,

$$C1 = C2$$

The low-end −3 dB point is given by the expression:

$$F_{Hz} = \frac{1}{2 \pi R C}$$

Where:

F_{Hz} is the −3 dB frequency (in hertz) for the frequency response curve
R is the resistance (R4 or R5 in Fig. 7-5A), in ohms
C is the capacitance in farads (C1 or C2 in Fig. 7-5A)

Example:

Assuming that R4 = R5 = 10 megohms (for sake of input impedance), calculate the values of the capacitors (C1 and C2) which will result in a low-end −3 dB frequency of 10 Hz.

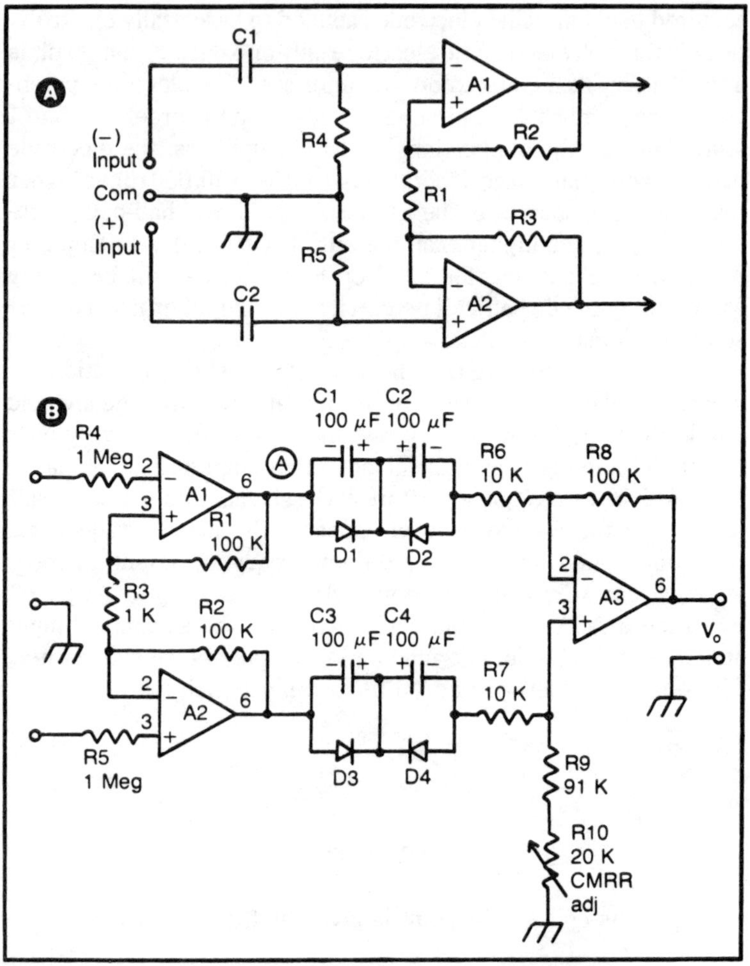

Fig. 7-5. A) Ac coupling at input of IA, B) inter-stage ac coupling.

$$F_{Hz} = \frac{1}{(2)(3.14)(R)(C)}$$

$$10_{Hz} = \frac{1}{(2)(3.14)(10^7 \text{ ohms})(C)}$$

$$C = \frac{1}{(2)(3.14)(10^7 \text{ ohms})(10 \text{ Hz})}$$

$$C = 1/6.28 \times 10^8$$

$$C = 1.59 \times 10^{-9} \text{ farads}$$

$$C = 1.59 \times 10^{-3}\ \mu F = 0.00159\ \mu F = 0.0016\ \mu F$$

It is essential that the input components be matched in order to prevent deterioration of the common mode rejection ratio. The capacitors are a little more difficult to match, but an attempt should be made. If you cannot match the components using instruments, then buy precision components. The resistors should also be matched closely, at least to 1 percent or so. Resistors are available to 0.05 percent, if you want to pay the cost. Unfortunately for the small builder, these resistors are a little difficult to obtain in quantities of less than fifty pieces each.

Another approach, common in biomedical instruments, is shown in Fig. 7-5B. This amplifier uses the ac-coupling between the stages and the output stage. The input lines may or may not use the method of Fig. 7-5A; the input resistors are used when the operational amplifiers are not terribly high impedance types (e.g., 741 amplifiers). If BiMOS or BiFET amplifiers are used, then the 1-megohm input resistors may not be needed. Those newer operational amplifiers have tremendous input impedances, so provide a great deal of isolation between the power supplies and the input lines.

Note: the biomedical amplifiers presented in this book are for example only, and are not intended to be used with human subjects. There are severe patient safety problems associated with medical electronics, and one should be cognizant of them before pursuing any design or development on instrumentation intended to be used on humans. There is also a regulatory problem, since the U.S. Food and Drug Administration takes an interest in medical devices offered for sale.

The ac-coupling circuits used in Fig. 7-5B may look a little odd at first. Note that the diodes shunt the capacitors. The purpose of these diodes is to short out the unused capacitor on the "wrong" half of the ac waveform. Keep in mind that the electrolytic capacitors used in these applications are polarity sensitive. Consider the signal at point "A" in Fig. 7-5B. When this signal is positive, then diode D1 will be forward biased, and therefore shorts out capacitor C1. Diode D2 is reverse biased, so does not affect C2; the signal passes through C2 to the output amplifier. On the negative half of the signal, diode D1 will be reverse biased, so the signal passes through C1. But diode D2 is forward biased, thereby shorting out C2. Thus, we find that only one of the capacitors is in the circuit at a time. Consequently, the capacitance seen by the circuit is not the series capacitance of the C1/C2 combination, as one might initially suspect, but the capacitance is $C1 = C2 = C = 100\ \mu F$. The diodes used in Fig. 7-5B are 1N914, or equal, signal diodes. In some older equipment, the 1N60 type of diode was used.

IC INSTRUMENTATION AMPLIFIERS

The instrumentation amplifier (IA) has proven to be such an impressive device for so many different analog circuit applications that it has been made into a special-purpose integrated circuit in its own right. Several different manufacturers offer these devices, but we have space here for only a couple of the most common and better known devices.

The first device which we will consider is the Burr-Brown Model INA-101 device. This instrumentation amplifier is packaged in the 10-pin metal TO-100 IC can. The device is built to have an ultra-low voltage drift, a specified value on the order of $0.25\ \mu V/°C$ (i.e., microvolts per degree centigrade); the offset voltage is $25\ \mu V$ while the nonlinearity is a low 0.002 percent. The device is also characterized as "low noise," and has a spec of $13\ nV/(Hz)^{1/2}$ at a frequency of 10 kHz. The common mode rejection ratio (CMRR) is 106 dB at 60 Hz (which is where the CMRR rejection is needed by most users!). The input impedance is 10^{10} ohms, so this device can easily be used in biomedical applications and others where low amplitude signals must be acquired from sources that have high output impedances. The INA-101 is also low cost, at least relative to other IC and hybrid instrumentation amplifiers. The dc power supplies operate to ±20 volts dc, and dissipates 600 milliwatts. The input signal is allowed an excursion equal to the power supply voltages, a factor that is a little unusual for operational amplifiers. The operating temperature range of the INA-101 is the standard industrial range, 0 to 70° C, while a version in military temperature range (−55 to +124° C) is available at higher cost.

Figure 7-6 shows a sample circuit based on the Burr-Brown INA-101 IC instrumentation amplifier. This is a differential amplifier that can be programmed to have gains up to ×1000 with a single external resistor, R2.

In the circuit of Fig. 7-6, the common lead is grounded, and the power supply terminals are connected to V+ and V− (dual polarity supplies). Note that the power supply lines are bypassed each with two capacitors. The $0.1\ \mu F$ capacitor in each line is used for high frequency decoupling, while the $4.7\ \mu F$ capacitor is for low frequen-

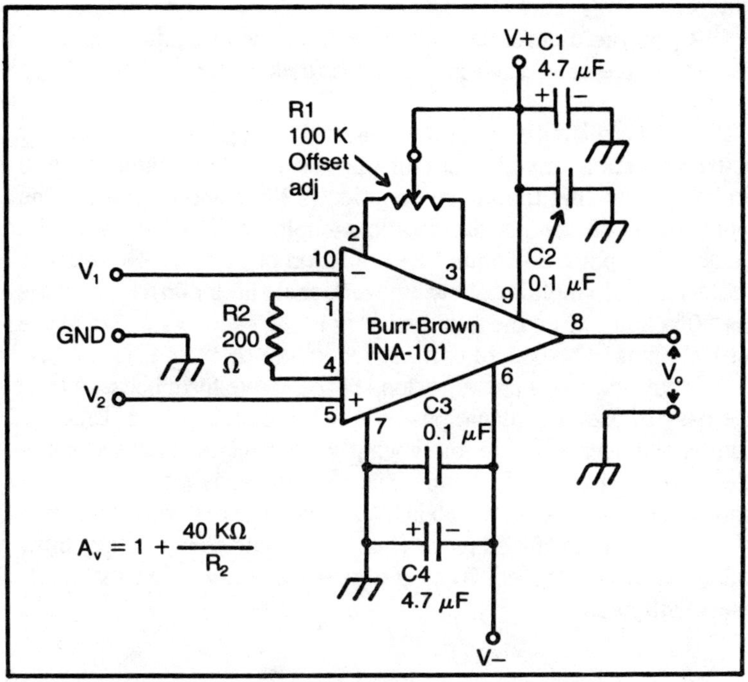

Fig. 7-6. Burr-Brown INA-101 instrumentation amplifier.

cies. These capacitors should be mounted as close as possible to the body of the ICIA. The 0.1 μF capacitors are the most critical and should be mounted even closer to the amplifier than the 4.7 μF devices.

The inputs (pins 5 and 10) and output (pin 8) terminals are not remarkable in this circuit, and are standard. If we want this device to be an ac-coupled amplifier, we can place a capacitor in series with each input line, and then place a 10-megohm resistor from each input terminal to common/ground. The input impedance on the INA-101 is so high that the resistors may prove to be optional for many applications. The purpose of the resistor is to keep the input bias currents from charging the capacitors, and thereby causing a dc input signal that will eventually (perhaps in a very short time in cheaper devices!) cause the amplifier to latch up.

The gain is set by resistor R2, according to the following relationship:

$$A_{vd} = \frac{40\ K}{R_2} + 1$$

Where:

A_{vd} is the differential voltage gain of the amplifier.
R_2 is the resistance in kohms of resistor R2.

If you are familiar with the classic three-operational-amplifier instrumentation amplifier circuit discussed earlier, then you will probably note that the above equation is the same as that for the input stage of the instrumentation amplifier if resistors in the feedback loop are 20 kohms. Examination of the Burr-Brown data sheet for that amplifier show, indeed, that the feedback resistors are 20 kohms, while the four resistors in the second stage (A3) are all 10 kohms. The gain of that second stage, therefore, is one (1).

Most textbooks use equations of the above form because they are easy to use in explanations of how the circuit works. Unfortunately, the user wants to know what value resistor (R2, for example) will yield a specified gain. You see, the gain will probably be known, and the designer's job is to calculate the resistor(s) that will yield that predetermined gain. Thus, we should do a little algebraic manipulation on the equation and produce the version that most users will need:

$$R_2 = (40\ \text{K})/(A_{vd} - 1)$$

Where:

The terms are exactly as given above

The output offset null function is provided by potentiometer R1. This arrangement is pretty standard for IC amplifiers, except that the wiper of the potentiometer is connected to V+ instead of V−, as is the usual case.

The project shown in Fig. 7-6 is a gain of X200 differential amplifier because of the value of R2 (200 ohms). We can adjust this circuit for other gains by changing the value of R2 according to the above equation. We can also make it a variable gain by making R2 a series combination of a fixed resistor and a potentiometer. This might be done in order to obtain variable gain under front panel control, or, it might be done in order to trim gain to some precise value. We can also select from several different resistors in the R2 position using either DIP switches, a front panel switch, or, an electronic switch. In the latter case, we can control the CMOS switch from a microcomputer and make a computer settable gain device.

National Semiconductor makes a series of four integrated circuit instrumentation amplifiers of two different types. The generic number for all four devices is LM-363. In Fig. 7-7, we see a circuit involving three fixed gain versions, namely:

LM-363-10
LM-363-100
LM-363-500

These devices can be designated generically as LM-363-xxx, in which the "xxx" is the gain. These devices cannot be easily programmed for variable gain, and are fixed into gains of ×10, ×100, or ×500 (three very popular gains).

The basic circuit of Fig. 7-7 can be used with all three devices, so is reasonably universal in that regard. The two inputs, IN1 and IN2, are inverting and noninverting respectively. There is a ground connection (i.e., a common) on the LM-363-xxx device. These devices are packaged in the 8-pin metal IC can. The gain can be varied, but only with difficulty. The frequency response, however,

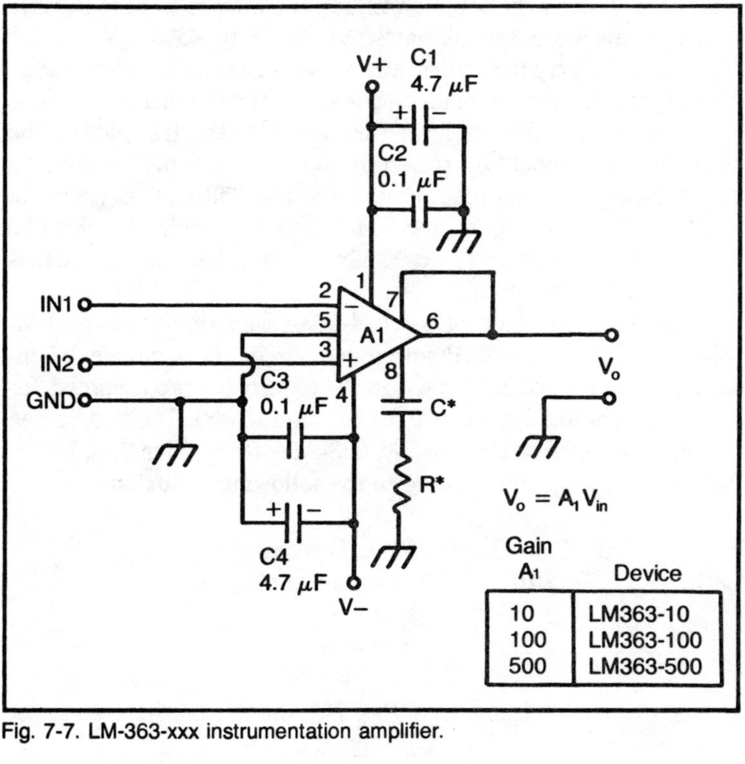

Fig. 7-7. LM-363-xxx instrumentation amplifier.

can be varied with an external lag network consisting of a series combination of a resistor and a capacitor.

The use of small gain blocks open a number of applications where circuits had previously been a little more complex. There are several variations of a circuit which I have published in TAB Books publications before. The original project was a transducer (resistive Wheatstone bridge) preamplifier used to help a physiologist friend of mine with a signals acquisition problem which he was suffering through. That preamplifier is a lot easier to build, and is therefore somewhat cheaper in production versions, if an LM-363-10 or LM-363-100 device was used instead of the 741-type of device used originally. There is also a possibility that the LM-363-xx device can be obtained in chip die form by transducer manufacturers, and built into the transducer. Such a device would have a much higher sensitivity that present resistive strain gauges, and would thus be easier to apply. A typical fluid pressure transducer, for example, has a sensitivity of 5 μV/V/mm Hg (i.e., 5 μV output per volt of excitation of the transducer per mm Hg of pressure), so are almost always very low-voltage output devices. If the excitation potential is 5 Vdc (presently the common value), and the pressure is 100 mm Hg, then the output potential would be only 2500 μV, or 2.5 millivolts. If we try to acquire such a signal through a 10-foot cable, then we find that the 60 Hz pick-up is so substantial that the signal is obscured. But, if the amplifier was built into the transducer, the output potential could be raised to 25 mV or 250 mV, depending upon which gain is selected. Even if the LM-363-xxx device is not available in chip die form, the LM-363-xxx is small enough to be mounted on the transducer externally (a project based on this idea is found later in this book).

The fourth version of the LM-363 series device is the LM-363-AD. This 16-pin DIP integrated circuit is a unique IC instrumentation amplifier, and can be externally programmed for three different fixed gains: ×10, ×100, and ×1000. There are three gain-setting terminals on the LM-363-AD device: pins 2, 3, and 4. We can set the gain according to the following protocol:

Gain Level	*Short Together*
10	(none-leave all open)
100	3 to 4
1000	2 to 4

An external switch is shown in Fig. 7-8, and this can be used to set

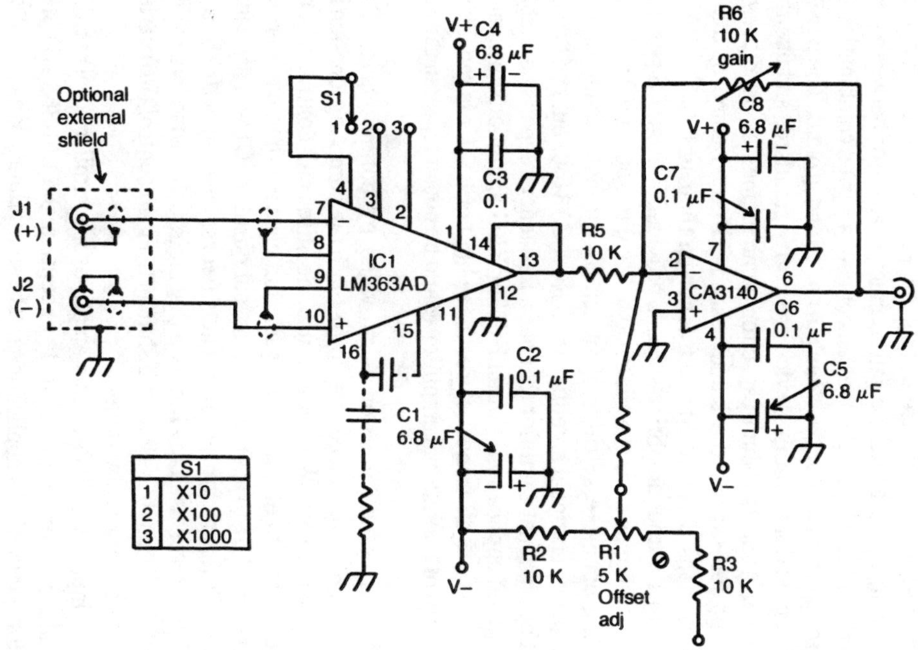

S1	
1	X10
2	X100
3	X1000

Fig. 7-8. LM-363-AD gain-of-10-100-1000 amplifier.

the gain to whichever is needed. Alternatively, we can use printed circuit board jumpers; can be used to set the gain on a custom basis. We can also use CMOS electronic switches in order to set the gain under microcomputer control. The gain setting techniques are wide, and that makes this device a candidate for many different applications.

The frequency response of this amplifier can be set in one of two different ways. First, we can connect a capacitor between pins 15 and 16, or, alternatively, we can connect a series combination of a resistor and a capacitor between pin no. 16 and ground. The power supply connections on this device are pretty standard, and use the same 0.1 μF and tantalum capacitor combinations as used previously, and for exactly the same reasons. Reread some of those earlier sections if you need a refresher on their use.

A 0 to 1 output amplifier stage is provided in Fig. 7-8 permits us to vary the gain over the ranges 0 to 10, 0 to 100, and 0 to 1000, depending upon the settings of switch S1 and potentiometer R6. The output operational amplifier is an RCA CA3140, but can be any form of operational amplifier that has a frequency response suitable for the given application. For an ECG or EEG amplifier, for example, an upper −3 dB frequency response limit of 100 Hz is required, so any form of operational amplifier would be possible, even the old 741 device.

The offset null function is provided by potentiometer R1, which is used to inject a counter-offset into the summing junction of the output amplifier. We can also use the offset null control as a dc balance control to eliminate the gain shift caused when we go from one position of S1 to another. The output offset of IC1 will probably be a little different for different gains (worst case on ×1000), and this circuit can be used to eliminate that shift. Otherwise, we would find that the output base line will shift as the gain control (R6) varies from zero to maximum value.

The input terminals of this amplifier are similar to other differential amplifiers with one little exception: the LM-363-AD provides for *guard shield* applications. These methods, which will be dealt with shortly, are used in low signal level applications where the common-mode rejection ratio must be maintained at any cost. Such examples include professional-grade EEG amplifiers, certain transducer applications or other cases where a low-level signal must pass through a region of extreme interfering signals (such as 60 Hz fields that are present in most locations, especially industry). For further information, see the *Guard-Shield Section* given below.

GUARD-SHIELD THEORY

Figure 7-9A shows the usual situation for a differential amplifier which uses shielded input lines, especially where low-level signals are processed. Voltages V1 and V2 are the components which form the differential input voltage (V2-V1), which is the voltage that the amplifier wants to see. These voltages are applied to the differential inputs of the amplifier through lengths of shielded cables. There is also one additional voltage present on the system, V_{cm}. The V_{cm} voltage is the common-mode potential, and is applied equally to both inputs. If the differential amplifier has a high CMRR, then this voltage will not significantly affect the output potential. For example, if the CMRR was 100 dB, we find that the output potential from this voltage will be 100 dB below the output that would be produced if the same voltage was a differential signal. If this signal is, for example, the 60 Hz interference picked up from the ac power mains in the room, then the interference is suppressed and is not a problem.

Resistor R in Fig. 7-9A is the source impedance of the signal source. This value might be quite low, as in the case where the driver is another voltage amplifier which has a low (e.g., 100 ohms) output impedance, or, quite high in the case of biophysical amplifiers. In a moment, we will use this resistor to show a potential problem.

The shielded cable consists of a coaxial arrangement in which there is an inner conductor separated from the outer shield by an insulator material. There is a capacitance formed between the inner conductor and the shield. This capacitance is due to the fact that these two conductors are in close proximity to each other.

Figure 7-9B shows a model of the circuit rearranged to present the source resistance and the capacitances in a different light. Resistors R1 and R2 are components of R (in Fig. 7-9A), and are

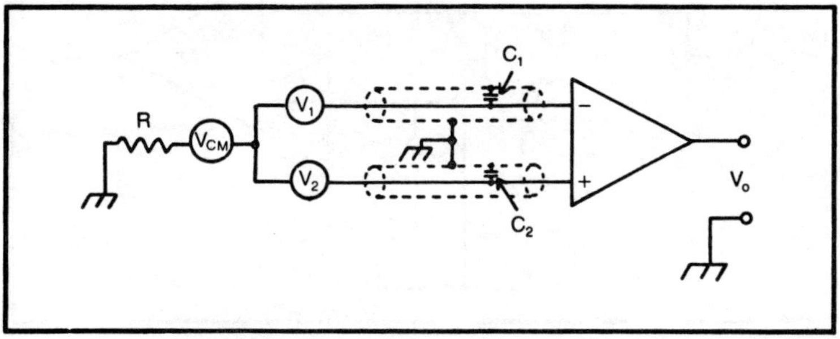

Fig. 7-9A. Standard shielding arrangement.

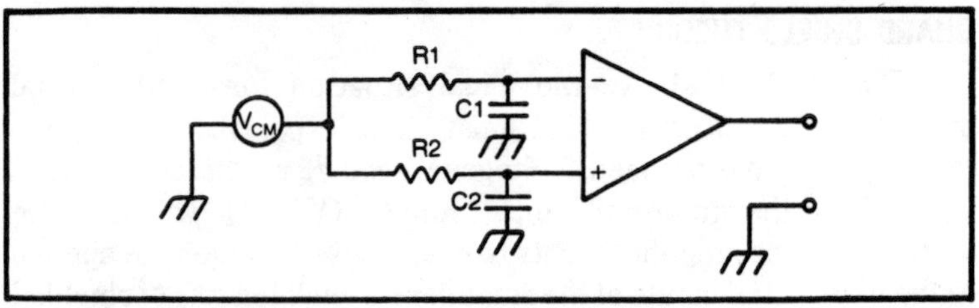

Fig. 7-9B. Equivalent circuit.

each equal to approximately R/2. When this circuit is drawn we see that it forms a low-pass filter RC network consisting of a series resistance and a parallel capacitance. There is one of these low-pass filters for each input. The −3 dB limit caused by these networks is given by:

$$F_{Hz} = \frac{1}{6.28\ R\ C}$$

Where:

F_{Hz} is the frequency in hertz of the −3 dB point
R is the resistance of either R1 or R2, in ohms
C is the capacitance of either C1 or C2 in farads

Although the −3 dB frequency might prove to be a problem, that problem can be overcome by additional frequency-shaping, or, other tactics. We have another problem, however, that is somewhat more sinister. Suppose that we do not have both of the following equalities: R1 = R2 and C1 = C2. If either R1 ≠ R2, or C1 ≠ C2,

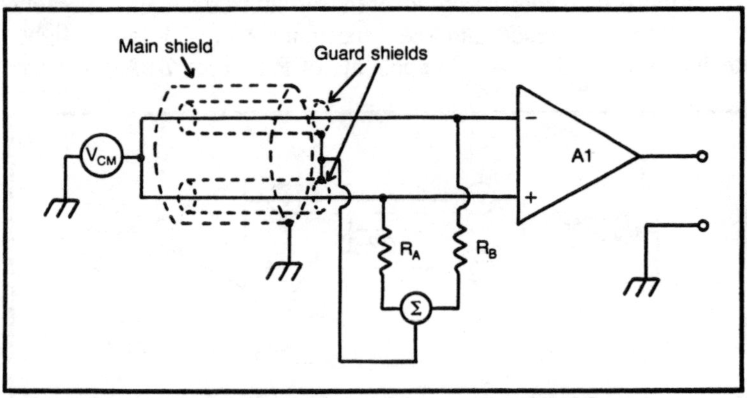

Fig. 7-9C. Guard-shield circuit.

then there is an unbalance in the network. Under that condition, the voltage across the capacitors from the common-mode signal source will be unequal. Thus, we have a situation in which a common-mode voltage is made into a differential signal and therefore produces an offset potential at the output, *sigh*.

In some cases, designers will go to extreme lengths in order to keep these two factors equal. But perhaps the best method is to use *guard-shield* methods, as shown in Fig. 7-9C.

In a guard-shield system, the two signals from the two input lines (or some point inside of the amplifier) are summed together in a resistor network, or an amplifier (usually with unity or very low gain), and then applied to the shields on the wires. This places the shields at the same potential as the signal conductors, so effectively neutralizes the capacitance, or at least its effect. The main shield surrounds the guard shields, and is insulated from them. This main shield can be grounded, preferably at one end only. In certain medical amplifiers, notably the ECG amplifier, the guard-shield method is used, but the guard amplifier is called a "right leg amplifier."

ISOLATION AMPLIFIERS

There are many applications for differential amplifiers which require a high degree of isolation between the inputs and the power supply and/or the output terminal. In some industrial situations, for example, there are tremendous potentials available at the signal acquisition site, but the instrumentation operates at lower, or "normal" potentials. Another application is biomedical amplifiers in which there is a patient safety question. In the late sixties, there was a lot of concern over some archaic medical equipment that was apparently dangerous to certain patients. Those patients who had indwelling catheters, or, surgical wounds, may have been at risk from leakage currents from ac-powered equipment. Patients with certain types of cardiac catheters, either in the catheterization laboratory or in certain intensive care units, are at special risk because the catheter represents a direct line to the sensitive heart muscle. In those cases, we now connect the patient to the input of an isolation amplifier that provides 10^{12} or so ohms of isolation between the differential input terminals and the ac power mains.

There are several different ways to accomplish isolation. In many devices, the signal amplitude modulates a 50 kHz to 500 kHz carrier that is passed to the nonisolated section through a transformer that is practically an infinite impedance to 60 Hz. Power to the isolated section is also produced by a carrier of the same frequency, in sort of a dc-to-dc converter circuit. Other cases, such as the cardiac output computer which used the catheter discussed above, may use battery power for absolute isolation . . . the application *is* critical.

Another form of isolation amplifier uses optoisolators to couple the signal between the isolated and the nonisolated sides of the amplifier. The signal will either frequency modulate or amplitude modulate the carrier and then apply the carrier to the optoisolator.

Still another method, used by one of the more popular medical bedside monitors used in hospitals, was called ***current starvation***. As the signal amplitude varied, the current drawn from the carrier power supply (dc-to-dc converter) also varied, and the variation was reflected on the nonisolated side of the power supply.

Figure 7-10 shows a Burr-Brown hybrid isolation amplifier circuit. This particular device, is one of a series of low-cost isolation amplifiers which are offered by Burr-Brown. The 3652 and 3650 are cousins, with the 3650 being a current-input device, and the 3652 being optionally either a current-input (identical to the

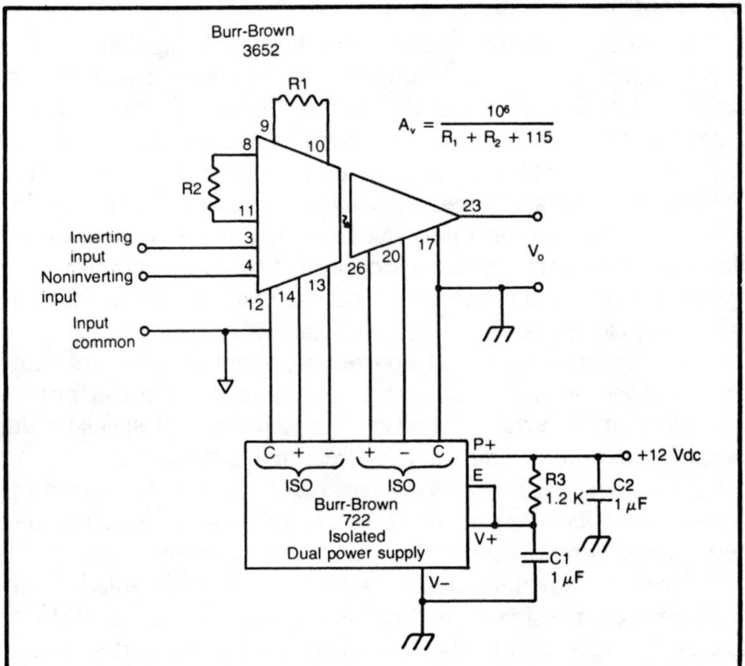

Fig. 7-10. Isolation amplifier.

3650) device or a voltage input device, depending upon the wiring of the input stages. These devices are especially designed to withstand large common-mode voltages (i.e., they offer 140 dB of rejection at ±2000 volts continuous applied potential). This feature means that the 3650 and 3652 can be used in applications such as medical electronics, where the defibrillator used in cardiac resuscitation may apply a couple of kilovolts for 5 to 10 milliseconds), or industrial applications where large potentials may be present at the signals acquisition site (a steel-mill example was given by one reader). The devices offer 0.25 microamperes of leakage current at 240 Vac (RMS) at a frequency of 60 Hz (where it is important); the leakage capacitance is 1.8 picofarads. The 0.05 percent linearity has excellent stability, on the order of 0.05 percent per 1000 hours. The 15 kHz −3 dB bandwidth and 1.2 V/μS slew-rate are more than sufficient for most instrumentation applications (except those with very fast rise-time signal components).

Notice in Fig. 7-10 that there are two independent ground connections. One is on the nonisolated side of the amplifier, while the other, indicated by an open triangle, is on the isolated side of the circuit. For the isolation integrity to remain high, it is imperative that these grounds are not connected together. Given that the ac power line is connected to ground on one side, it is wise to chassis-ground the nonisolated common connection (indicated by the conventional three-bar ground symbol), and float the isolated ground. The input signals are taken with respect to this floating, isolated ground. The output connections are taken with respect to the nonisolated ground.

The power connections are separated on the isolated and nonisolated sides of the 3650/3652 devices. Pins 12, 13, and 14 form the common, V− and V+ terminals for the isolated side of the amplifier, while pins 17, 20, and 26 form the common, V− and V+ terminals for the nonisolated side of the amplifier. The nonisolated power can be the same ±V power source as is used for the rest of the instrument, while the isolated supply must come from a source with high isolation of its own, or else the isolation will be deteriorated. The possible power sources include batteries or dc-to-dc converters that are based on a high frequency oscillator. In Fig. 7-10, we are using a dual dc-to-dc converter, the Burr-Brown 722 device (*B-B* also makes a quad device, the 724, which offers four low-current isolated dual-polarity dc sources). Since there is an extra "port" available on the 722, we will use it to power the nonisolated side of the 3652 device.

The gain of the circuit is set by a pair of resistors which are external to the case. The resistors, R1 and R2 in Fig. 7-10, are set to be equal to each other in most cases, even though that is not always a strict requirement. The gain expression is:

$$A_v = \frac{10^6}{R1 + R2 + 115 \text{ ohms}}$$

Where:

A_v is the gain
$R1$ is the resistance of R1 in ohms
$R2$ is the resistance of R2 in ohms
115 ohms is the nominal value of internal resistances which serve to affect gain of the device

As is usually the case, textbook equations rarely are published in the form which most readers need them. In the case of the above expression, we will usually set R1 = R2, so we may want to rewrite the equation algebraically to find this value, *R: see below.*

$$A_v = \frac{10^6}{R + R + 115}$$

$$R + R + 115 = 10^6/A_v$$

$$2R = (10^6/A_v) - 115 \text{ ohms}$$

$$R = \frac{(5 \times 10^5)}{A_v} - 57.5 \text{ ohms}$$

Example

Assuming that R1 = R2 = R, as in the expression above, calculate the value of R which will yield an overall gain of ×1000.

$$R = (5 \times 10^5/A_v) - 57.5 \text{ ohms}$$
$$R = (5 \times 10^5/1000) - 57.5 \text{ ohms}$$
$$R = (500) - (57.5 \text{ ohms})$$
$$R = 442.5 \text{ ohms}$$

Normally, unless gain error was of profound importance, the practice is to use a 440 ohm or 450 ohm resistor, which are the nearest standard precision resistor values.

Decoupling capacitors are normally used on the power supply

lines of the nonisolated side of the amplifier. In most cases, the recommendation would be 0.1 μF on each line for the high frequencies, and 1.0 to 5.0 μF on each line for the lower frequencies. Keep in mind that the polarities are different on those two power lines, and be careful to observe capacitor polarity.

LF152/LF352 INSTRUMENTATION AMPLIFIERS

Before leaving the subject of amplifiers altogether, let's take a look at one additional instrumentation amplifier which will prove popular with many readers: the National Semiconductor LF-152/LF-352 devices. The main difference between the '152 and '352 versions is the matter of temperature range. The '152 is a premium-grade device which operates to full specifications over the so-called "military" temperature ranges of −55° C to +125° C, while the '352 only operates over the "commercial" temperature range of 0° C to + 70° C. Another difference is the power supply potentials. The premium-grade LF-152 operates to potentials of ±22 volts, while the '352 device will only tolerate ±18 Vdc.

These devices are BiFET operational instrumentation amplifiers that have high-impedance junction field-effect transistors (JFET) in the input stages. The input impedance is, therefore, extremely high, on the order of 2×10^{12} ohms, with input leakage currents of 3 picoamperes. The LF-152/352 comes in a 16-pin DIP package, and features single-resistor gain setting, should you want it (actually, two resistors set gain, but we can leave one of them constant and vary the other to form gains over the range 1 to 1000). This latter feature makes it possible to vary gains over a wide range without having to first worry about deteriorating the common-mode rejection ratio (CMRR) in the process. It is also possible to drive guard-shields, as is done in Fig. 7-11.

The LF-152/352 devices feature a high CMRR, on the order of 110 dB, or more, at a gain of ×100. It is easy to adjust for output offsets, and to program the frequency response through the use of an external resistor-capacitor network.

The circuit in Fig. 7-11 is a high input-impedance instrumentation amplifier for use in a wide variety of applications in the physical and life-sciences, and in engineering. This device can easily be built into existing equipment, or, on its own. Many builders, who already have either *Tektronix* or *Hewlett-Packard* oscilloscopes or strip-chart recorders (or other brands which accept plug-ins), will buy the blank plug-in module kits from the instrument manufacturer (usually at a reasonably low price) and then build a circuit such as Fig.

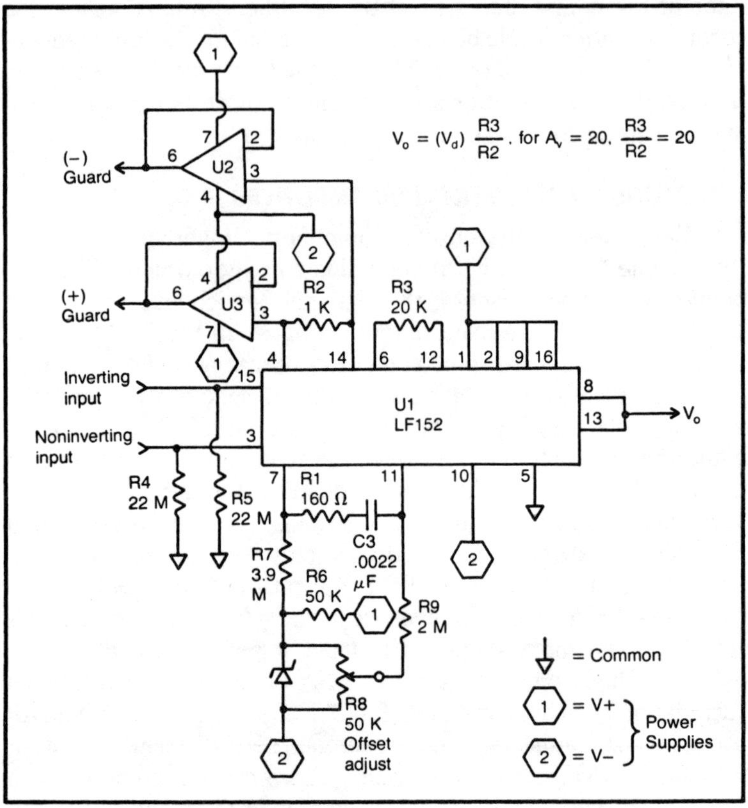

Fig. 7-11. LF-152 amplifier circuit.

7-11 inside. The result is a plug-in that is compatible with an existing instrument, looks real good (assuming that you don't botch the front panel!) and is custom-designed for your own application. You will also find that it is considerably cheaper in most cases to roll your own. A case in point is a project which I designed for a physiologist friend. He needed a medium-gain amplifier (i.e., ×100 or so), but lacked the $600 which the strip-chart recorder manufacturer wanted. We bought a blank plug-in for $65, and some operational amplifiers and other components for around $35, and invested 5-hours of electronic lab time. The result was a medium-gain amplifier which more than met the scientist's needs, at a cost which was one heckuva lot less than the cost of a new plug-in for the recorder. It could also be modified easily and without tears . . . something that one might have a little trouble convincing the powers-that-be on a $600 unit!

The guard-shield driver amplifiers, U2 and U3, are ordinary 741-family devices (as the pinouts show), or, they can be RCA BiMOS or some other form of premium operational amplifier. One modification which I recommend over the design as I originally built it (i.e., Fig. 7-11) is to make U2/U3 part of a dual op-amp package like the 1458 or RCA CA3240 devices. That way, the differences in thermal environment won't be a problem.

The gain of the amplifier shown in Fig. 7-11 is set at 20, which was needed in the original project. But that gain can be varied at will by using the expression:

$$A_v = R_3 / R_2$$

Where:

A_v is the gain
R_2 is the resistance of R2*
R_3 is the resistance of R3*
(* It is important that R2 and R3 be expressed in the same units, e.g., ohms, kohms, or megohms)

One variation on the theme used by many designers is to make R3 a fixed resistance, and then vary R2 to set the gain. It is usually the case that we will set R3 = 1 megohm (i.e., 1000 kohms), and then vary R2 to find the gain. If we want to express resistances in kohms, then the following will apply:

$$A_v = 1000 \text{ kohms}/R2$$

or, if we want to use ohms,

$$A_v = 10^6 \text{ohms}/R2$$

Solving these equations for the value of R2 when the required gain is known (the usual case):

$$R2_{(kohms)} = 1000/A_v$$

and,

$$R2_{(ohms)} = 10^6/A_v$$

Example

Find the value of R2, in ohms, which will yield a gain of ×500.

$$R2 = 10^6/A_v$$

$$R2 = 10^6/500$$
$$R2 = 2000 \text{ ohms}$$

If we want to make the gain variable, then we can make either R2 or R3 a potentiometer, or, alternatively, a combination of a potentiometer and a fixed resistor in series. In some cases, we might want to provide switch selectable resistances at R2, or a potentiometer, and a fixed resistor at R3. We could also use electronic CMOS switches to control the gain with a microcomputer or some other instrument. There are a lot of possibilities.

If we want the amplifier to be ac-coupled, which might be necessary in biomedical applications and some chemistry applications, then we may place equal valued capacitors in series with each input line. If these capacitors, C, are equal, then the frequency response would be given at the low-end −3 dB point by:

$$F_{Hz} = \frac{1}{2 \pi R C}$$

Where:

F_{Hz} is the low-end −3 dB point in the frequency response curve, in hertz

R is the resistance of R4/R5

C is the capacitance of the input capacitors

Now that we know the expression which the theory texts would have us use, let's rewrite it in a form which we all can use. We know the value of *pi*, and R in Fig. 7-11 is 22 megohms (2.2×10^7 ohms), so let's rewrite the expression above given these constants when we know the desired frequency response (again, the usual situation).

$$C = \frac{7.23 \times 10^{-9}}{F_{Hz}}$$

Example

Find the value of input capacitors (C) for the circuit of Fig. 7-11, when the frequency response must have a low-end −3 dB point of 0.05 hertz.

$$C = \frac{7.23 \times 10^{-9}}{F_{Hz}}$$

$$C = \frac{7.23 \times 10^{-9}}{0.05}$$

$$C = 1.446 \times 10^{-7} \text{ farads}$$

or, since microfarads are more useful:

$$C = 0.1446\ \mu F, \text{ or, } 0.15\ \mu F \text{ in standard values}$$

MORE BIOELECTRIC AMPLIFIERS

Most readers who have bought more than one of my books will note a distinct bioelectronic flavor to some of the projects. Part of this fact is due to my own involvement in medical electronics for almost a decade. Another factor is that these amplifiers also find wide application in a variety of instrumentation applications, not only by life-sciences people, but also by engineering, hobbyists, science fair entrants and ham radio operators (at least those who build some of their own instruments).

Note that these projects are intended mostly for animal experimentation except by those who know enough about electrical safety in the medical environment to safely use them on real people.

The amplifiers which I will present in this section are basically adaptations of earlier circuits for the specific purpose of electrocardiograph and electroencephalograph applications. The signal parameters for most such instruments are given below:

Electrocardiograph (ECG):
Signal Amplitude: 1 to 1.5 millivolts
Source Impedance: 1 to 10 kohms, depending upon type of electrodes used
Frequency Response: 0.05 Hz to 100 Hz for "diagnostic" types, and 0.05 Hz to 50 Hz for monitoring types of instruments

Electroencephalograph (EEG):
Signal Amplitude: 20 – 100 μV
Source Impedance: 1 to 10 kohms
Frequency Response: 0.05 Hz to 40 Hz

You will note that the low-end −3 dB point in the frequency response curve is indicated as 0.05 Hz. You might rightly ask why not dc? This frequency is so low as to be near dc, so why mess with the capacitors and resistors needed to so-limit the frequency response?

The answer is that ECG, EEG, and other bioelectric amplifiers must be ac coupled, but also have low-frequency components near dc. The reason for this is that the metal electrodes are connected to the conductive and electrolytic human skin using an electrolytic paste or jelly. This condition creates a battery potential, called a half-cell potential, that can approach a volt or so. In other words, the dc offset created by the electrodes will be 1000 times higher than the signal potential. But, if we ac couple the instrument, then the dc component is sloffed off and no longer bothers us. The low frequency components are salvaged by placing the frequency response very near dc, e.g., 0.05 Hz.

In most cases, the gain of a bioelectrical amplifier is based on the assumption that the display device, usually a strip-chart recorder, an oscilloscope or the input of an A/D converter that feeds data to a computer, will use a 1-volt input range. If other input ranges are used, then adjust the gain accordingly. The gain will be 1000 mV/1 mV, or ×1000 if a 1-volt output is assumed. If the display device is an oscilloscope with a variable input sensitivity, then we can adjust the gain and the oscilloscope sensitivity according to our own requirements. For example, we could set the oscilloscope for a deflection factor of 100 mV/cm. If we want the peak of the waveform to occupy 4 cm on the vertical scale, then we will want a signal of 4 cm × (100 mV/cm), or 400 mV. The gain in that case will only be 400 mV/1 mV, or 400. Some oscilloscopes that use plug-in vertical modules can be directly driven. Some models have available blank-plug modules in which you can create your own circuitry. The deflection factor of the mainframe is typically 50 mV, in which case, the ECG amplifier will need a gain of only 50.

EEG amplifiers, are a little more difficult to construct, but are not beyond your ability with modern components. The primary problem is that the signal is so low in amplitude. Typical waveform amplitudes will be around 50 microvolts, so much higher gains are needed. In most cases, we find that for the same display device, an EEG requires 20 to 30 times the gain of an ECG amplifier. The input impedance should be very high, but ECG-level impedances will suffice. Noise is also a very important factor for EEG amplifiers. The low-level signals can be buried in the noise normally produced by cheap operational amplifiers, so one is usually better off using a low-noise model in the "front-end" of an EEG amplifier.

Figure 7-12A shows one of the optional circuits which we will provide in the next few pages. This circuit is called "Option-1" in our text. The project is little more than merely an ECG amplifier, but also contains a circuit called a QRS-detector. The "QRS-complex" of the ECG waveform is the sharp spike that is created when the heart contracts. This spike is used to trigger light or audible indicators, and is also the source of the signal used to count the heart rate in heart rate meters. The QRS-detector circuit is what produces the dramatic "beep-beep-beep" in medical movies on TV. In reality, most users of those monitors find the beep so disconcerting that they *turn it off* except when transporting the patient and cannot see the monitor screen. There are high and low alarms that will alert them if the beep-beep-beep stops.

The input amplifier is the main analog low-noise gain stage. This device is the Burr-Brown Model INA-101 low-noise integrated circuit instrumentation amplifier. This amplifier produces a gain that is determined solely by the resistance between pin nos. 1 and 4, according to the equation:

$$A_v = \frac{40\text{ K}}{R} + 1$$

Where:

R is the resistance between pins 1 and 4

In the case of the INA-101 stage shown in Fig. 7-12A, then, the gain can be set either high or low by use of two different resistors. If the jumper is in place, the gain will be:

$$A_v = \frac{40\text{ K}}{2.2\text{ K}} + 1$$

$$A_v = 18 + 1$$

$$A_v = 19$$

If the jumper is removed, then the total resistance is higher and the gain will be lower:

$$A_v = \frac{40\text{ K}}{R4 + R5} + 1$$

$$A_v = \frac{40\text{ K}}{(2.4 + 2.2)\text{K}} + 1$$

$$A_v = \frac{40\text{ K}}{4.6\text{ K}} + 1$$

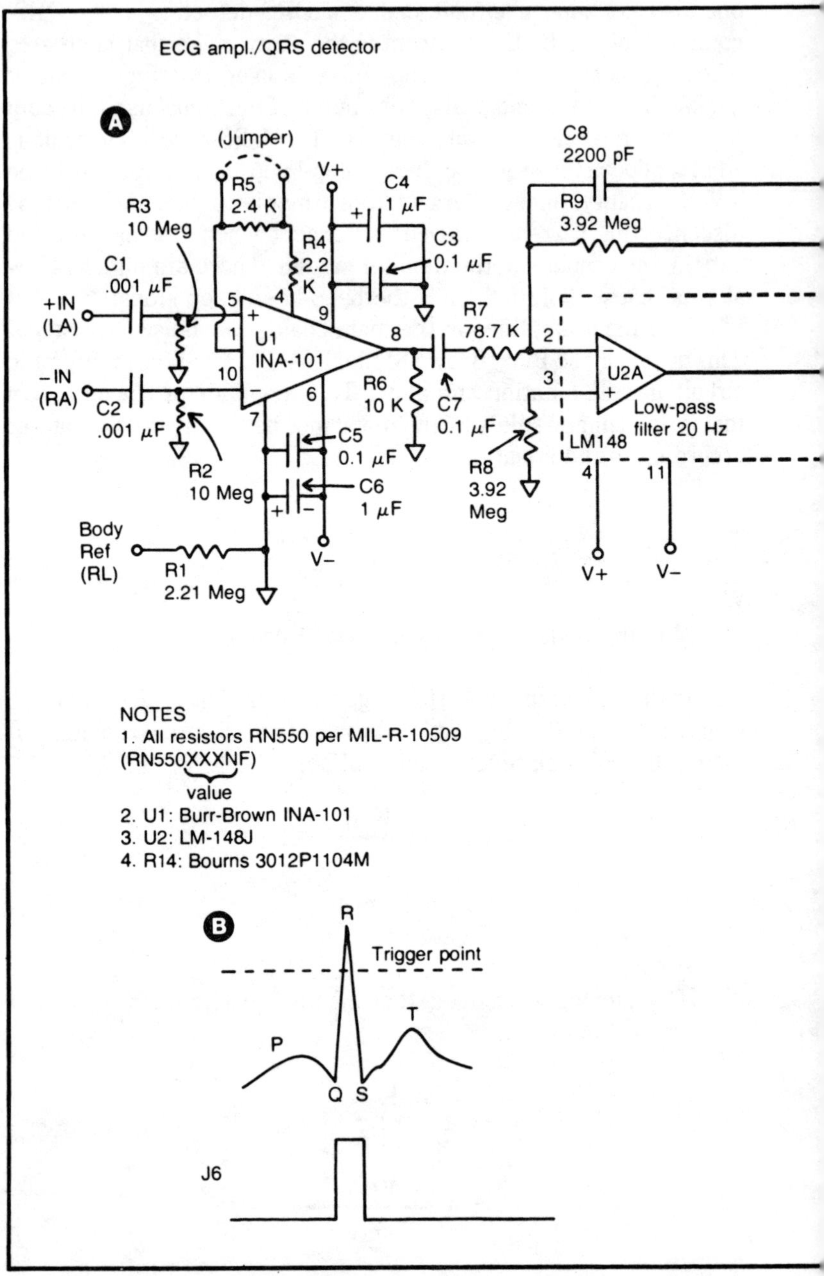

Fig. 7-12. Bioelectronic amplifiers A) Option-1, B) ECG trigger signal, C) example of double-triggering.

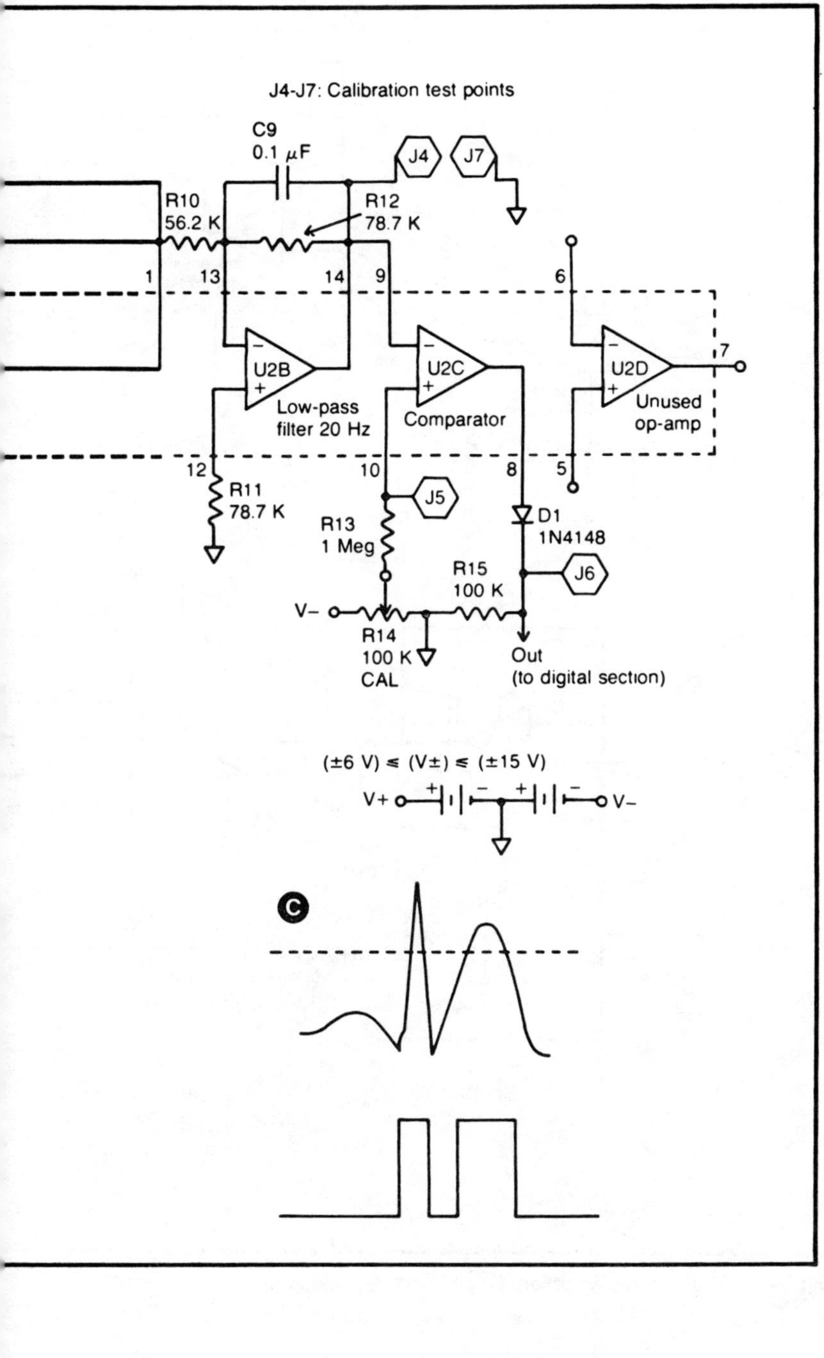
J4-J7: Calibration test points
C9
0.1 μF
J4
J7
R10
56.2 K
R12
78.7 K
1
13
14
9
6
U2B
Low-pass
filter 20 Hz
U2C
Comparator
U2D
Unused
op-amp
7
12
10
8
5
R11
78.7 K
J5
R13
1 Meg
D1
1N4148
R15
100 K
J6
V−
R14
100 K
CAL
Out
(to digital section)
(±6 V) ≤ (V±) ≤ (±15 V)
V+
V−
C

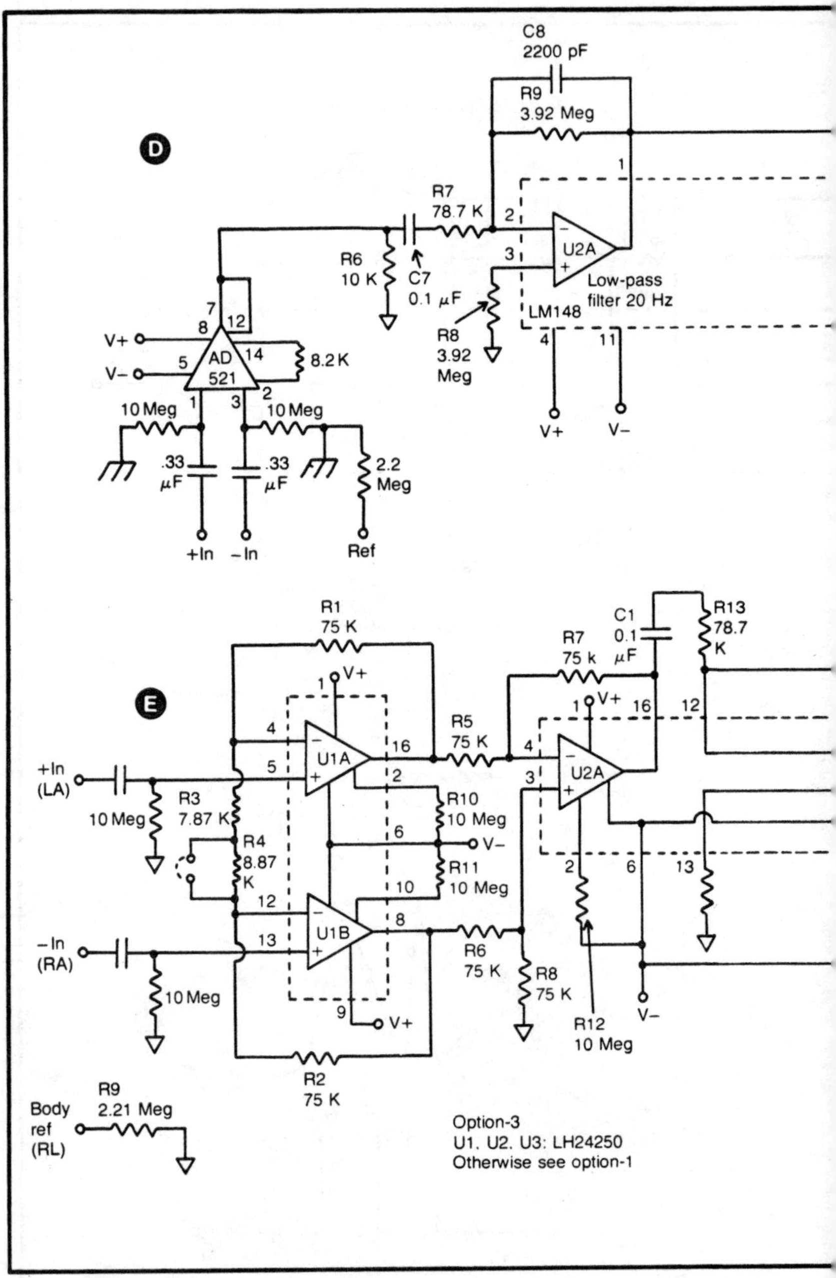

Fig. 7-12. Bioelectronic amplifiers D) Option-2, E) Option-3.

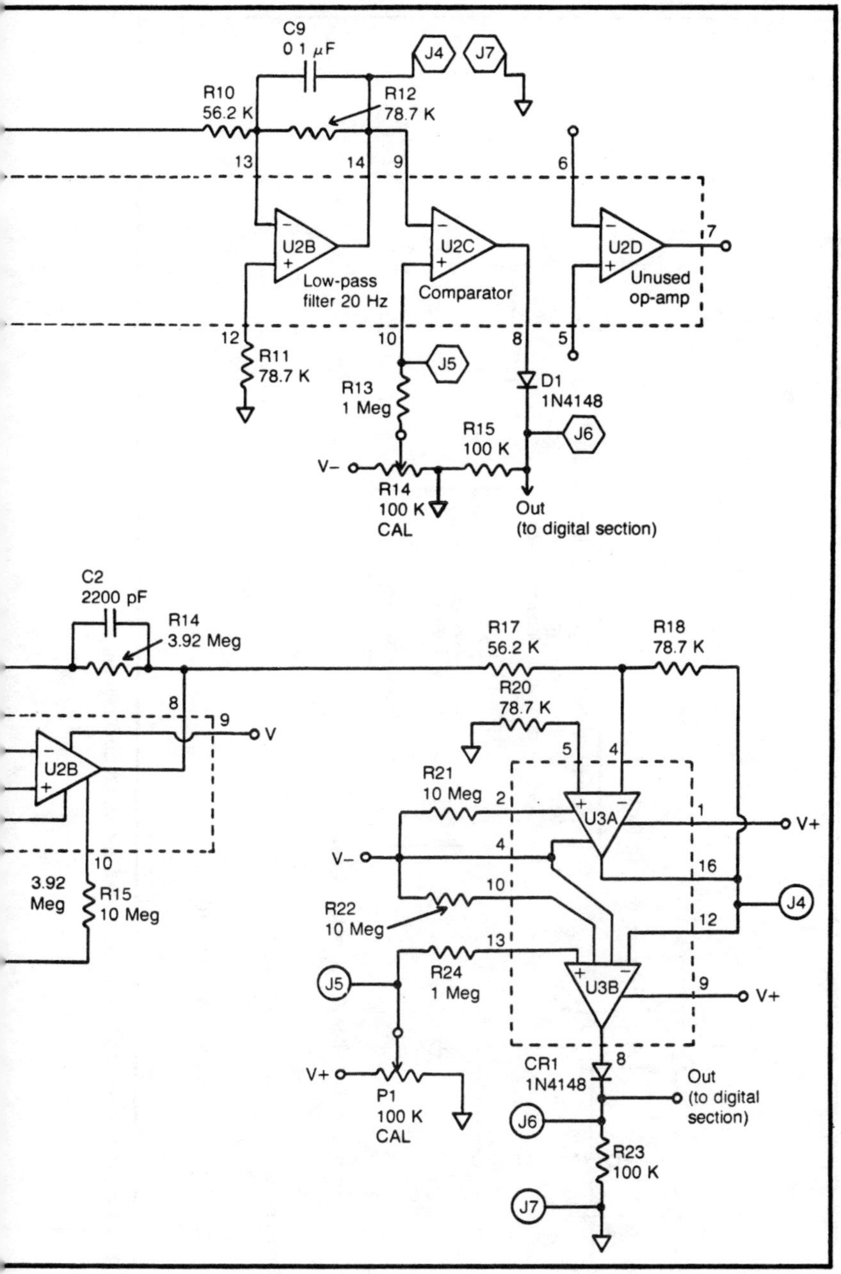

C9
0 1 μF
J4
J7
R10
56.2 K
R12
78.7 K
13
14
9
6
U2B
U2C
U2D
7
Low-pass
filter 20 Hz
Comparator
Unused
op-amp
12
10
8
5
R11
78.7 K
J5
R13
1 Meg
D1
1N4148
R15
100 K
J6
V–
R14
100 K
CAL
Out
(to digital section)
C2
2200 pF
R14
3.92 Meg
R17
56.2 K
R18
78.7 K
R20
78.7 K
8
9
V
U2B
5
4
R21
10 Meg
2
U3A
1
V+
V–
4
16
10
J4
R22
10 Meg
12
3.92
Meg
R15
10 Meg
13
R24
1 Meg
J5
U3B
9
V+
V+
P1
100 K
CAL
CR1
1N4148
8
Out
(to digital
section)
J6
R23
100 K
J7

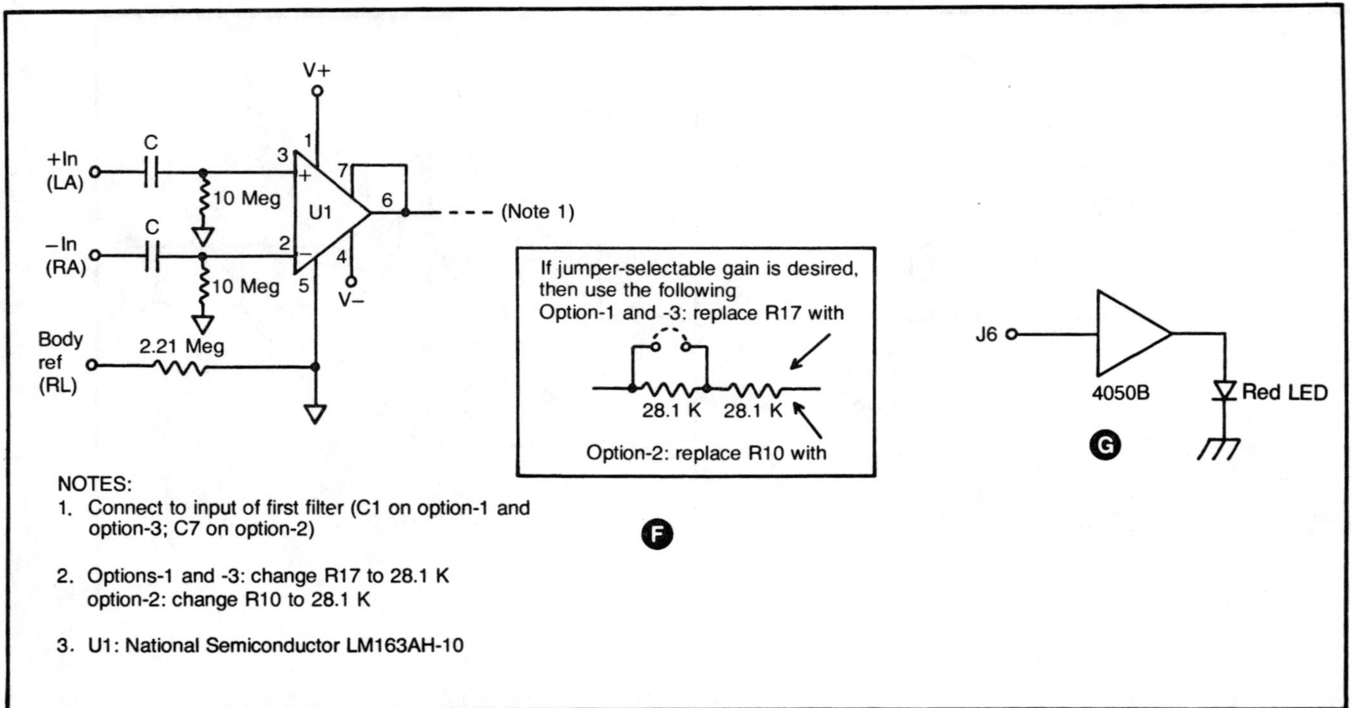

Fig. 7-12. Bioelectronic amplifiers F) alternate input circuit, G) LED blinker.

$$A_v = 8.7 + 1 = 9.7$$

Ac-coupling is provided by capacitors C1 and C2 in conjunction with resistors R3 and R2, respectively. It is important that these components be closely matched, if at all possible. The resistors can be bought as 1-percent or better precision resistors, while the capacitors may have to be hand matched on a bridge or other similar circuit (or, using one of those new digital capacitance meters now on the market). The frequency response here is:

$$F_{Hz} = \frac{1}{6.28\,R\,C}$$

$$F_{Hz} = \frac{1}{(6.28)(10^7\text{ ohms})(10^{-9}\text{ farads})}$$

$$F_{Hz} = 1/6.28 \times 10^{-2}$$

$$F_{Hz} = 16\text{ Hz}$$

The reason for this higher-than-normal frequency for the −3 dB point is that the QRS detector does not want to be fooled by the low frequency components of the waveform. If it is necessary to actually display the ECG waveform, then modify the circuit by replacing C1 and C2 with 0.33 μF capacitors.

The 2.2 megohm resistor in series with the "body reference" electrode is used for the sake of safety. In the case of a catastrophic failure in which some 120-volt power supply comes across the electrodes, the current to the patient will be limited to 120/2,200,000, or 5.5 microamperes ac. This resistor should be a 1-watt or larger if it is to survive the 120-volt line (which, hopefully, will never happen!).

The three body electrodes will be: RA, LA, and RL. For a Lead-I ECG signal, the RA is connected to the inverting input of the amplifier, the LA is connected to the noninverting input of the amplifier while the RL is connected to the *body reference* or "common" input of the amplifier. RA means "Right Arm," LA means "Left Arm, " while RL means "Right Leg."

Gain selection is by a jumper on the printed circuit wiring board. For low gain, the jumper is left open, while for high gain the jumper is in place (effectively shorting R5). The jumper could be replaced with either mechanical or electronic (CMOS) switch in order to allow external selection of the gain.

The capacitors connected to the INA-101 are placed as close as possible to the body of the amplifier, and are intended to decouple the circuit. If these capacitors are not used, then it is possible that the amplifier will oscillate.

If you want a displayable ECG signal, then change the capacitors as indicated above, and connect a gain-of-100 amplifier to the output of the INA-101. This step will provide the needed 1000 millivolts output with no restriction on the frequency response. If the capacitors are not changed, then the waveform will be distorted by smoothing of the low-frequency components. If the gain is not raised, then the signal can only be displayed on the most sensitive positions of your oscilloscope.

The QRS detector consists of U2 and the associated circuitry. The need for this circuit is shown in Fig. 7-12B and 7-12C. The ideal situation for the Lead-I ECG is shown in Fig. 7-12B. Only the QRS-complex passes over the threshold point where a pulse is generated; the P-wave and T-wave are not capable of passing this point, so will not create an output pulse. The output at point J6 will be one and only one square pulse per heart beat, as indicated by the QRS complex.

Another situation, less ideal, is shown in Fig. 7-12C. Here the T-wave will also cross the threshold and will therefore create an extra pulse. The extra pulse will cause an extra count. In my years as a Biomedical Equipment Repair Technician, and later as a Certified Clinical Engineer, I found numerous cases where the nurse would complain that the heart rate on the meter was too high for the patient's real heart rate, as measured the normal way with a sweep second hand. The usual case was that the heart rate indicated was *twice* as high as it ought to be. The problem was that the ECG amplifier gain was turned up too high, so created the double-pulse situation shown in Fig. 7-12C. This usually only occurs on patients whose T-wave or P-wave are almost as high as the QRS complex.

The solution might seem to be reduction of the gain range, but that is not useful for medical monitors because of the wide range of signal amplitudes the user will face. The better solution is to differentiate the ECG signal such that only high frequencies come through. Note that the slope and duration of the QRS complex are different from the other features. On the QRS complex, the slope is very high, and the duration short compared with features of the other two main waves. If we differentiate the ECG signal, therefore, we will suppress the low-frequency P-wave and T-wave features. Capacitor C7 and resistor R7 form a differentiating network with a low-end frequency response of 20 Hz or so. The frequency response will therefore roll off at approximately −6 dB per octave below 20 Hz. Additional roll-off is provided by stages U2A and U2B, which are 20 Hz low-pass filters. In this circuit, only a narrow band of frequencies around 20 Hz will pass, and this will heighten the QRS with respect to the other features, thereby making it more difficult to false trigger the detector stage (U2C).

The signal at J4 will be the trigger for the threshold detector, and is a much-distorted QRS waveform. The trigger stage is nothing but a comparator in which one voltage is the ECG-differentiated waveform, and the other is selected from a potentiometer. The output at J6 will be a pulse with a width a little narrower than the QRS complex.

A two-stage low-pass filter with a −3 dB frequency of 20 Hz is formed with two of the stages of integrated U2. This device is an LM-148J, which is a quad 741-family device. Such an IC contains four independent (except for power supply connections) operational amplifiers of the 741 class. "741-family" devices are frequency compensated and are considered unconditionally stable. Stages U2A and U2B form the low-pass filter. The filtering action is not difficult to see in this circuit. The gain of a normal inverting follower is given by the ratio of the feedback resistor to the input resistor (e.g., R9/R7 for U2A and R12/R10 for USB. But this gain is modified when a capacitor is placed across the feedback resistor. At frequencies below the −3 dB cut-off frequency, the gain is given solely (well, . . . *almost*) by the ratio of the resistors. But the gain will roll above the −3 dB frequency at a rate determined by:

$$F_{Hz} = \frac{1}{2 \pi R_f C_f}$$

Where:

- F_{Hz} is the −3 dB frequency, in hertz
- R_f is the feedback resistance (e.g., R9 and R12 in Fig. 7-12A) in ohms
- C_f is the feedback capacitance shunting the feedback resistance (e.g., C8 and C9), in farads

Stage U2C is a threshold detector, and is used to generate the output pulse. This stage is used as a voltage comparator. The calibration potentiometer (R14) is used to set the noninverting input of U2C at some negative voltage, which can be measured on test point J5. The ECG signal is applied to the inverting input. If both voltages are equal, then the output of U2C is zero. If, on the other hand, the voltage on the inverting input is more positive than the voltage on the noninverting input (*note:* more positive could also mean "less negative"), then the output of U2C will be saturated high negative. Since diode D1 will be reverse biased, however, this pulse will not appear on the output (J6). If, on the other hand, the voltage (ECG signal) on the inverting input is more negative than the noninverting input, then the operational amplifier (U2C) thinks that it is seeing a negative input signal, so the output will be high positive (saturated). This signal is propagated to the output terminal J6.

The dc power supply for this circuit can be any voltage between ±6 volts dc and ±15 volts dc. The power supply lines probably do not need bypassing in the vicinity of the LM-148 device, but bypassing *is* advisable near the INA-101. Each power supply lead is bypassed with 0.1 μF and 1 μF, and these should be placed as close as possible to the body of the operational amplifier. The 0.1 μF unit should be closer than the 1 μF, because it is used to decouple high frequencies.

Figure 7-12D shows the same circuit with the LM-148 replaced with an *Analog Devices, Inc.* AD-521 instrumentation amplifier. The gain of this amplifier is set approximately by the 8.2 kohm resistor between pins 2 and 14. Otherwise, the circuit is the same.

The last circuit of this type which we will consider is shown in Fig. 7-12E. Here we have replaced both integrated circuits of the past two circuits with a series of three dual-operational amplifiers, the LH-24250. These amplifiers can be replaced with the RCA CA3240 device, incidentally, or any other dual operational amplifier, should you care to use a more commonly available op-amp pair. The pinouts would have to be modified on the drawing wherever necessary.

U1A and U1B for a pair of input amplifiers in a differential-input/differential-output circuit. Taken with U2A, these two amplifiers form a classical three-amplifier instrumentation amplifier, a circuit which is described earlier in this book. The gain of U1 is:

$$A_{v(U1)} = \frac{(2)(75\ K)}{R3 + R4} + 1$$

Thus, when the jumper is in place, the gain of the stage will be

$$A_{v(U1)} = \frac{150\ K}{7.87\ K} + 1$$

$$A_{v(U1)} = 19.1 + 1 = 20.1$$

When the jumper is removed, the gain is reduced to:

$$A_{v(U1)} = \frac{150\ K}{(7.87 + 8.87)K} + 1$$

$$A_{v(U1)} = \frac{150\ K}{16.74\ K} + 1$$

$$A_{v(U1)} = 8.96 + 1$$
$$A_{v(U1)} = 9.96 \approx 10$$

For the remainder of the stages, the gains are given by the ratios of the feedback to input resistances, as modified by the capacitance above the −3 dB points. The low-pass filter stage is contained fully within U2. The input stage, U2A, will convert the differential output of U1 to a single-ended output as required by the rest of the stages. The gain of this stage is unity because all four resistors in this differential amplifier are equal to each other. The gain of U2B, however, is set by the ratio R14/R13, which is approximately 50. The gain of the final amplifier (U3A) stage is R18/R17, which is 1.4. The total gain of the amplifier is the product of the stage gains. In this case, this product is:

$$A_{v(t)} = A_{v(U1)} \times A_{v(U2)} \times A_{v(U3)}$$

when the jumper is in place, therefore, the gain will be

$$A_v = (20)(50)(1.4)$$
$$A_v = 1400$$

and when the jumper is removed:

$$A_v = (10)(50)(1.4)$$
$$A_v = 700$$

If you want to make the gain of this amplifier variable between two limits, then we can place a potentiometer in the circuit instead of R3 and R4. I recommend using a 6.8 kohm resistor in series with a 10 kohm potentiometer. The 10 kohm potentiometer can be either panel-mounted or trimpot style, depending upon application (i.e., whether or not you want a gain control available to the user).

Figure 7-12F shows a new front-end for the previous circuits based on the *National Semiconductor, Inc.* LM-363-10 amplifier. This fixed gain instrumentation amplifier has been described earlier in this book. It functions as a differential gain block with a very high input impedance. The gain is fixed, and will be ×10 (LM-363-10), ×100 (LM-363-100), or ×500 (LM-363-500) depending upon the type number ordered. In our case, the gain-of-10 LM-363-10 device is selected for this application. The LM-163AH-10 device called out in the schematic drawing is a military temperature range model (−55° C to +125° C) of the basic LM-363 device.

The other circuits have to be changed if this amplifier is used. The gain will not be maintained unless one of the stages following the "front-end" have increased gain.

Also, in the differential stage just discussed above, we find that it is necessary to convert this stage into a simple inverting follower. In Fig. 7-12E, therefore, remove resistors R6 and R8, and ground no. 2 of U2A.

The output of the threshold detector in all versions of this circuit is a short duration pulse, slightly less in duration than the QRS complex of the ECG waveform. This signal was intended for inputting to the cardiotachometer (heart rate meter) circuit. The circuit can also be used to create a blinking light that will light up once per heart beat. Figure 7-12E shows a typical example. Here we use a 4050B CMOS noninverting buffer to drive a light-emitting diode. For LEDs with 10 to 15 milliampere currents, this circuit will work well. The buffer will drive the LED without the need for a current limiting resistor. If an open-collector TTL inverter is used, then connect the LED in series with a 330 ohm resistor, and then connect the series pair from the output of the inverter to the ±5 volt dc power supply.

You will find that this LED will not blink bright enough if the threshold detector is adjusted too high on the QRS complex. We will need to use a lower threshold, which will often increase the probability of a false trigger situation. Otherwise, place a one-shot multivibrator between the output of the QRS detector and the LED blinker. The monostable ("one-shot") multivibrator will have an output pulse duration that is long enough to turn on the LED, but not so long that it would interfere with the operation of the circuit. If we select an upper limit to the meter of 300 beats per minute, then we will have a duration of 0.2 seconds between QRS complexes. Two-hundred milliseconds is a relatively long time, a 50 millisecond one-shot would do the trick rather nicely.

SIGNAL-ACQUISITION AMPLIFIERS

It is commonly believed that the microcomputer is the be-all and end-all of electronics, and that analog electronics is a dead field. Such could not be further from the truth! Even though the field of digital electronics, especially the microcomputer area, is growing immensely fast, it is also true that analog electronics is healthy. In fact, although some analog electronic functions have been taken over by digital logic and/or software, there is still a tremendous need for analog circuitry. Some realignment of roles and assignments has taken place, of course, but in the main the analog designer is still very much needed. The basic fact is that its an analog world! The types of signals most commonly available from sensors, transducers, electrodes, (etc.) are analog in nature, i.e., voltages or currents. In fact, we can generalize even further and make the statement that most such signals are *time-varying* voltages or currents.

There is also the fact that sometimes the best display method for a given signal is the analog waveform. We can represent a waveform digitally, i.e., in the form of a table of values, but that is difficult for the human to comprehend. A waveform on a cathode-ray tube or chart-recorder paper is so much easier to read than a digital presentation. Thus, we sometimes require analog electronic circuits not only on the input of the system but on the output as well.

In this section, we will continue our earlier theme and discuss some analog electronic circuits that may be used for signals acquisition in a wide variety of applications. In these cases, the amplifiers presented are dc coupled. Should you require ac coupling for any reason (biological, physiological, medical and many chemical instruments must be ac coupled because of half-cell potentials), then use the same capacitor-resistor schemes as were developed in earlier circuits with these circuits.

Figure 7-13A shows an integrated circuit instrumentation amplifier that is available from National Semiconductor. This device is available in several different model types depending upon the required temperature range. The LF-152 device will provide its full performance over the standard military temperature range of −55° C to +125° C. The LF-352, on the other hand, is guaranteed only over the standard commercial temperature range of 0° C to +70° C. For most readers, however, the LF-352 is more than sufficient.

The gain of the LF-352 amplifier is set by the ratio of two resistors (although this is not analogous to an operational amplifier

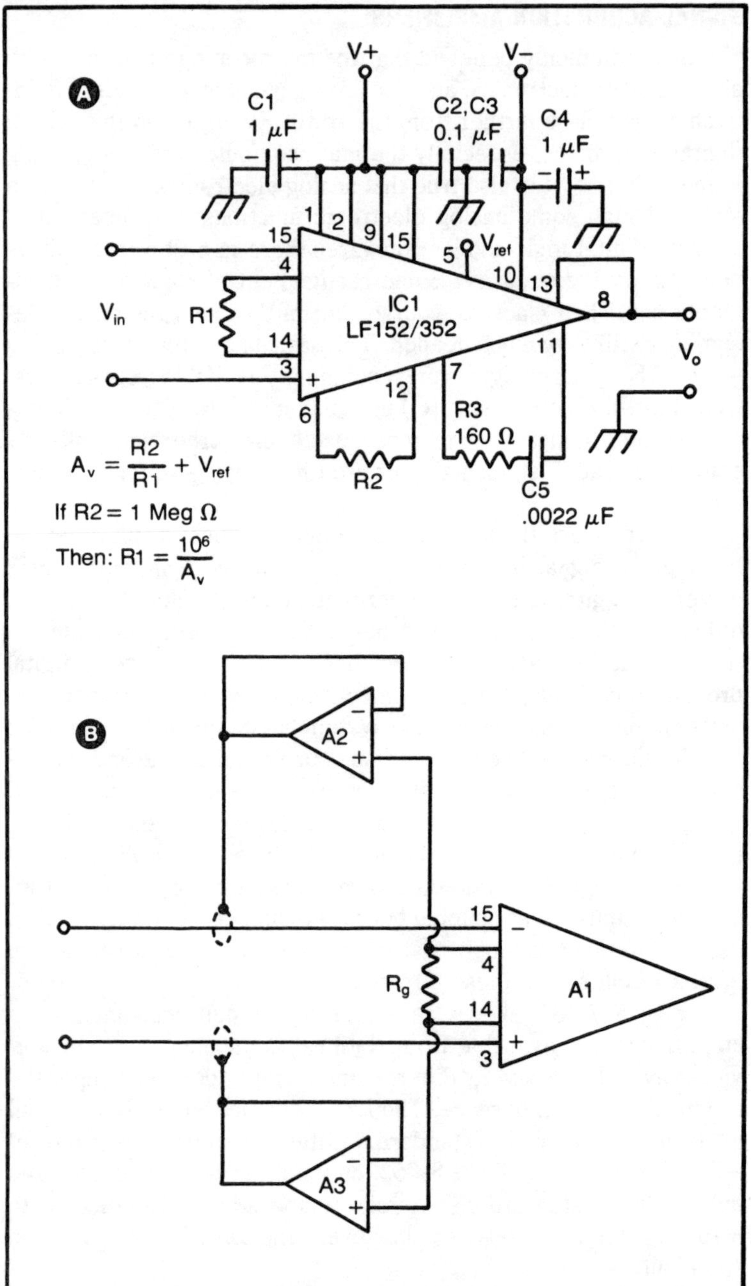

$$A_v = \frac{R2}{R1} + V_{ref}$$

If R2 = 1 Meg Ω

$$\text{Then: } R1 = \frac{10^6}{A_v}$$

Fig. 7-13. A) LF-152/352 amplifier circuit, B) with input guarding.

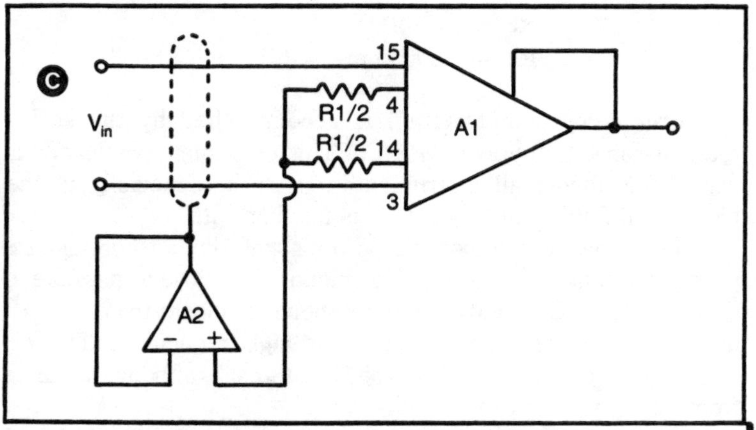

Fig. 7-13. C) alternate guarding.

gain circuit!), shown in Fig. 7-13A as R1 and R2. The gain is given by:

$$A_v = \frac{R2}{R1} + V_{ref}$$

Where:

A_v is the voltage gain
$R2$ is the resistance of R2, in ohms
$R1$ is the resistance of R1, in ohms
V_{ref} is the reference voltage, in volts (i.e., the pin no. 5)

Or, if R2 is set to 1 megohms, and V_{ref} is made zero (i.e., grounded), the gain equation becomes:

$$R2 = \frac{10^6}{A_v}$$

Where:

$R2$ is the resistance of R2, in ohms
A_v is the gain

Example

Calculate resistance of R2 in the above case for a gain of 400.

$$R2 = \frac{10^6}{A_v}$$

$$R2 = 10^6/400$$
$$R2 = 2500 \text{ ohms} = 2.5 \text{ kohms}$$

Frequency compensation is accomplished by the series resistor-capacitor network, R3/C5. This particular combination is useful for almost all instrumentation work, especially in the biomedical fields that this author is familiar with.

The capacitors across the power supply lines to ground are used for decoupling, and should be mounted as close as possible to the pins of the IC. The 0.1 μF units should be closer than the 1 μF units, as these are intended to take out high frequencies. The V+ and V− voltages are intended to be 12 volts each, although anything from ±9 volts to ±15 volts will work.

A modification of the circuit is shown in Fig. 7-13B which will permit guard-shield applications. The guard shields of the two input lines are separate from each other, and are each driven by their own unity gain isolation amplifier, A2 and A3. These amplifiers can be almost any operational amplifier, even though it is probably wise to use a premium-grade device if the signal levels are low. The RCA CA3240 is a good choice as are any of the dual operational amplifiers in the National Semiconductor catalog. In this figure, the input signals for the guard-shield drivers are taken from the two pins that connect to the gain-setting resistor, R_g (i.e., R1 in Fig. 7-13 A).

A slightly different version is shown in Fig. 7-13C. In this version, there is only one guard shield surrounding both wires of the differential input to the LF-352 device. Again, a guard-shield driver amplifier is used. In this case, the gain setting resistor is divided in half, and the guard-shield signal is taken from the junction of the two resistors (each of which has a value of R1/2). In both Figs. 7-13B and 7-13C, only the input circuitry is shown—all other circuitry is the same as in Fig. 7-13A.

PROGRAMMABLE-GAIN AMPLIFIERS

A programmable-gain amplifier is designed such that the gain of the circuit is set by a binary input word. The number of possible gain combinations depends upon the number of bits in the binary control circuit. For a bit-length N, the possible number of combinations is 2^N. In Fig. 7-14, there are two binary inputs (B0 and B1) so the possible combinations are 2^2 or 4.

The control mechanism in Fig. 7-14 is the gain setting resistor. In this case, we are dealing with a National Semiconductor LF-352 device (the same idea can be applied to almost any amplifier in

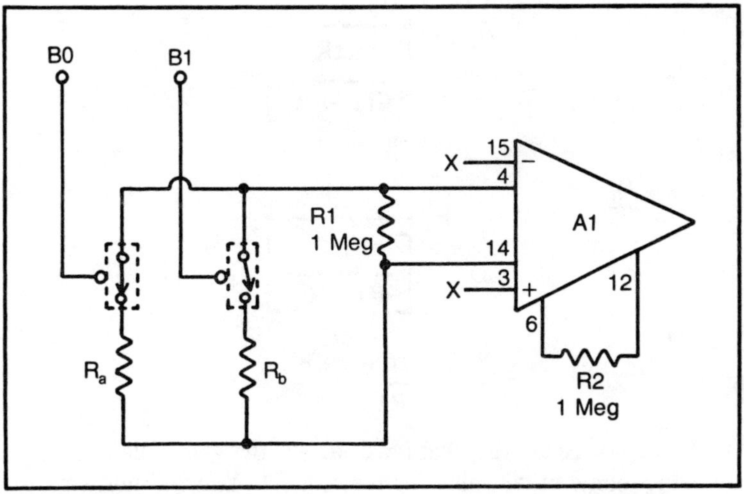

Fig. 7-14. Programmable gain amplifier.

which the gain is set by one or more resistors). Let's assume that the other gain resistor is set to 1 megohm also. If all switches are open, then the total gain would be unity ($10^6/10^6 = 1$).

The gain is set by varying the total resistance across pins 4 and 14 of the LF-352 ICIA. The resistance is changed by connecting one or more resistors in parallel with R1. The possible combinations are given for the admissible digital codes below:

B0	*B1*	*Resistors used for gain selection*
0	0	R1, R_a, R_b
0	1	R1, R_a
1	0	R1, R_b
1	1	R1

The following gain equations apply, assuming that $V_{ref} = 0$:

Code 00

$$A_v = \left[\frac{10^6}{\dfrac{1}{\dfrac{1}{R1} + \dfrac{1}{R_a} + \dfrac{1}{R_a}}} \right]$$

Code 01

$$A_v = \frac{10^6}{\left[\dfrac{R1R_a}{R1 + R_a}\right]}$$

Code 10

$$A_v = \frac{10^6}{\left[\dfrac{R1\,R_b}{R1 + R_b}\right]}$$

Code 11

$$A_v = \frac{10^6}{R1}$$

In the last case, i.e., that for Code 11, the gain is unity unless you select some other value for resistor R1. You will not want to leave R1 out of the circuit, as might be the inclination of some, because the amplifier gain will try to go to infinity when the two control inputs are both open. This situation could occur either by design, or, by a programming error, or, during the transition period between two other codes. In the latter case, the amplifier output will see spikes as the gain goes extremely high for a few microseconds between valid codes.

DIFFERENTIAL LINE-DRIVER AMPLIFIER

A line-driver amplifier is one that will drive a 600-ohm balanced audio line of the type normally used in broadcast, recording and high quality public-address work. These amplifiers typically use a pair of operational amplifiers in a differential configuration such as Fig. 7-15. This amplifier differs from the earlier line-drive amplifier in that it uses a differential input circuit.

The circuit shown in Fig. 7-15 uses a pair of ordinary operational amplifiers. If a full "hi-fi" range is not needed, then these amplifiers can be of the 741-family of devices. A wise selection might be the 1458 device, which contains two 741-family operational amplifiers within a single 8-pin miniDIP integrated circuit package. The gain of this configuration is computed a little differently from other amplifiers because this circuit is basically a pair of noninverting followers with gain. The gain is:

$$A_{vd} = \frac{2\,R_2}{R_1} + 1 \quad \text{Iff } R_2 = R_3$$

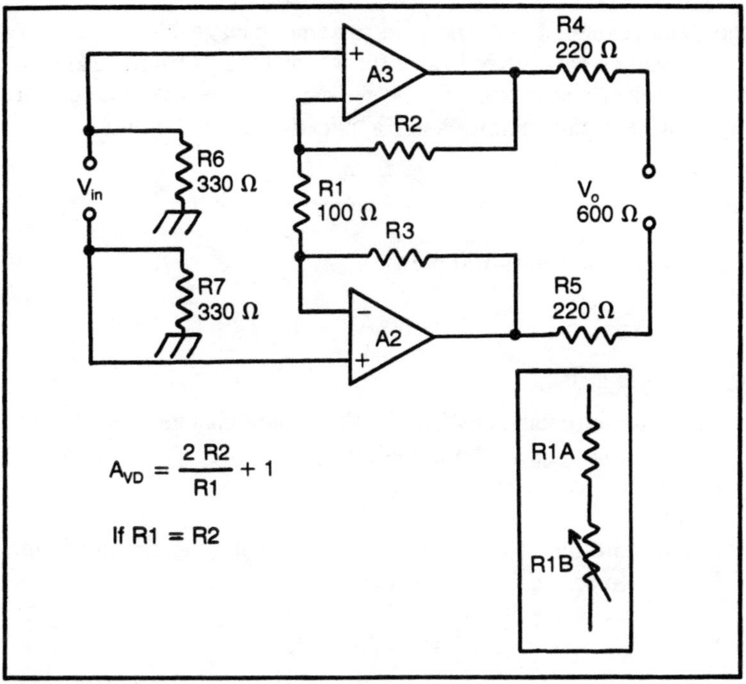

Fig. 7-15. 600-ohm line driver amplifier.

Where:

R_1 is the resistance of R1
R_2 is the resistance of R2, and is expressed in the same units (i.e., ohms, kohms, or megohms) as R_1
A_{vd} is the differential voltage gain of the circuit

It is common practice to make the common resistor (R1) fixed at some low value and then select a value for R2 and R3 that will yield the required gain. It is also common practice, when gain control is desired, to make R1 a combination of a fixed resistor and a potentiometer (see insert for Fig. 7-15). The variable resistance of the potentiometer makes it possible to adjust the gain of the circuit, while the fixed resistor limits the maximum permissible gain of the amplifier.

Note in the equation above that R1 appears in the denominator. If it were allowed to go to zero, then the gain would try to go to infinity . . . and will become extremely high as the resistance of R1 drops low. If we make the resistor a combination of a fixed resistor and a potentiometer, however, this problem is overcome because the fixed resistor limits the gain to some manageable amount.

Since we have a fixed value for R1, and the gain will be known from other circuit requirements, we can set a new version of our equation that will permit easier application of the circuit:

$$R_2 = \frac{R_1(A_{vd} - 1)}{2}$$

Or, since R1 is set at 100 ohms:

$$R_2 = 50(A_{vd} - 1)$$

Where:

R_2 is the resistance of R2 in ohms (note change from above)
A_{vd} is the gain of the amplifier

Example

Calculate the value of R2 required to make the circuit of Fig. 7-15 operate as a gain of 50 amplifier.

$$R_2 = 50(A_{vd} - 1)$$
$$R_2 = (50)(50 - 1)$$
$$R_2 = (50)(49)$$
$$R_2 = 2450 \text{ ohms}$$

If the above resistance is obtained in a precision resistor, then the gain will be very nearly perfect. But, if we use a "standard" 2.4 kohms, i.e., 2400 ohms, then an error of only 2 percent is obtained:

$$\%Error = (2450 - 2400)(100\%)/2450$$
$$\%Error = (50)(100\%)/(2450)$$
$$\%Error = 5000\%/2450$$
$$\%Error = 2.04 \text{ percent}$$

Of course, it is necessary that both R2 and R3 of Fig. 7-15 be of the exact same value. Otherwise, we will have not only a gain error but a common-mode error that will prove difficult to deal with . . . it kind of obviates the need for a differential amplifier!

The output network of the circuit in Fig. 7-15 is made to approximate an impedance of 600 ohms. The normal output impedance of most operational amplifiers is on the order of 50 to 100 ohms (75 ohms is a frequently quoted number). If we have a 220-ohm resistor in series with each output line, therefore, the apparent impedance as far as the line is concerned will be (220 ohms + 75 ohms) × 2 or 590 ohms. This is a very good match for the 600-ohm line impedance.

The input circuit can be open-ended (i.e., without R6 and R7) in many cases, but if it received audio from another 600-ohm line, then we will have to use the load resistors to terminate the line. The 330-ohm resistors are used because the combination appears to form an impedance of nearly 600 ohms, and is balanced with respect to ground. If the input signal source is some balanced source, such as a microphone or electronic instrument pick-up, then it may not be necessary to use any resistors. If the input signal source is single-ended, such that a differential input is not needed, then simply ground one of the inputs and use the other.

The operational amplifiers to use for a "high fidelity" or good quality public-address amplifier will have a frequency response up to 15 kHz closed loop. One must divide the gain-bandwidth product of the prospective amplifier by 15,000 in order to find a suitable device. The RCA CA3140 is a nice choice.

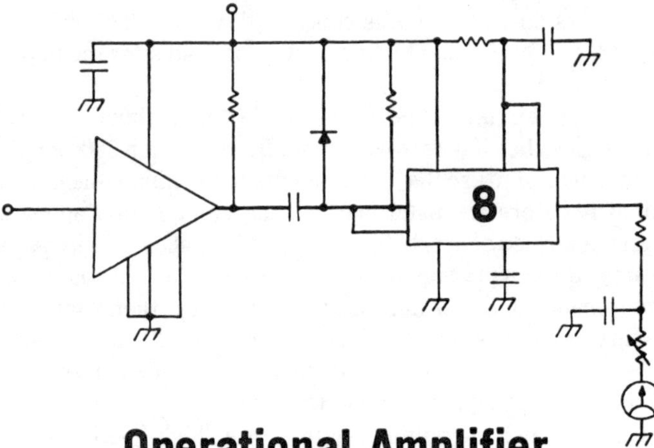

Operational Amplifier Instrumentation Circuits

Operational amplifiers are often used for purposes other than straight amplification. In many electronic instruments, for example, the operational amplifier will perform such mathematical operations as integration and differentiation, comparison and so forth. In this chapter, we will examine some of these applications so that you can eventually use them as building blocks in some project or another. Keep in mind that many analog electronic projects are little more than old-fashioned analog computers permanently "programmed" by the design to perform one and only one job.

PRECISE DIODE

The ordinary PN junction semiconductor diode is linear only at junction potentials larger than its V_g junction potential. For germanium diodes (which are all but obsolete) this potential is 200 millivolts, while for the more modern silicon diodes this potential is 600 to 700 millivolts (0.6 to 0.7 volts). Operation in the range from 0 to 700 millivolts sees a substantial nonlinearity that can affect the quality of an instrument circuit, or produce distortion of one type or another. We can, however, linearize the diode by making use of the operational amplifier's feedback loop. An example of such a circuit is shown in Fig. 8-1. This circuit is usually called a precise diode, or precision diode circuit.

If the input voltage applied to this circuit is negative, then the voltage appearing at the output terminal of operational amplifier A1

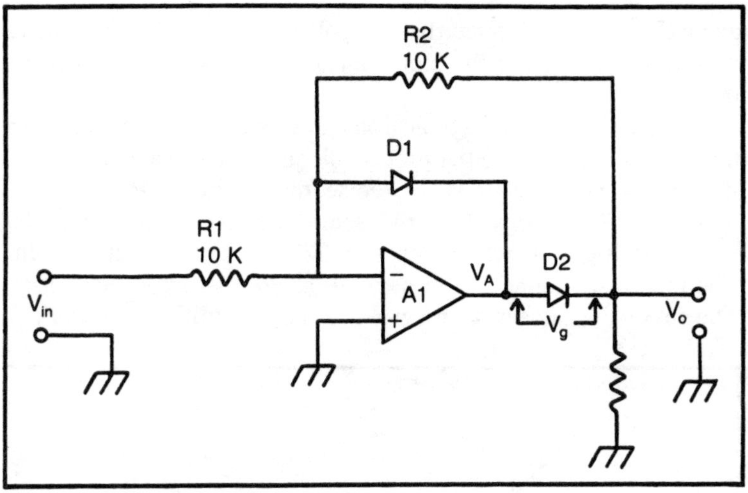

Fig. 8-1. Precise diode circuit.

is positive. This potential will reverse bias diode D1 and forward bias D2. The feedback current that nulls the input current, in the usual manner of inverting follower operational amplifier circuits, will be generated by a voltage that contains the voltage drop across D2. Thus, the voltage drop is nulled out by the feedback loop of the amplifier ($V_A - V_g$).

In the event that the applied input voltage is positive, we find that diode D1 is forward biased, and will effectively clamp the input voltage to near zero (less V_g for D1) through the operational amplifier output terminal. Since the circuit of Fig. 8-1 is both linear and unidirectional, it is an ideal rectifier circuit.

Precise diodes are used in instrumentation circuits that require operation at low values of input voltage, and cannot tolerate nonlinearity. Examples of typical circuits include those used in quadrature detectors and in ac rectification for ac voltmeters.

ABSOLUTE-VALUE AMPLIFIER

An absolute value amplifier is one that will produce an output that is the absolute value of the input voltage. Recall from junior high school arithmetic that the absolute value of a number is the number without regard for sign. Thus, an absolute value amplifier will always produce a positive output voltage regardless of the polarity of the input signal voltage. A transfer function of an absolute value amplifier is shown in Fig. 8-2A. The output voltage (V_o) is graphed along the vertical axis, while the input voltage is graphed along the horizontal. Note that the polarity of V_o is always positive regardless of whether the input signal voltage is positive or negative.

Figure 8-2B shows a typical absolute value amplifier. Note that this circuit is similar to the precise diode discussed above. In fact, the circuit consists of two precise diode circuits of opposing polarities. Amplifier A1 is the same negative-input circuit described in Fig. 8-1, while amplifier A2 is a positive-input circuit. The inputs are connected in parallel, as are the output terminals. The circuit operates in such a way that these amplifiers will work on

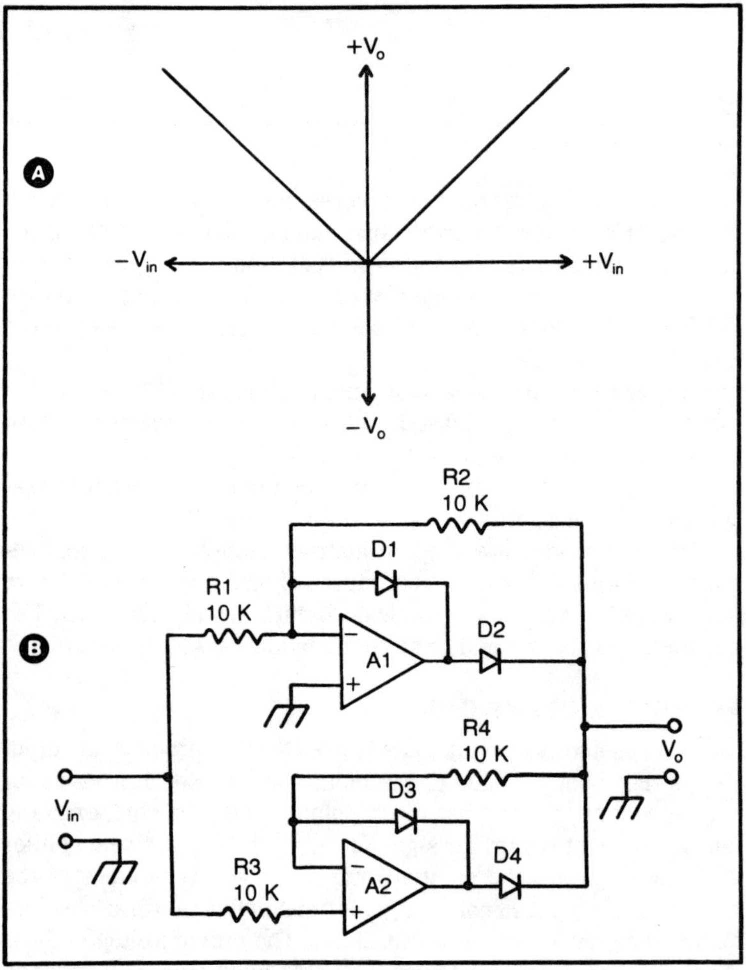

Fig. 8-2. Absolute value amplifier A) transfer function, B) circuit.

opposite polarity signals, so will not interfere with each other. The gain of this amplifier, like that of Fig. 8-1, is unity for both positive and negative input signals.

MULTI-FUNCTION AMPLIFIER

There is at least one hybrid electronic device, made by Burr-Brown Corporation, which will perform a large number of analog circuit functions. The basic circuit for the B-B 4302 device is shown in Figs. 8-3A through D. The transfer function of the B-B 4302 is of the form:

$$V_o = V_y \left[\frac{V_z}{V_x}\right]^m$$

Where:

- V_o is the output potential
- V_y is the potential applied to the "Y" input
- V_z is the potential applied to the "Z" input
- V_x is the potential applied to the "X" input
- m is an exponent between 0.2 and 5, and is set by an external resistor network

The basic circuit is shown in Fig. 8-3A. The terminals marked "A," "B," and "C" are used to connect the external resistor network that will set the "m" exponent. The power supply inputs are generously decoupled with 4.7 μF electrolytic (tantalum) capacitors for low frequencies and 0.1 μF capacitors for high frequencies. The 0.1 μF capacitors should be mounted as close as possible to the body of the device. Of the three input voltage lines, the V_y has a compensating input resistor.

The external resistor networks for values of m from 0.2 to 5 are shown in Fig. 8-3B, C, and D. In Fig. 8-3B we see the network required for values of m from 1 to 5. The value of m is given by the equation:

$$m = (R1 + R2)/R2$$

In most cases, we will find that the value of *m* is known, so we will select one of the resistor values and then calculate the value of the other resistor in order to yield the required *m* value. The use of exponent values higher than unity will permit transfer functions such as squarer (i.e., m = 2).

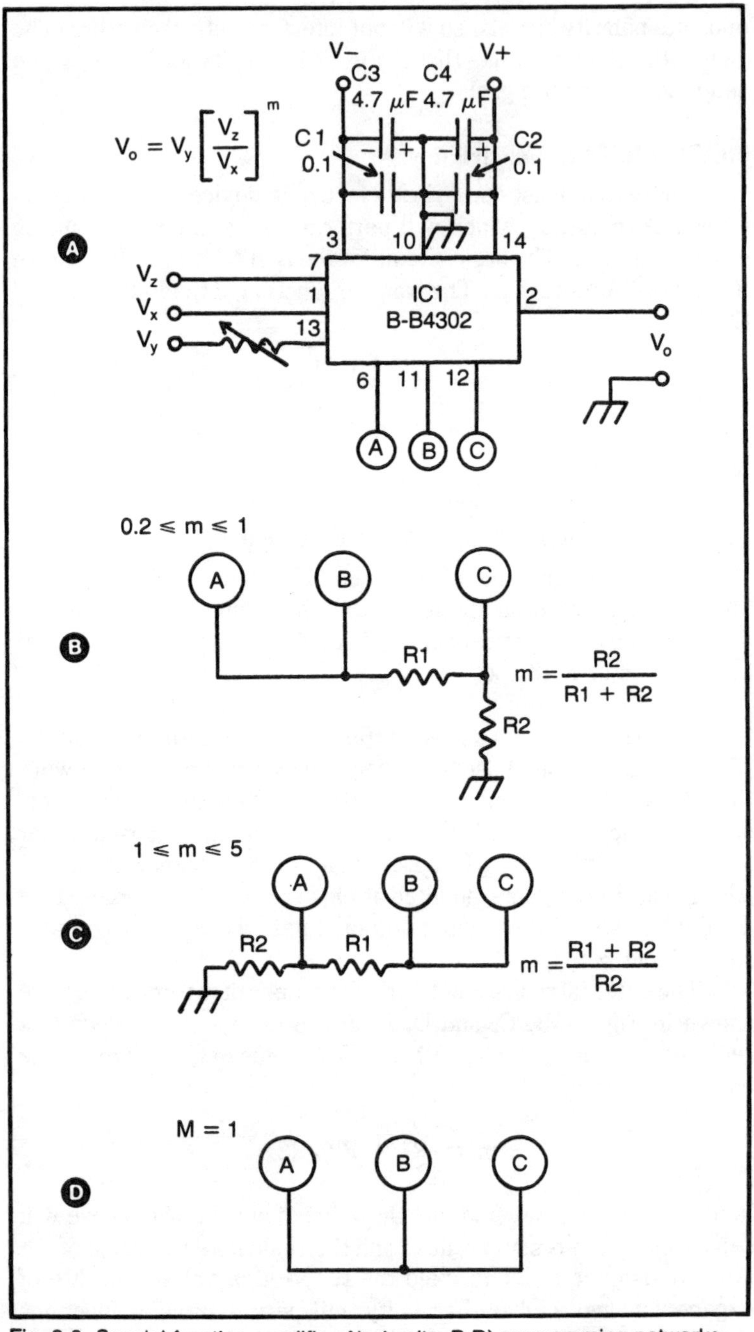

Fig. 8-3. Special function amplifier A) circuits, B-D) programming networks.

The circuit to produce exponents of 0.2 to 1 is shown in Fig. 8-3C. The value of *m* is given by:

$$m = R2/(R1 + R2)$$

and for m = 1, there are no resistors at all; terminals A, B, and C are simply tied together (see Fig. 8-3D).

BUILD A LOW-COST pH METER

Perhaps the most common electronic instrument in the scientific and medical laboratory is the pH meter. These instruments are used in medicine to measure the acidity/alkalinity of human blood, gastric juices, and other fluids. In the chemistry laboratory they are used for a wide variety of measurements of acidity/alkalinity, as well as similar usage in a wide variety of scientific laboratories. Homeowners, gardeners, and others will use a pH meter or some similar measurement system to analyze soil.

The instrument might be a stand-alone device sitting on a counter, or, it might be embedded into some other instrument. In all cases, however, the pH meter is merely a millivoltmeter (with appropriate scale calibration) and a special electrode (of which, more in a moment).

The term "pH" relates to the measurement of relative acid-base (i.e., acidity/alkalinity) balance in a liquid. This figure is related to the concentration of the *hydrogen-plus* (H^+) ion in the solution being tested. More specifically,

$$pH = -\mathrm{Log}_{10}(H^+)$$

A neutral solution is one that is neither acid nor base, and it is assigned a pH of exactly 7.00. Acidic solutions have a pH lower than 7, while basic solutions have a pH that is greater than 7. The numerical scale on the modern pH meter normally covers the range of pH from 1 to 14.

An electronic pH meter is little more than the aforementioned millivoltmeter with a special scale. The signal for the millivolter is developed by a special *pH electrode* that can be immersed in the solution being examined. The electronics portion of the typical pH meter is a high-stability millivoltmeter, while the display can be either an analog meter calibrated with a 1 to 14 pH scale, or, an appropriately scaled digital voltmeter (digital panel meters are often the only thing "digital" in many "digital pH meters!"). There are several pH meters on the market that reflect their millivoltmeter heart by providing a millivolt measurement function that can be selected using a front panel switch. Although almost any reasonable millivoltmeter circuit can be used, only one will be used in this project.

There are several different pH electrode configurations (although they all work the same . . . with the exception of the old but now being revived antimony electrode). The version shown in Fig. 8-4, however, is probably the most common in instruments made for medical and life-sciences use. This system uses a *glass* electrode whose output voltage is measured relative to that of a *calomel* (i.e., mercurous chloride) *reference electrode*. Electrodes used to measure pH depend upon the development of electrical potentials across a barrier separating two materials of the same phase, or, across the interface between materials of different phase (e.g., metal-to-electrolyte interface). In the calomel reference electrode, this potential is generated by an interface of pure mercury (Hg) and a solution of a potassium chloride (KCL) saturated with calomel.

One of the problems with some reference electrode designs is that a second potential might develop at the interface between the solution being tested and the KCL in the electrode. This problem is eliminated, or at least reduced, by the use of a *salt bridge* in the pipe from the calomel reference electrode (see Fig. 8-4). The calomel reference electrode is preferred because it possesses a reasonable long-term stability. That is to say, its half-cell potential does not deteriorate markedly over long periods of time.

The second electrode used in the pH configuration is the *glass electrode,* also shown in Fig. 8-4. This electrode produces a potential across a thin membrane, in this case a special type of very thin glass. The glass bulb shown in Fig. 8-4 is the membrane, and it is immersed in the solution being tested. The solution on the inside of the glass bulb is usually a chloride with a pH of 1. A silver-silver chloride (Ag-AgCl) electrode connected to the output wire is immersed in the solution inside the bulb. The electrical potential developed across the glass membrane is approximately 60 millivolts per pH unit (mV/pH) at room termperature (i.e., 30° C), but it will vary with temperature from approximately 30 mV/pH at 0° C to around 90 mV/pH at 100° C. One manufacturer presents a graph of the pH electrode output showing it to have a negative slope (see Fig. 8-5) and is bipolar. Note that the transfer functions for three different temperatures are shown. For any actual pH electrode, at any given temperature, the actual slope will be within the ranges

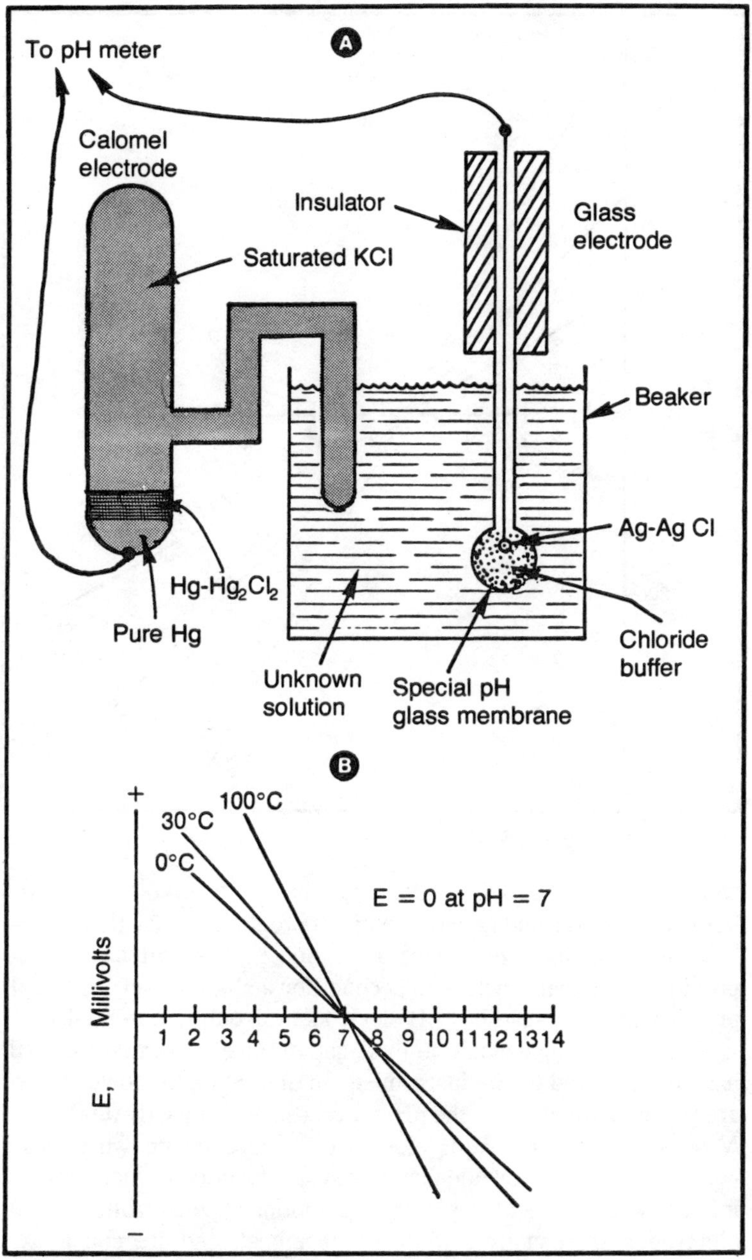

Fig. 8-4. (A) Construction of calomel and glass electrodes used in the electronic measurement of pH. (B) output curve of the pH electrode set. Temperature affects the slope, but it is a nominal 57 mV/pH in most common probes.

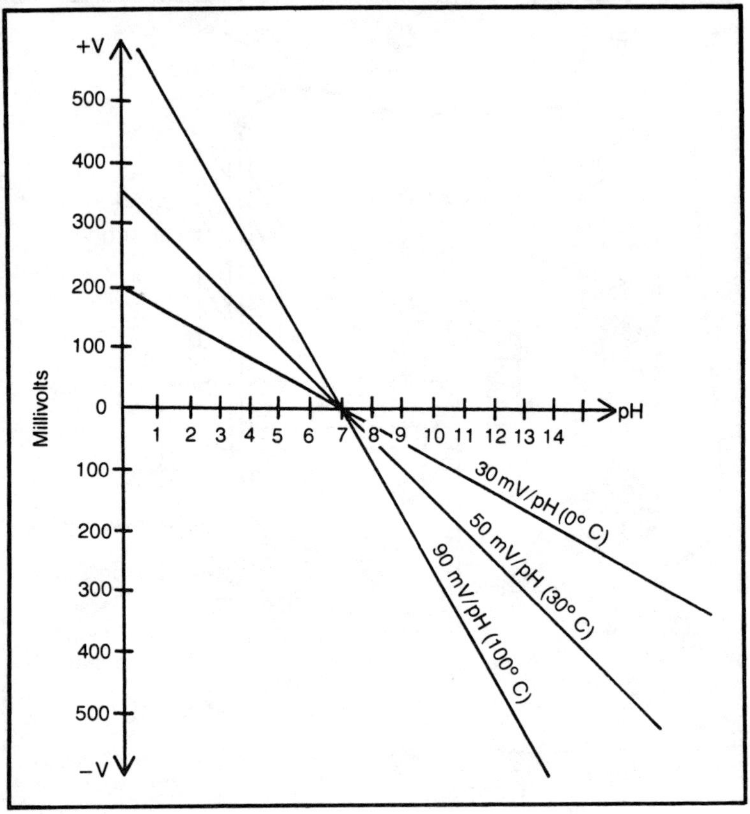

Fig. 8-5. pH electrode transfer function.

shown for 0° C and 100° C. Also, note that the zero-volts crossover between positive and negative output potentials is exactly at pH = 7. This is an idealized case, and some means will have to be provided in actual practice to account for a slight offset. In a real, practical kind-you-can-buy pH electrode, the output potential at pH = 7 may be a slight positive or negative offset potential that will have to be nulled by the instrument. In this example, however, we are going to assume that the pH = 7 crossover is exactly 0.00 volts. *Note:* some makers of pH electrodes have been known to hand select electrodes for idealized transfer functions, including 50 mV/pH @ 30° C and a pH = 7 potential of zero volts. These electrodes are usually sold at a higher cost, and are considered premium. As long as the transfer function is stable, however, these "errors" can be designed-out by the instrument designer. For our present purposes, then, the transfer functions shown in Fig. 8-5 are considered actual rather than merely ideal, and we will base our design on them.

The glass electrode is preferred for most applications over the competitors (for example the antimony electrode) because it will work consistently, with little fuss and muss, under a variety of trying conditions. The glass electrode is generally preferred for all *in vitro* (lit. "in glass") measurements (see Fig. 8-4 again). There are some specialized glass electrode designs for special applications such as *in vivo* (lit. "in life") measurements such as stomach acidity or on-line blood pH. In those cases, which are common in gastroenterology and cardiac catheterization laboratories, they may well use the old-fashioned antimony electrode.

The cell potential across the glass electrode's membrane is produced through the migration of a tiny number of ions across the barrier. Therefore, the glass electrode has a very high source impedance, on the order of 200 to 300 megohms. In order to prevent loading of the electrode by the amplifier input impedance, therefore, it becomes necessary to specify an extremely high input impedance. In the past, it has been policy to specify not less than 1000 megohms (according to an approximate "5X" rule), but today it is easy to provide input impedances of 10^{12} ohms or more (MOSFET and some JFET-input operational amplifiers are able to boast this kind of impedance, as are so-called "electrometer" amplifiers).

A TYPICAL pH METER

There are several approaches to pH meter design; here we will describe three of them. Two of the designs are essentially analog, while the third is "digital." All of them are essentially millivoltmeters.

The two analog versions are essentially the same, and are shown in Figs. 8-6A and 8-6B. Note that Fig. 8-6B is equipped with a digital voltmeter output display, so may be called a "digital pH meter" by the wonderfully creative dudes in the advertising department. In reality, it is no more digital than the first—it only appears to be so because of the seven-segment LED or LCD output display (*sigh!*).

Although the two designs of Fig. 8-6 are essentially similar, the design of Fig. 8-6A is probably a little easier (not much) because it doesn't matter much what polarity the output potential takes . . . we can always connect the meter leads to deflect downscale for acidic solutions and upscale for basic solutions. It also doesn't matter much the scale factor of output potential V_o, as long as we can

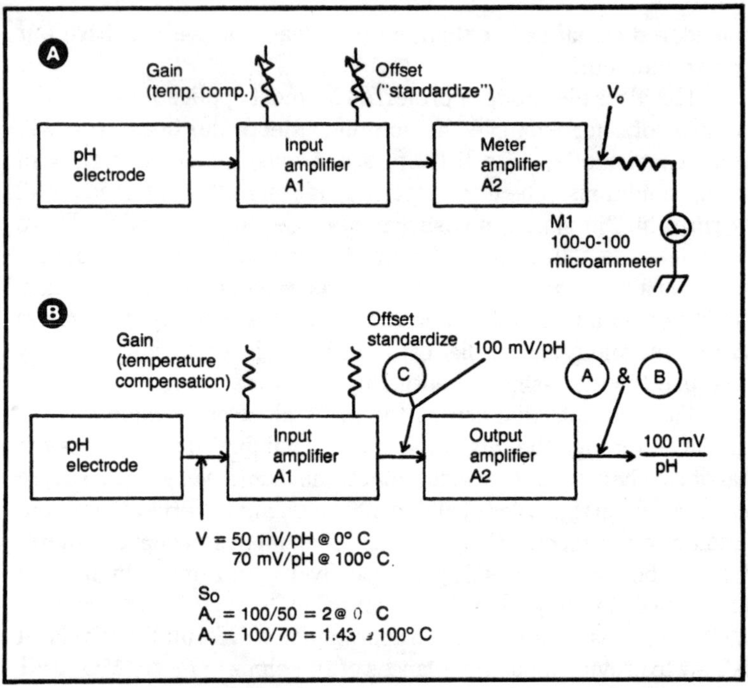

Fig. 8-6. pH meter block diagram A) analog output, B) digital output.

select a meter/resistor combination that will deflect full-scale positive when pH = 14 solutions are measured, and full-scale negative when pH = 1 solutions are encountered.

There are two controls shown in Fig. 8-6A: temperature compensation (i.e., amplifier gain), and "standardization" (i.e., amplifier offset). The temperature compensation varies the gain of one of the amplifiers in the chain in order to change the slope of the signal according to the temperature. The tempco control is usually a front panel control that the operator sets, although in some cases of more recently designed instruments the tempco control is set automatically in response to the measured temperature of the solution.

The meter amplifier may or may not be needed, and is used to scale the output of amplifier A1 to whatever factor is needed to drive the meter. In some cases, this amplifier will house the standardization control in order to optimize our control over all contributing amplifier/electrode offset potentials. The standardize control, incidentally, is used to null out offsets and is often adjusted with either a shorted electrode input to A1, or, a pH = 7 calibration solution immersing the electrode (the preferable method).

The pseudo-digital pH meter is shown in Fig. 8-6A. In this case, as mentioned earlier, the rubric "digital" (which is so magical for instrument sales today) is conferred merely from the use of a digital voltmeter to replace the analog voltmeter of Fig. 8-6A. The stages of this instrument are essentially the same as in Fig. 8-6A. There will be a pH electrode driving a set of amplifiers (two or more stages) that finally drive a 0 to 1999 mV digital voltmeter. Let's consider a real design example. In order to follow this discussion, please refer to Figs. 8-6B, 8-7, and 8-8.

Note: in any design project it helps to make a block diagram of the circuit functions needed prior to making any circuit decisions. This is only common sense. The mark of the novice designer is the all too eager willingness to hop right into the design process and start committing circuitry to paper before really understanding the functions the circuitry is to perform. The block diagram is to the design process what an outline is to a book or magazine article. It is also the place where you can make certain calculations and most easily see combinations that can make the overall circuit one heckuva lot simpler.

In Fig. 8-6B we are showing a slightly different electrode from before. In this case, the slopes will vary from 50 mV/pH at 0° C to 70 mV/pH at 100° C, Fig. 8-7 shows the different signals at critical points in Figs. 8-6B and 8-8. Slope "A" in Fig. 8-7 is the desired final output voltage function. We are going to use a 3½-digit digital voltmeter for the display. For reasons of economy (i.e., the most common range), a 0 to 1999 millivoltmeter is selected. Thus, we have a hint as to the appropriate scale factor. Since we cannot paste a new scale onto the face of the digital voltmeter (as we can in an analog instrument), we must appropriately scale the output signal voltage to produce a meaningful reading on the output display. If we select an output scale factor for slope "A" of 100 mV/pH, then we will make the best use of the output display instrument. At pH = 14, the output voltage will be (14)(100 mV/pH) = 1400 millivolts. By manipulating the decimal point (easily done mechanically on DVMs), the reading will be "14.00" when the pH is 14.

Amplifier A2 is an inverting follower operational amplifier, so the output slope will be inverted relative to the input slope. The input slope of A2 is the output slope of amplifier A1, while the input of A1 is the pH electrode signal. Figure 8-8 shows the actual circuit.

The input amplifier A1 must have an input impedance of 1000 megohms or more (i.e., 10^9 ohms). If we use the RCA CA3140

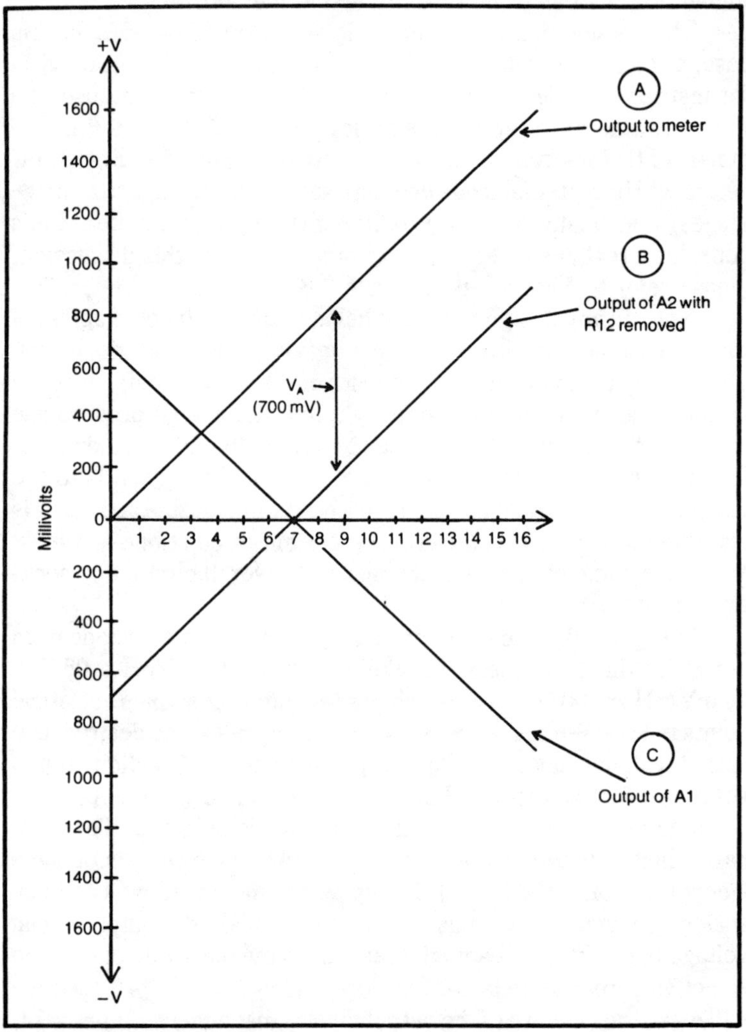

Fig. 8-7. Transfer functions.

device, we can approach an input impedance of 10^{12} ohms, which is more than sufficient. This input impedance must be maintained in the actual construction, however, and that is not always so easy. The amplifier must be built on only the highest grades of printed circuit material, be dust and grease free and should not be handled with bare hands. When this amplifier is assembled, it should be handled with cotton "finger cots" rather than bare hands in order to keep skin oils from deteriorating the input impedance. It is also necessary to keep the amplifier shielded, so the circuit is built inside of a metal container such as a *Pomona* box or die-case *Bud* box. Alternatively, the shielded container can be hand-built.

Another design criterion for this amplifier (A1) is low-drift. Temperature drift can be a real problem with this type of circuit, and for that reason early designs featured chopper-input amplifiers (which are very costly). But today, drift is sufficiently overcome by analog circuit designers that one does not always need the chopper-input amplifier. We must, however, keep this factor in mind. The way to do that with the RCA CA 3140-series of devices is to keep the $V+$ and $V-$ potentials low. In this circuit, the potentials are limited to ±5 volts dc, thereby setting the stage for a better drift factor. It is also necessary that amplifier A1 be equipped with a top-hat heatsink (the type normally used for TO-5 transistors is OK, provided that they are flexible. The inflexible varieties may not fit the IC metal can).

The gain of amplifier A1 is the temperature compensation mechanism, so some means must be provided for gain variation. In almost all cases, the gain of an amplifier will be V_o/V_{in}, or, the output scale factor divided by the input scale factor. Since amplifier A2 has a gain of −1, we must use the overall output scale factor of 100 mV/pH for the output of A1. Also, since the input scale factor for A1 is the pH electrode signal, it will have a variable scale factor that depends upon the temperature of the solution being measured (which is the main reason why a variable gain is needed in the first place!). Thus, we have two end points to consider, the gain at 0° C and the gain at 100° C. These calculations are made below and in Fig. 8-6B:

At 0° C

The slope is 50 mV/pH, so the gain must be:

$$A_v = V_{of} / V_{if}$$
$$A_v = (100\ \text{mV/pH})/(50\ \text{mV/pH})$$
$$A_v = 100/50 = 2$$

(Note: here and below, V_{of} is the output scale factor in mV/pH, while V_{if} is the input scale factor, also in mV/pH.)

At 100° C

The slope is 70 mV/pH, so the gain must be

$$A_v = V_{of} / V_{if}$$

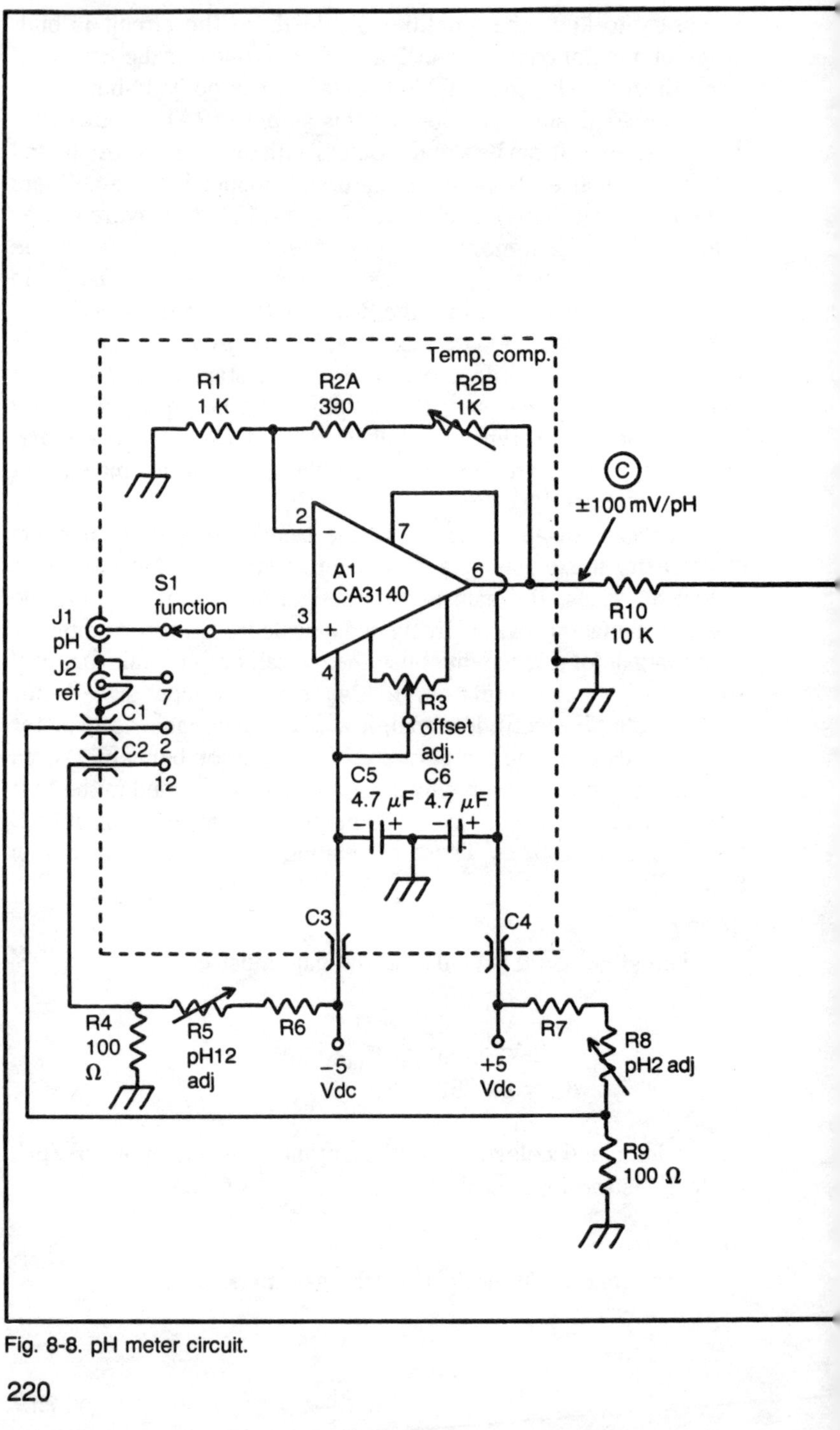

Fig. 8-8. pH meter circuit.

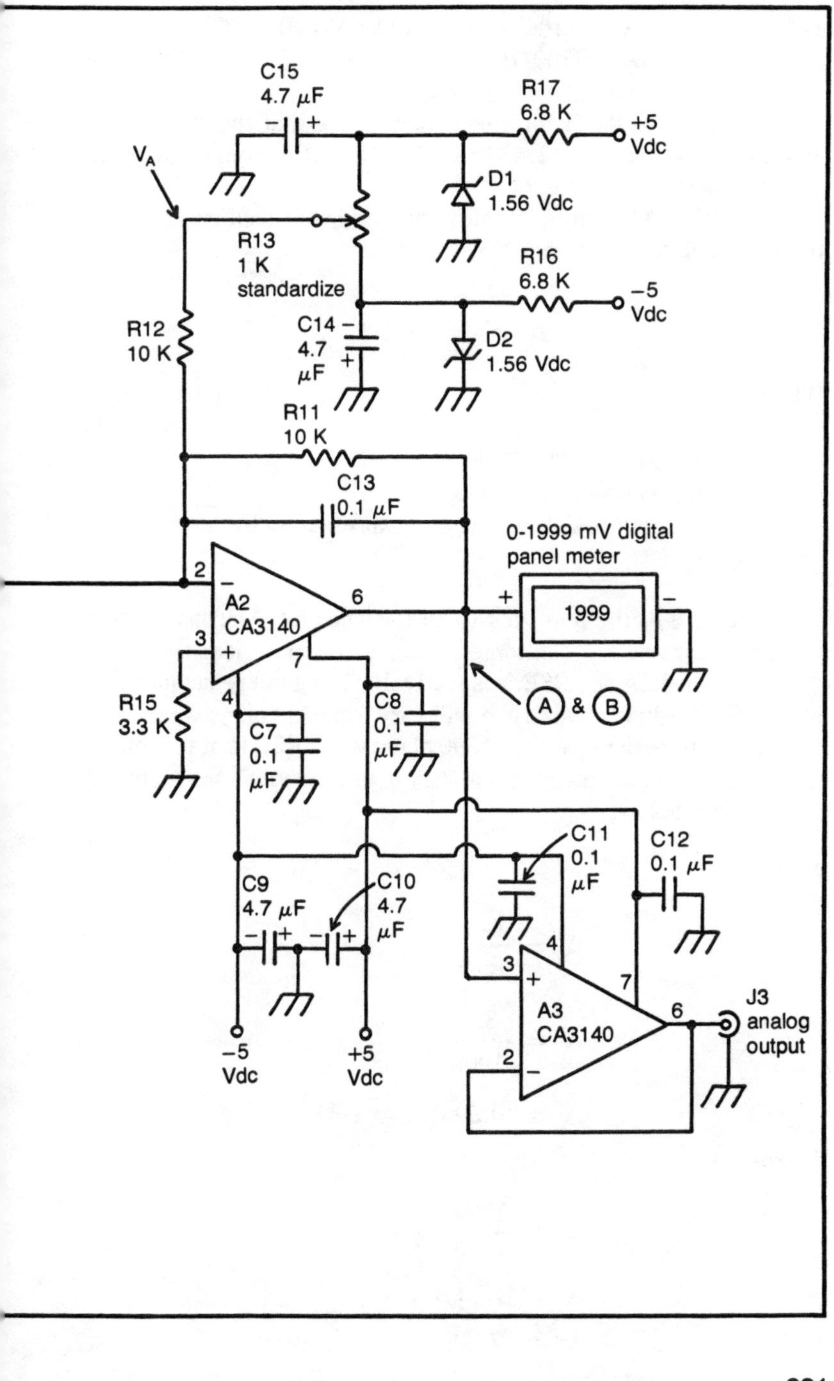
C15
4.7 µF
R17
6.8 K
+5
Vdc
VA
D1
1.56 Vdc
R13
1 K
standardize
R16
6.8 K
−5
Vdc
R12
10 K
C14
4.7
µF
D2
1.56 Vdc
R11
10 K
C13
0.1 µF
0-1999 mV digital
panel meter
A2
CA3140
1999
R15
3.3 K
C7
0.1
µF
C8
0.1
µF
A & B
C11
0.1
µF
C12
0.1 µF
C9
4.7 µF
C10
4.7
µF
A3
CA3140
J3
analog
output
−5
Vdc
+5
Vdc

$A_v = (100 \text{ mV/pH})/(70 \text{ mV/pH})$
$A_v = 100/70 = 1.43$

Thus, our scale factor calculations show that the gain of the amplifier must vary from 1.43 to 2.00, with a little room on the ends of the range for error factor.

Amplifier A1 is in the noninverting follower configuration, so the gain will be:

$$A_v = \frac{R_f}{R_i} + 1$$

Where:

A_v is the gain (dimensionless)
R_f is the feedback resistance
R_i is the input resistance, and is expressed in the same terms as R_f

In Fig. 8-8, the feedback (R_f) resistance for A1 consists of a fixed resistor and a potentiometer connected in series; these are designated as R2A and R2B, respectively. The input resistance (R_i) is R1. The feedback resistance will vary from 390 ohms when R2B is a minimum resistance, and 1390 ohms when R2B is at maximum gain position. Thus, the gain will vary over the ranges shown in the calculations below:

Minimum

$$A_v = \frac{R_f}{R_i} + 1$$

$$A_v = \frac{390}{1000} + 1$$

$$A_v = 0.39 + 1 = 1.39$$

Maximum

$$A_v = \frac{1390}{1000} + 1$$

$$A_v = 1.39 + 1$$

$$A_v = 2.39$$

Thus, with a gain range of 1.39 to 2.39, we satisfy the criteria of 1.43 to 2. We might want to consider reducing R2A to 330 ohms, so that the gain will be 1.33 to 2.33, which provides a little more cushion at the bottom of the range at the expense of some of the upper-end range.

Some anti-noise capacitors, used to decouple the dc power supply lines are used in Fig. 8-8, amplifier A1. Low-frequency noise problems are addressed by capacitors C5 and C6, which are 4.7 μF tantalum units. These capacitors are mounted right inside of the shielded box housing A1. The high-frequency noise is decoupled by using a pair of feedthrough 1000 pF capacitors, designated C3 and C4 (also used for the calibration lines, capacitors C1 and C2). These capacitors are designed to feed through a chassis or other bulkhead, and will mount with either a threaded nut or solder to the bulkhead (solder types are only used where the bulkhead is solderable!). Normally, capacitors for high-frequency decoupling are mounted as close as possible to the op amp, but that may not be strictly necessary in this case because of the intentionally small dimensions of the shielded case of A1.

The input connectors for amplifier A1 must be low-loss, high impedance types. If you cannot obtain genuine pH connectors from the electrode maker, then substitute some other high impedance connector: *do not use phenolic* based connectors unless certified by the manufacturer as high impedance or (alternative method of specifying) low loss. The UHF SO-239 connector can sometimes be used (unless some dope is likely to mistake your pH meter for a C.B. set), as are the various forms of high-voltage connectors on the market. The important thing is that both J1 and J2 be coaxial types that will accommodate the shielded cable normally used with pH electrodes.

The "function" switch shown in this circuit is not strictly necessary, but does add a certain operating ease. Since pH is related to a specified number of millivolts, we can calibrate the meter using millivolt sources . . . or almost so. It must be recognized that this simple circuit does not take into consideration the effects of temperature, so makes the assumption of room temperature solutions. We have four positions on the function switch: pH, 7, 2, and 12. The pH position measures the voltage coming from the pH electrode, so is the active position. The "7" position causes the input of the amplifier to be grounded, so will measure 0 volts . . . the potential for pH = 7 in our idealized electrode. The "2" and "12" positions provide voltages that will be equivalent to pH = 2 and pH

= 12 (acid and base calibration points). Note that the adjustment potentiometers, R5 and R8, are internal trimpots and are thus not available to the operator. NEVER give the operator a control that will screw up the calibration or functioning of ANY instrument unless it is ***absolutely necessary*** to do so. Shortly we will discuss calibration of these potentiometers.

Please be aware that all forms of internal calibration are a poor second to the use of well prepared, accurate pH calibration grade buffer solutions. The utility of the internal calibration of the function switch is that it allows reasonably accurate (for many uses) measurement without the need for costly solutions, or the errors that operators inattention to solution Q.A. can cause.

Amplifier A2 is the driver for the digital voltmeter. It inverts the A1 output signal from point "C," and produces the signal required for the amplifier input (refer again to Fig. 8-7). Curve "A" in Fig. 8-7 was the ideal curve, or the one with the 100 mV/pH slope and an origin ("Y-intercept") of 0, 0. But the output of amplifier A1 (i.e., curve "C") does not have an origin which, when inverted by A2 to produce curve "B" will yield the ideal curve. There is an offset potential of 700 millivolts between curves "A" and "B." This offset represents 50 percent of full-scale, so must be eliminated. The *Standardize* control (R13) provides the offset voltage V_a that will correct this error. The gain of the standardize circuit is R11/R12, or unity (1). Thus, V_a must be equal to the offset potential, or 700 millivolts under ideal conditions. The circuit shown in Fig. 8-8 allows V_a to be varied over the range ±1560 mV, so will more than accommodate the needs of the circuit. The standardize control (R13) is an operator adjustment, as is the temperature compensation (R2B).

The 0- to 1999- mV digital panel meter used as the output indicator is connected directly to the output of amplifier A2. In order to reduce the effects of signal noise (which can really mess up the A/D conversion in some DVMs), a low-pass capacitor is connected across R11 in amplifier A2. Also provided is a buffered output of the same scale factor (i.e., 100 mV/pH) so that an analog signal can be provided to a strip-chart recorder, oscilloscope, or computer A/D converter input. In the latter case, incidentally, it might be wise to use the BCD outputs provided on some digital panel meters instead of a separate A/D converter for the computer. The reason for this method is to make the A/D output and DVM/DPM reading harmonious. You could otherwise arrive at a situation where conversion differences provide inconsistent readings (e.g., 6.4 on one and

6.2 on the other). There is always some measurement and (in digital instruments) quantization error, so the design goal is to make the readings consistent with each other as well as accurate as possible within budget constraints.

Initial Calibration

The pH meter will be calibrated initially by the maker, and then (using a lesser procedure) by the user every time it is used. The following procedure will yield results:

1. Turn on the circuit with the *Function* switch set to "7" and allow to warm up for a period of 15 minutes. Disregard the output meter reading.
2. Set the *temperature coefficient* control at 30° C, or about ⅓ way between limits (0 and 100° C).
3. Adjust *standardize* (R13) for a meter reading of "7.000" (the last digit may "bobble" between "0" and "1." This is normal and is not a sign of trouble).
4. Set the *Function* switch for "2," and then adjust R8 for a reading of "2.00" on the meter.
5. Set the *Function* switch for "12" and adjust potentiometer R5 for a reading of "12.00" on the meter.
6. Seal potentiometers R5 and R8 with a dab of rubber cement on each adjustment point.

Operator Adjustment (without Buffer Solutions)

1. Set the *Temperature Compensation* control (R2B) for the appropriate temperature (30° C if room temperature).
2. Set the *Function* switch to "7."
3. Adjust *Standardize* (R13) for a reading of "7.000" on the meter.
4. Set *Function* to "2" and check reading for 1.9 to 2.1.
5. Set *Function* to "12" and check reading for 11.90 to 12.10.
6. The meter is ready for use.

Operator Adjustment (With Buffer Solutions)

1. Set the *Temperature Compensation* control (R2B) for the appropriate temperature (30° C if room temperature).
2. Immerse the glass electrode in a solution that is neutral (pH = 7.00), such as well-distilled water.
3. Adjust *Standardize* for a reading of 7.00.
4. Use buffer solutions in the pH 2-4 and pH 11-13 range to further check calibration.
5. The meter is ready for use.

Alterations to the Circuit

There are a couple of things which you can do to make this circuit a little "nicer" for the user. For example, you could "digitize" the temperature compensation function by adding another position on function switch S1. A signal from a temperature source can be measured here.

MICROCOMPUTER-CONTROLLED "DIGITAL" pH METER

These days most electronic instruments newly made for the physical and life-sciences are microcomputer controlled. There are a couple of ways which will make the pH meter "microcomputer-controlled." The simplest trick for an existing pH meter is to simply connect the analog output of Fig. 8-8 to the A/D converter input of a microcomputer. Alternatively, we can use the BCD output lines of the panel meter (not always available at low cost). If an existing instrument lacks an analog output, then it is still possible in many cases to provide one by adding a unity gain noninverting follower buffer. If the scale factor is other than 100 mV/pH, then it will be necessary to also provide some software scaling to account for the difference.

If the A/D converter used on this project is BCD and 3½ digit, then little scaling will have to be done . . . the binary output will be consistent with the people-orienting readings. But if a straight binary A/D is used (probably the case!), then some scaling is needed. For the normal range, the output voltage of Fig. 8-8 will be 0 to 1.5 volts (pH up to 15), or 0 to 1500 millivolts. We will, therefore, represent pH = 15 with the binary word 11111111_2 or FF_{16}.

Another method is shown in Fig. 8-9. Here we keep the analog circuitry at a minimum, and replace some circuit functions with computer software. This circuit also includes, in addition to pH measurement, automatic setting of the temperature compensation. A thermistor or other form of electronic thermometer is immersed into the fluid being measured along with the pH probe. Some pH probes, incidentally, already contain the temperature sensor so are ready to go. The pH electrode will produce pretty much the same scale factors as the previous examples, so will range as shown in Fig. 8-7. By using an amplifier with a gain of −10, and an offset scale factor, we can make the output of A1 on the order of 7 volts at full-scale. We measure pH by using a digital-to-analog converter and a voltage comparator (COMP) to potentiometrically measure V1. The voltage at the output of the DAC is made maximum when

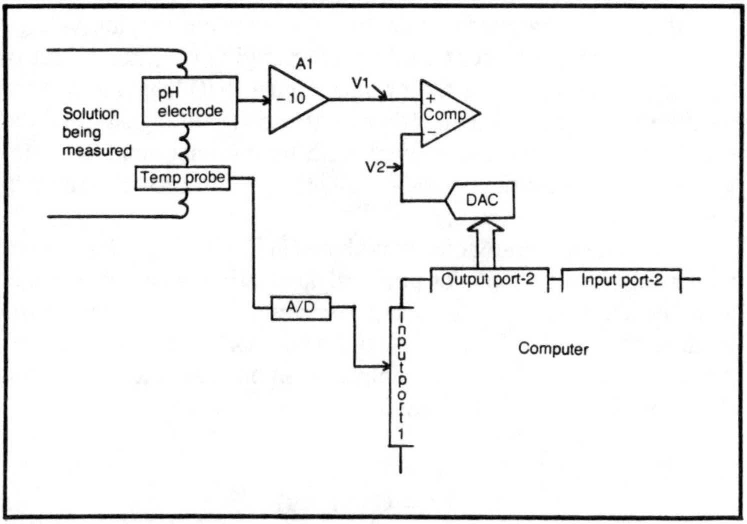

Fig. 8-9. Block diagram for a digital pH meter.

the routine is initialized, and then decremented each step until a potential is found that is equal to, or immediately below V1. The voltage comparator output will drop LOW at that time signaling the computer that the measurement is finished. A look-up table can be provided to equate DAC states with pH readings.

The pH meters described in this section can be built in a variety of different configurations. The only critical layout problem regards the input amplifier, A1. The input amplifier should be constructed in such a way that the distance between it and the input connectors is minimized. In fact, it should be built on the connector, if possible. Also, conspire to keep heat sources away from the input amplifier, less drift becomes a problem.

INTEGRATORS AND DIFFERENTIATORS

Integration and differentiation are two related mathematical processes which are often used in electronic instrumentation. Electronic integrators and differentiators are easily made with operational and other forms of amplifier. The naive, but none the less useful, definition of integration is that it allows us to find the area under a curve, even if irregular. Similarly, the process of differentiation is such that we are able to find the instantaneous rate of change of a curve. In both cases, we are assuming that the curves are functions and are at least continuous over the range and domain of interest to the circuit.

If you integrate a function V(t) (i.e., a time-varying voltage signal), and then differentiate the integrated signal, the output of the differentiator will be the original signal, V(t). Similarly, if we differentiate V(t), and then integrate that dV(t)/dt signal, then the result is V(t). If we experiment with operational amplifier integrators and differentiators we will see this result, provided the gain constants are set correctly.

A simple integrator circuit is shown in Fig. 8-10A. This circuit, which may be too simple for practical application (as we shall see in a moment), uses a series input resistor and a capacitor in the feedback network. We can use the same sort of analysis on this circuit as we used before on simple dc amplifiers. First, we write the expressions for I1 and I2:

$$I1 = V_{in}/R1 \qquad \textbf{(a)}$$
$$I2 = C_1(dV_o/dt) \qquad \textbf{(b)}$$

By Kirchhoff's Current Law (KCL), we also know that I1 and I2 are equal to each other:

$$I2 = -I1 \qquad \textbf{(c)}$$

Now, substituting (a) and (b) into (c):

$$\frac{C_1 dV_o}{dt} = \frac{-V_{in}}{R1} \qquad \textbf{(d)}$$

Now, lets integrate both sides of equation (d) with respect to time:

$$\int \frac{C_1 dV_o}{dt}\, dt = \int \frac{-V_{in}}{R1}\, dt \qquad \textbf{(e)}$$

$$C_1 V_o = \int \frac{-V_{in}}{R1}\, dt \qquad \textbf{(f)}$$

Collecting terms and solving for V_o:

$$V_o = \frac{-1}{R_1 C_1} \int V_{in}\, dt \qquad \textbf{(g)}$$

Where:

V_o is the output potential in volts

V_{in} is the input signal potential in volts
R_1 is the resistance of R1 in ohms
C_1 is the capacitance of C1 in farads

The term R_1C_1 in equation (g) is the *time constant* of the integrator, in seconds (provided that R is in ohms and C is in farads). It also expresses the gain of the integrator, which is:

$$A_x = -1/R_1C_1 \quad \textbf{(h)}$$

Let's consider some of the implications of Expression (h). A rule of using integrators is that the time constant R1C1 be long (i.e., 5X or more) relative to the period of the applied signal; otherwise, some low-pass filtering might occur but no real integration. But that can provide us with an exciting little gain problem when that value is too small. Keep in mind that the time constant (R1C1) appears in the denominator of the gain expression. Thus, as it gets small the gain becomes very large in a quick hurry. Consider, for a minute, a time constant of one-second. This could be obtained with the gentlemanly values of R1 = 1 megohm, and C1 = 1 μF: R1 × C1 = (10^6 ohms)(10^{-6} farads) = 1 second. The gain is:

$$A_v = -1/R1C1$$
$$A_v = -1/(10^6)(10^{-6})$$
$$A_v = -1/1 \text{ second}$$
$$A_v = -1$$

The integrator described above would have a gain of −1, which means that the output potential would rise at a rate of *1-volt-per-second-per-input-volt.* But what happens with smaller time constants. For example, if we make C1 = 0.01 μF, and R1 = 10 kohms. The time constant is a lot smaller, so the gain is a lot higher:

$$A_v = -1/R_1C_1$$
$$A_v = -1/(10^4 \text{ ohms})(10^{-8} \text{ farads})$$
$$A_v = -1/10^{-4} \text{ seconds}$$
$$A_v = -10^4$$

In this case, the gain is 10,000. This means that the output voltage will rise at a rate of 10,000-volts-per-second-per-volt-of-input-signal! If we apply a 1-volt signal to that input, we will saturate the integrator very rapidly. Let's assume that the maximum permissible output voltage is ±10 volts (a reasonable value for ±12 Vdc power supplies and easy to calculate). If we apply a 1-volt input signal, then the output of the first integrator would take 10-seconds to saturate. But, in the second case, with the gain set at 10,000, the output will saturate at 1/10,000 the amount of time, or, 0.001 seconds (i.e., 1 millisecond). To the outside world, that integrator will seem to simply latch-up on power-on condition.

The reason why that circuit of Fig. 8-10A is too-simple is that

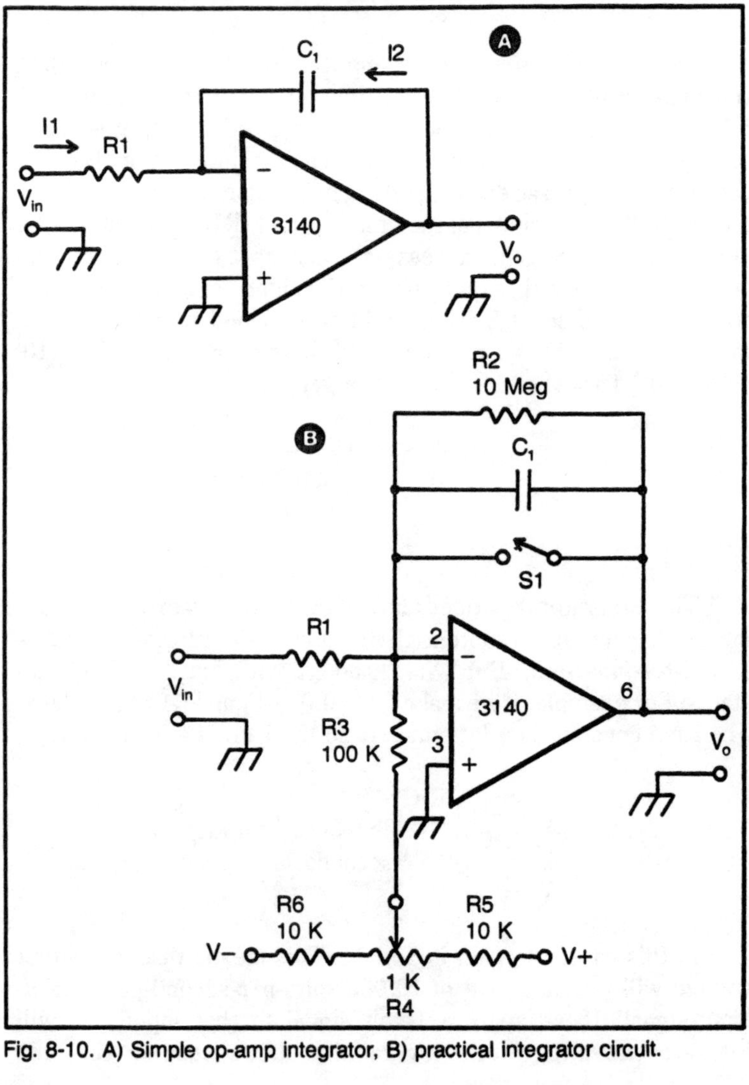

Fig. 8-10. A) Simple op-amp integrator, B) practical integrator circuit.

bias currents will tend to cause an output offset signal that will charge the capacitor in a short period of time that is dependent upon the time-constant of the integrator. Also, if there is a dc-offset potential on the input signal, as is often the case, then the integrator will also latch up. The trick is to add a couple of small components and make a more complex integrator than that of Fig. 8-10A. An example of a minimum-components integrator is shown in Fig. 8-10B.

The operational amplifier selected is the RCA CA3140 device. This device was chosen because it has an extremely high input impedance, thus only a tiny input bias current. Those bias currents tend to charge the integrator capacitor and produce a rising error term. I tried several different operational amplifiers while experimenting with the integrator circuit and found this to be among the best op amps for the application. The 741-family devices were all but useless in this regard. They had such tremendous offsets and bias currents that the capacitor charged sufficiently to saturate the amplifier output within a couple seconds after turning the power on. Certain then-premium devices, such as the μA-725 device, cost plenty of money compared with the 741 (and the CA3140!), but fared only a little better than the 741 device. The 725 would saturate in a few seconds, but that was still unacceptable. The CA3140 device was the only one of those tested which filled the bill. Therefore, I recommend BiMOS (like the CA3140) or BiFET operational amplifiers only when designing integrators.

The switch in parallel with the integration capacitor is used to dump the charge at the beginning of an operation. This switch might be a relay, a mechanical switch, or a CMOS electronic switch. It should be closed momentarily immediately before the integrator is expected to operate in order to dump any charge in the capacitor. This charge would be seen as a constant error term in the final result.

Resistor R2 is used to cancel the effects of erroneous dc offset signals on the input signal. If this resistor was not used, then slight amounts of dc level would cause the output waveform base line to creep upscale until the amplifier was saturated. The value of this resistor must be very high with respect to the value of R1, so most designers use 10 megohms as a minimum value. If we make this resistor too low in value, then the circuit acts more like an amplifier with a gain of R2/R1, and a frequency response roll-off of −6 dB per octave above a −3 dB point determined by R2/C1; that's *not* what we want!

The purpose of potentiometer R4 is to provide a means for canceling the drift of the integrator, which even the CA3140 will exhibit to a limited degree. The adjust procedure is as follows:

1. Short the input end of R1, making V_{in} zero.
2. Apply power and wait 5 mins. for the integrator to stabilize. Connect a voltmeter to V_o.
3. Close S1 for several seconds.
4. Adjust R4 for zero volts output and no drift.
5. Repeat steps 3 and 4 until no further improvement is possible.
6. If it proves impossible to obtain both zero output and zero drift, then add either a second circuit like R4, or, a unity gain inverting amplifier with an offset control to follow the integrator. Of course, this will invert the polarity of the integrator output signal.
7. Remove the input short, the integrator is ready for use.

The values of R1 and C1 must be selected carefully. Most authorities recommend a time-constant of 5X or 10X the period of the waveform being integrated.

The mathematical process of differentiation is exactly opposite the process of integration. It should not surprise us, therefore, to find that the differentiator is made simply by reversing the roles and positions of the capacitor and resistor: compare Fig. 8-11A with Fig. 8-10A. In the differentiator circuit (Fig. 8-11A), the capacitor is in series with the input signal, while the resistor is in the feedback loop around the amplifier. Using a method of circuit analysis similar to that used before we can find the transfer function of the differentiator:

From Kirchhoff's Current Law (KCL):

$$I2 = -I1 \qquad \textbf{(a)}$$

We know that

$$I1 = \frac{C_1 dV_{in}}{dt} \qquad \textbf{(b)}$$

and,

$$I2 \; \frac{V_o}{R_1} \qquad \textbf{(c)}$$

Substituting (b) and (c) into (a):

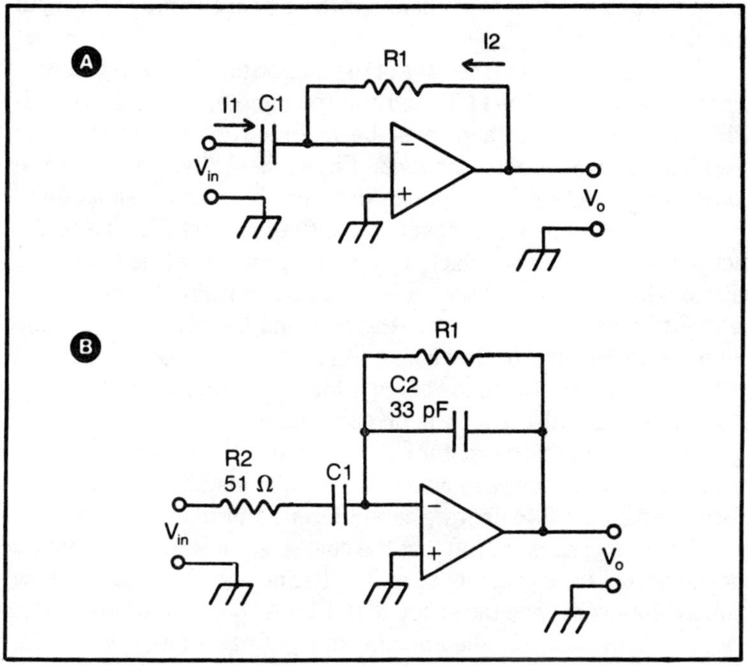

Fig. 8-11. A) Simple op-amp differentiator, B) practical differentiator circuit.

$$\frac{V_o}{R_1} = \frac{-C_1 dV_{in}}{dt} \qquad \text{(d)}$$

Now, let's solve Equation (d) for output voltage V_o:

$$V_o = \frac{-R_1 C_1 dV_{in}}{dt} \qquad \text{(e)}$$

or, rearranging to make a little more sense when read quickly:

$$V_o = -R_1 C_1 \frac{dV_{in}}{dt} \qquad \text{(f)}$$

Equation (f), then, is the transfer function of the simple operational amplifier differentiator. As in the case of the integrator, the "gain" of the circuit is given by R1C1. The "d/dt" used with V_{in} is calculus notation to let us know that we are looking at the rate of change of V_{in} with respect to time. The operational amplifier integrator, then, will produce an output that is proportional (by R1C1) to the instantaneous rate of change with respect to time of the input signal.

As was also true in the case of the integrator circuit, the simple differentiator of Fig. 8-11A is a little too simple. In most cases, it will not work nicely, and must be modified a little in order to overcome certain real problems. Figure 8-11B shows a typical compensated differentiator which will work in practical situations.

The problems which beset real differentiators are due to the fact that the differentiator acts like a high-pass filter (the integrator is also a low-pass filter) with a +6 dB/octave rising characteristic gain. At some point, however, the gain and the phase shift of the circuit will meet the Barkhausen criterion and the circuit will oscillate mercilessly. Also, being a high-pass filter, the circuit is also subject to noise which is on the line.

Resistor R2 and capacitor C2 are added to the circuit in order to reduce the frequency response in the area of instability, while still allowing the circuit to operate as a differentiator in lower frequency regions. It is generally true that the combination R2C2 must have a much shorter time constant than R1C1. The actual design of these components is beyond the scope of this book, but will be covered in another of my more detailed (forthcoming) TAB books, *The Analog Subsystem*.

WINDOW COMPARATOR

A voltage comparator will examine two voltages, and then indicate their relationship by producing output levels. Figure 8-12A shows a comparator made from an operational amplifier. The comparator is basically an amplifier with far too much gain. If we connect the operational amplifier with no feedback resistor, then the gain of the circuit is essentially the open-loop gain of the integrated circuits. For most operational amplifiers, the gain (A_{vol}) is 20,000 for garbage-grade devices, and more than 1,000,000 for some premium devices. In Fig. 8-12A we are using the operational amplifier comparator to compare V1 and V2. In this circuit, V1 is applied to the inverting input, while V2 is applied to the noninverting input of the operational amplifier. In the event V1 = V2, we see the output is zero:

$$V_o = (A_{vol})(V1 - V2)$$
$$V_o = (A_{vol})(0)$$
$$V_o = 0$$

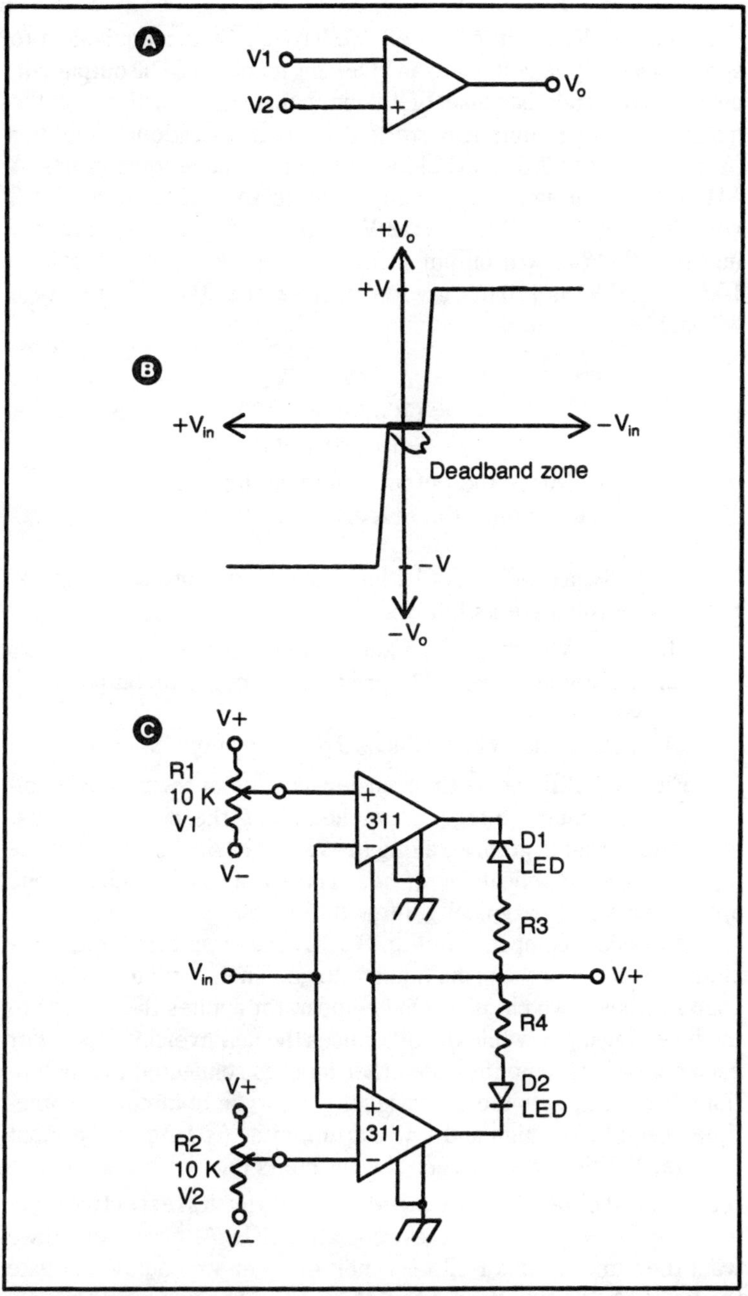

Fig. 8-12. A) Comparator is an op amp with no feedback network, B) transfer function, C) window comparator.

In the case of V1 greater than V2, we have a situation equivalent to applying a positive voltage to an inverting amplifier. The output will be positive, and, because of the immense gain, will be at the maximum voltage output permitted for that operational amplifier (from 0.5 volts to 3.5 volts below V+). Let's assume some values. A 741 operational amplifier usually permits an output voltage of 3 volts less than V+. With a ±12 Vdc power supply, therefore, we find that the 741 will output from −9 to +9 volts. If the gain is 100,000 and V1 is 100 mV greater than V2, then the output voltage will be:

$$V_o = (A_{vol})(V1 - V2)$$
$$V_o = (100,000)(-.100)$$
$$V_o = -10,000 \text{ volts*}$$

*Since the output voltage is limited to 9 volts, the output will be 9 volts.

Similarly, a condition of V1 less than V2, will produce a similar result. The rules are as follows:

1. V1 = V2, produces a zero output voltage.
2. V1 greater than V2, produces a negative output voltage.
3. V1 less than V2, produces a positive output voltage.

Figure 8-12B shows the transfer function of the voltage comparator. The input voltage is graphed along the horizontal axis, while the output is graphed along the vertical axis. The units of the input voltage are typically *millivolts.* There is a small deadband zone around zero that is typically a few millivolts.

A window comparator (Fig. 8-12C) uses two comparator circuits connected to the same input voltage. The alternate inputs are used for these two circuits. One comparator applies the voltage to the inverting input, while the other uses the noninverting input. For each operational amplifier the other input is connected to a potentiometer which sets the alarm level. This type of circuit is sometimes used to set high and low alarm limits. As long as the input voltage, V_{in}, remains between the set limits then both alarms keep turned off. If either limit is exceeded, however, its respective input will drop LOW and turn on the associated LED. Alternatively, if we want the same alarm for either condition, then we could AND-gate the two LM-311 outputs into one alarm line. The line would remain LOW until an alarm condition occurs, then it goes HIGH.

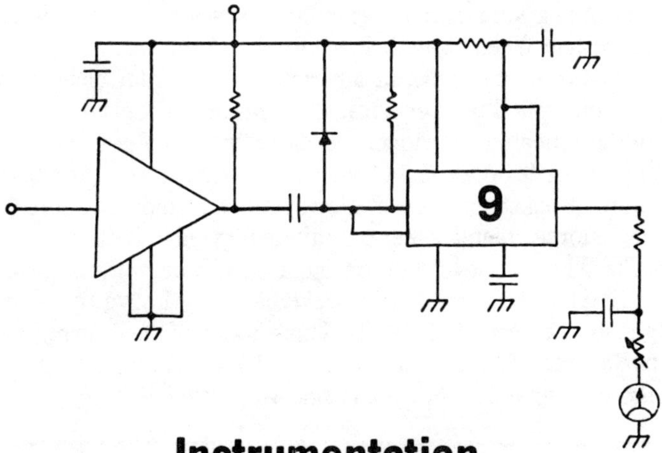

Instrumentation Techniques Using Electronics

In this chapter we will study some of the basic techniques of electronic instrumentation in measuring physical parameters. We will discuss such exotic topics as colorimetry, electronic temperature measurement (a frequently used and very popular electronic application), electrophoresis and photoplethysmography. Mind you, we are not under any illusions that there are hundreds of budding electrophoriologists in the readership. Our purpose here is to illustrate a technique. In the case at hand, a column cost a lot of money and the scientist could not afford to instrument it with the normal "store-bought" instruments used by other physiologists in that kind of experiment. A technician following my advice built an instrument package with relatively low-cost components and filled the need. The main thing here is not the specific application, but rather the principle. We will start our discussion with a look at the workhorse of the electronic measurement business . . . the Wheatstone bridge and strain gauges.

STRAIN GAUGES AND WHEATSTONE BRIDGES

The Wheatstone bridge has been around for more than a century, and is still one of the principal methods for making certain types of measurement. Despite its age, the ol' Wheatstone bridge hangs on continuously. Figure 9-1A shows the most usual version of the Wheatstone bridge. Notice that it consists of all resistances. Since many instrumentation transducers use variable resistances in order to indicate the parameter being measured, we can use those transducers in Wheatstone-bridge circuits. A thermistor, for example, is a ***thermal resistor,*** i.e., a resistor whose value changes with changes in applied temperature. If we place the thermistor in one arm of the Wheatstone bridge, therefore, we can produce an output voltage that is proportional to the temperature. Other examples of Wheatstone-based instrumentation systems include pressure (fluid and gas), force, displacement, and wind-speed/direction.

The Wheatstone bridge consists basically of two voltage dividers, called half-bridges. For the sake of clarity I have redrawn the bridge into the form of Fig. 9-1B. This version of the drawing shows the two voltage dividers: R1/R2 and R3/R4. From the elementary voltage divider equation we can find *V*1 and *V*2:

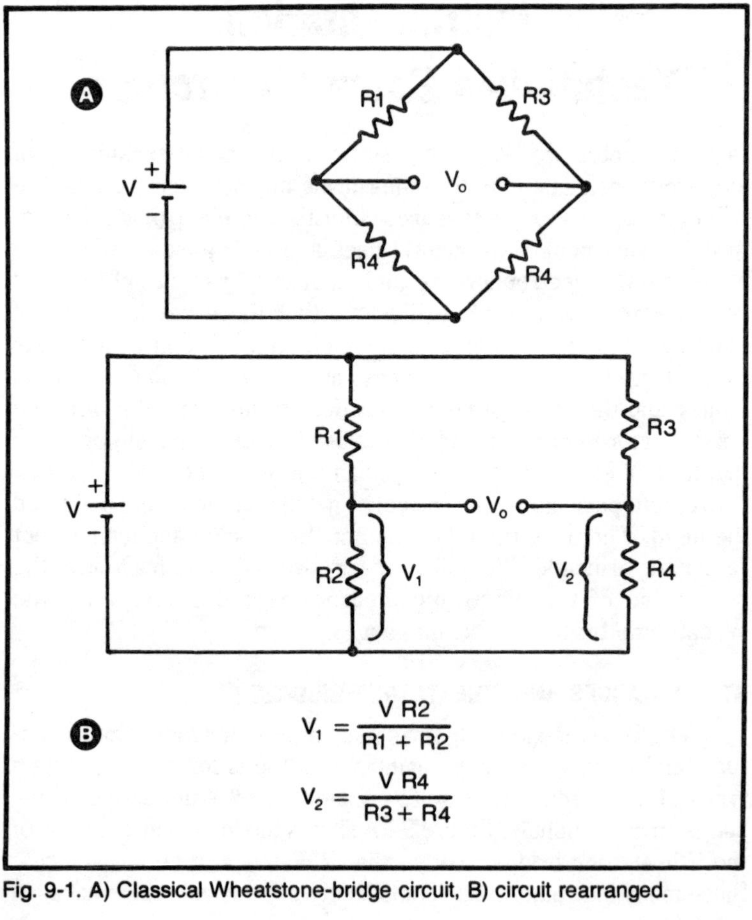

Fig. 9-1. A) Classical Wheatstone-bridge circuit, B) circuit rearranged.

$$V1 = \frac{VR_2}{R_1 + R_2} \quad \text{(a)}$$

and,

$$V2 = \frac{VR_4}{R_3 + R_4} \quad \text{(b)}$$

Output voltage V_o is the difference between V1 and V2 (V1-V2):

$$V_o = \frac{VR_2}{R_1 + R_2} - \frac{VR_4}{R_3 + R_4} \quad \text{(c)}$$

or, when V, the excitation voltage, is factored out:

$$V_o = (V) \quad \frac{R_2}{R_1 + R_2} - \frac{R_4}{R_3 + R_4} \quad \text{(d)}$$

In many instrumentation applications, all elements of the Wheatstone bridge will have the same resistance under "rest" conditions, so the output voltage will be given by:

$$V_o = (V) \quad \frac{R_2}{2\,R} - \frac{R_4}{2\,R} \quad \text{(e)}$$

We do not replace R2 and R4 with "R" here because these values will probably change as the applied stimulus changes.

An interesting effect occurs when the resistances are all equal, which in a transducer case means the applied stimulus is zero: all four resistances are equal, and the output voltage is zero. This is what makes the Wheatstone bridge so easily adapted to instrumentation applications. The null condition doesn't actually depend upon all of the resistance being equal, it is merely necessary that the ratios R1/R2 and R3/R4 be equal:

$$\frac{R1}{R2} = \frac{R3}{R4} \quad \text{(f)}$$

When using the Wheatstone bridge in an instrumentation circuit, we find that the circuit will have at least one, and possibly all four, arms of the bridge consisting of a transducer. This transducer will change resistance (ΔR) when the stimulating parameter changes. If we denote the change of resistance as *h*, then we can write the following expressions for the output voltages of bridges with one, two, or four active elements (i.e., arms).

1. If only one arm of the Wheatstone bridge has an active element, then the output voltage is:

$$V_o = \frac{V \times h}{4R} \quad \text{(g)}$$

Where:

V is the excitation voltage
R is the resistance of all elements, including the active element when the applied stimulus is zero
h is the change of *R* in the active element when stimulus is applied.

2. If two arms of the Wheatstone bridge contains active elements, then the output voltage is given by:

$$V_o = \frac{V \times h}{2R} \quad \text{(h)}$$

Where:
All terms are as defined above in (1)

3. If all four elements are active, then the output voltage will be:

$$V_o = \frac{V \times h}{R} \quad \text{(i)}$$

All of the above equations assume that, at rest when no stimulus is applied, or when the stimulus is some standard value declared by convention to be zero (e.g., 0° C is actually 273° K), all resistances in all four arms are equal.

A *strain gauge* is a remarkable little device which will allow us to measure such parameters as force, pressure, displacement, vibration, and so forth. The strain gauge operates on a principle known as *piezoresistivity,* that is, the change of resistance due to deformation of a conductor. Let's consider the basic expression for the resistance of a conductor (we assume a round wire for the sake of simplicity):

$$R = \rho \times \left[\frac{L}{A}\right] \quad \text{(j)}$$

Where:

- ρ is the *resistivity constant* of the material in ohms-meters (Ω-m)
- L is the length in meters (m)
- A is the cross-sectional area of the conductor in square meters (m^2)

The resistivity is a fixed constant, and is a property of the material making up the conductor. But the length (L) and the cross-sectional area (A) can be varied. If either of these factors is varied, obviously we will change the resistance. If the change of resistance is either linear, or varies nonlinearly in a predictable manner, then we can use the device as a strain gauge transducer.

Figure 9-2 shows how the basic strain gauge works. Figure 9-2A shows a conductor at rest, with no forces applied to it. The

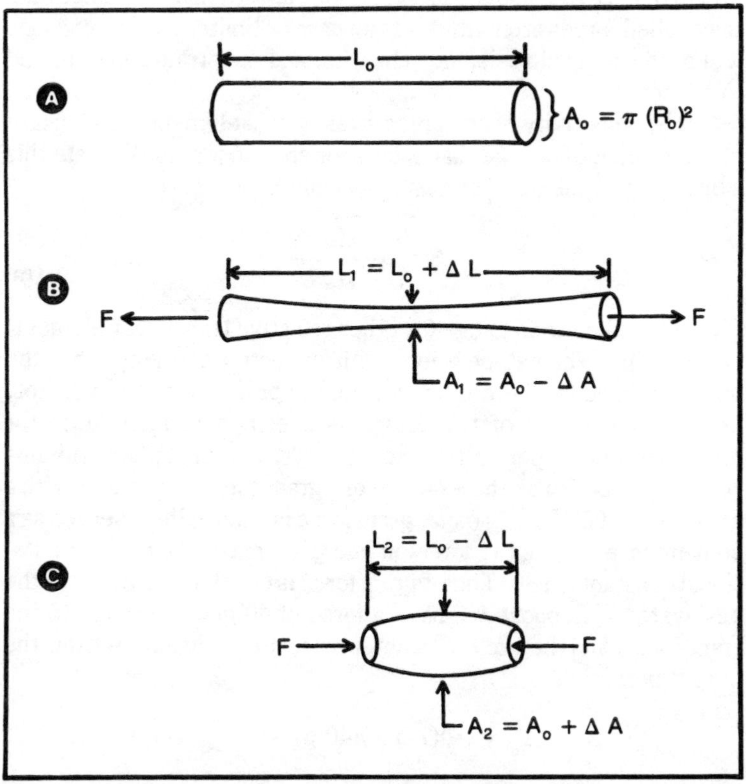

Fig. 9-2. Piezoresistivity example A) member at rest, B) tension force, and C) compression force.

area is A_o and the length is L_o. If we apply a *tension force* to the conductor, however, we have a situation as in Fig. 9-2B. The new length L_1 is

$$L_1 = L_o + \Delta L \tag{k}$$

and the new area is

$$A_1 = A_o - \Delta A \tag{l}$$

The dc resistance of this sample will increase a small amount, *h,* which is proportional to the applied force (as long as the force isn't too large!).

Similarly, in Fig. 9-2C, we have the situation where a *compression force* is applied. In this case, the resistance will decrease proportional to the applied force. As long as the change of length and the applied forces are within certain narrow limits, the two changes will produce equal resistance changes, and the strain gauge is said to be linear.

The sensitivity of any given strain gauge element is a measure of the amount of change that occurs for applied forces. We rate this sensitivity in the form of a *gauge factor,* S:

$$S = \frac{\Delta R/R}{\Delta L/L} \tag{m}$$

When the strain gauge (or other resistive transducer element) is placed in a Wheatstone bridge with four active elements, then the output voltage can be predicted from the bridge sensitivity factor, **Φ**. The sensitivity of the bridge is specified in microvolts (or millivolts) output per volt of excitation per unit of applied stimulus (μV/V/U). Let's use the example of a gram-force transducer with a sensitivity of 5 μV/V/g (*note:* gram force is used in life sciences as a convenience, and is the force on one gram mass when the gravitational constant is 980. Thus 1 gram force is equal to 980 dynes in the cgs system). Suppose we have a force of 40 grams applied to the transducer, and the excitation potential is +7.5 volts dc. What is the output potential?

$$V_o = \Phi(7.5\ \text{V})(40\ \text{g})$$

$$V_o = \frac{5\ \mu\text{V}}{\text{V-g}} \quad (7.5\ \text{V})(40\ \text{g})$$

$$V_o = (5\ \mu V)(7.5)(40)$$

$$V_o = 1500\ \mu V, \text{ or } 1.5\ mV$$

PRACTICAL WHEATSTONE-BRIDGE AMPLIFIERS AND EXCITATION SOURCES

The Wheatstone bridge has four junctions, two of which are for the excitation signal and two of which are output terminals. In this section, let's discuss what can be done with the Wheatstone, especially in an instrumentation frame of reference.

First, let's deal with the matter of the excitation potential. In the examples given previously, we have used batteries for the excitation potential. In many real instruments, we will also use batteries as the excitation, but more often it will be an electronic dc power supply. Certain instruments used in biomedical research and in clinical applications require battery excitation for the sake of safety. A cardiac output computer, for example, uses a thermistor in the tip of a catheter that is thread through a vein in the patient's arm into the right side of the heart and out the pulmonary artery. If the thermistor is placed nicely in the pulmonary artery, and the doctor or nurse does everything correctly, then this thermistor will measure temperature changes which are interpreted as correlatable to cardiac output in liters per minute. In that case, patient safety requires battery power.

In most other cases, however, we will use some form of electronic power supply to excite the Wheatstone bridge. It is essential, however, that the voltage must be *stable*—which means a voltage regulator. Recall the equation given earlier had a "V" term in it, and that means changes in *V* will change the output voltage. In most cases, we can use a three-terminal IC voltage regulator for the excitation potential.

We must be careful as to the value of the excitation potential. Some manufactured transducers which use a resistive Wheatstone bridge as their basis have maximum excitation potentials, usually under 10 volts. One very popular medical blood-pressure transducer requires not more than 7.5 volts of excitation, while others must have less than 5 volts. As a result of these values, it is often the case that the excitation potential will be supplied by a 5 Vdc three-terminal IC regulator. Of course, almost any Wheatstone-bridge transducer can use the LM-309K, LM-340K-5, LM-340T-5 or 7805. If the transducer has a high enough resistance, then the 100 milliampere LM-309H or LM-340LS-05 devices can be used. The load seen by the power supply is the so-called "looking back resistance" of the Wheatstone bridge. If you have a technical institute or engineering education you will recognize that term as being from Theveinin's theorem, but for the rest we can state that, for transducers in which all four resistance arms have the same value at rest the looking back resistance is the resistance of any one arm. If the voltage regulator output voltage divided by this resistance is less than 100 mA, then the low-power regulator can be used (it will result in a space savings).

It is also possible to use a simple zener diode voltage regulator for this application. The load represented by the Wheatstone bridge does not change appreciably during operation, so the zener is a reasonable choice—maybe. One problem with the zener is a small dependence on temperature, which could introduce error into the transducer Wheatstone bridge output potential. One solution is to use something like one of the *band-gap* IC zeners, or, a zener with a built-in heater (e.g., National Semiconductor LM-399 device) that keeps temperature constant.

The output terminals of the Wheatstone bridge are differential in nature. Thus, for proper operation we will want to provide a *differential amplifier* to process the signal at the output of the Wheatstone bridge. Figure 9-3 shows a typical circuit.

The example in Fig. 9-3 was originally intended for a low-cost fluid pressure transducer used in a medical research facility. The fluid was not blood, but was a relatively viscous protein material flowing in tubing. The tubing, incidentally, should be stiff-walled *Teflon®* (or some similar material) so that pulsations will not cause variations in the pressure waveform due to changes in the wall of the tubing. The connector pair can be almost anything that fits your own need, but in my case it was an Amphenol military-style MS connector in Size-14 shells and 5-pin inserts. Note that the usual practice is to place the male insert on the end with the transducer, with the female insert on the chassis mounted shell from the main amplifier unit. The reason for this is that the excitation potential is applied to one of those pins, and it is safer if that pin is a female type—guards against short circuits. Connector J1, therefore, is a 5-pin female chassis mounted type, while P1 is a mating male plug that is attached to the Wheatstone-bridge transducer via shielded wire.

Excitation for this amplifier is provided by a National Semiconductor LM-399 device, in which a 6.95 volt zener diode is packaged inside of an IC can with an electric heater. The heater is connected across the V+ power supply, and keeps the internal temperature of

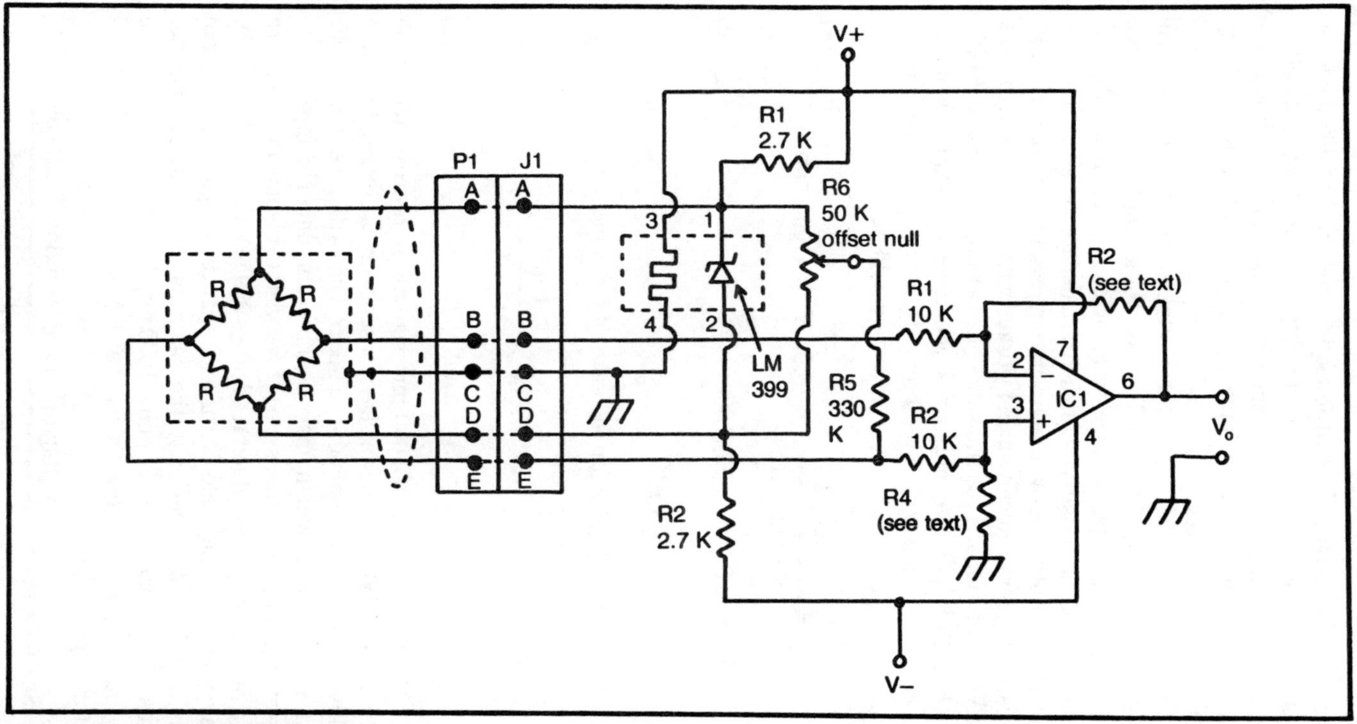

Fig. 9-3. Wheatstone-bridge amplifier.

the zener IC constant. This little tactic gives the LM-399 device a relatively high degree of voltage stability. If the 6.95 Vdc excitation potential is too high for some specific transducer, then it may be necessary to use a lower-voltage zener, or, an LM-309H (5 Vdc) regulator.

The amplifier, IC1, is a differentially-connected operational amplifier of the simplest type. The gain of the amplifier will be set according to our output voltage scale factor requirements. As discussed above, the sensitivity of such a transducer is given in terms of microvolts output per volt of excitation per unit of stimulation. In the case of the transducer which I worked with, the scale factor was 50 μV/V/mm Hg, and the transducer had a range of -200 to $+1200$ mm Hg. Only positive pressures would be encountered in the experiment. At full scale (1000 mm Hg in this case), the output voltage with the +6.95 volt excitation would be

$$V_o = \frac{50\ \mu V}{V - \text{mm Hg}} \times (6.95\ V) \times (1000\ \text{mm Hg}) \qquad \textbf{(a)}$$

$$V_o = (50\ \mu V)(6.95)(1000) \qquad \textbf{(b)}$$

$$V_o = 347{,}500\ \mu V$$

Converting to millivolts for the sake of convenience:

$$V_o = 347{,}500 \times \frac{1\ mV}{10^3\ \mu V} \qquad \textbf{(c)}$$

$$V_o = 347.5\ \text{millivolts} \qquad \textbf{(d)}$$

Thus, a pressure of 1000 mm Hg will be represented at the output of this Wheatstone-bridge transducer by a potential of 347.5 millivolts. How much gain do we need? That will depend upon our gain-scale factor, which may be different under different circumstances. For example, suppose we want to display this pressure on a digital voltmeter which reads 0 - 1999 mV (i.e., 1.999 volts). We could make the scale factor such that the output voltage from amplifier IC1 is numerically the same as the pressure reading. Thus, if we make 1000 mV equal 1000 mm Hg, then we could read the output pressure *directly* on the digital voltmeter. The scale factor is given by:

$$\text{Scale Factor} = \frac{\text{Voltage } (V_o) \text{ representing Full-Scale}}{\text{Bridge Output for Full-Scale}}$$

Or, when 1000 mV is to represent 1000 mm Hg, and the bridge output at that same pressure is 347.5 mV:

$$\text{Scale Factor} = \frac{1000\ \text{mV}}{347.5\ \text{mV}}$$

$$\text{Scale Factor} = 2.878$$

Of course, the "scale factor" is the gain required of the differential amplifier. Given the nature of the circuit used in Fig. 9-3, we would calculate the gain as:

$$A_v = R2/10\ K$$

or,

$$R2 = A_v(10\ K)$$

(assuming R2 = R4),

$$R2 = (2.878)(10\ K)$$

$$R2 = 28.78\ \text{kohms}$$

If the exact resistance value cannot be found, then make the gain of IC1 a little less than required by the scale factor, and then add a stage to follow which has a variable gain that will swing the total gain above and below the "correct" gain.

The 1000 mV example was used for the sake of convenience, but you may be faced with other cases. For example, computers use A/D converters to collect data from analog transducers and other sources. These A/D converters will have input ranges of 0 to 10 volts, 0 to 5 volts, −5 to +5 volts, 0 to +2.56 volts, and so forth. If we have a 0 to +2.55-volt (i.e., 2550 mV) A/D converter, and we want 2.55 volts to represent fullscale (1000 mm Hg), then the scale factor would be:

$$\text{Scale Factor} = \frac{2550\ \text{mV}}{347.5\ \text{mV}}$$

$$\text{Scale Factor} = 7.338$$

Another case might be the manufacture of a special plug-in transducer amplifier for your oscilloscope or strip-chart recorder.

One popular medium-priced oscilloscope which accepts plug-ins has a mainframe deflection factor of 50 mV/cm (vertical). We can buy blank plug-in assemblies with perforated boards inside to build our own circuitry. In that case, the total full-scale deflection would be 8 cm for 1000 mm Hg, and at a 50 mV/cm deflection factor 1000 mm Hg will be represented by a voltage of (8 cm)(50 mV/cm), or, 400 mV. In that case, the scale factor would be:

$$\text{Scale Factor} = \frac{400\ \text{mV}}{347.5\ \text{mV}}$$

$$\text{Scale Factor} = 1.151$$

Of course, in that case, we might want to eliminate the amplifier, and output the Wheatstone bridge directly to the differential input of the oscilloscope and sacrifice the upper end of the voltage range between 347.5 and 400 mV.

A scientist friend of mine was using a small, very popular force-displacement transducer on some medical experiments in an animal lab. The transducer used a Wheatstone bridge which, among several configurations, had a sensitivity of 5 μV/V/mg-force. He had a Tektronix oscillsocope (an elderly instrument) that did not have true differential inputs, but used two channels in an A-B type of pseudo-differential capability. Because of the low common-mode rejection of this instrument, and because of the lack of amplification between the oscilloscope and the transducer, and the 10-feet of shielded wire, there was a tremendous problem with 60 Hz interference on the small forces that he was measuring. This transducer was excited by a 6-volt lantern battery in a small chassis box, which further complicated the common-mode rejection problem. Figure 9-4 shows the solution, or at least an up-dated version using modern ICs (741 and 1458 devices were used in the originals).

The maximum force that this guy was going to measure was around 100 mg, so the output voltage at maximum will be:

$$V_o = \frac{5\ \mu\text{V}}{\text{V-mg}} \times (6\ \text{V}) \times (100\ \text{mg})$$

$$V_o = (5\ \mu\text{V})(6)(100)$$

$$V_o = 3000\ \mu\text{V} = 3\ \text{mV}$$

The output voltage at maximum stimulus (100 mg-force) will be 3 mV (which is only 0.003 volts!), and most readings will be a lot

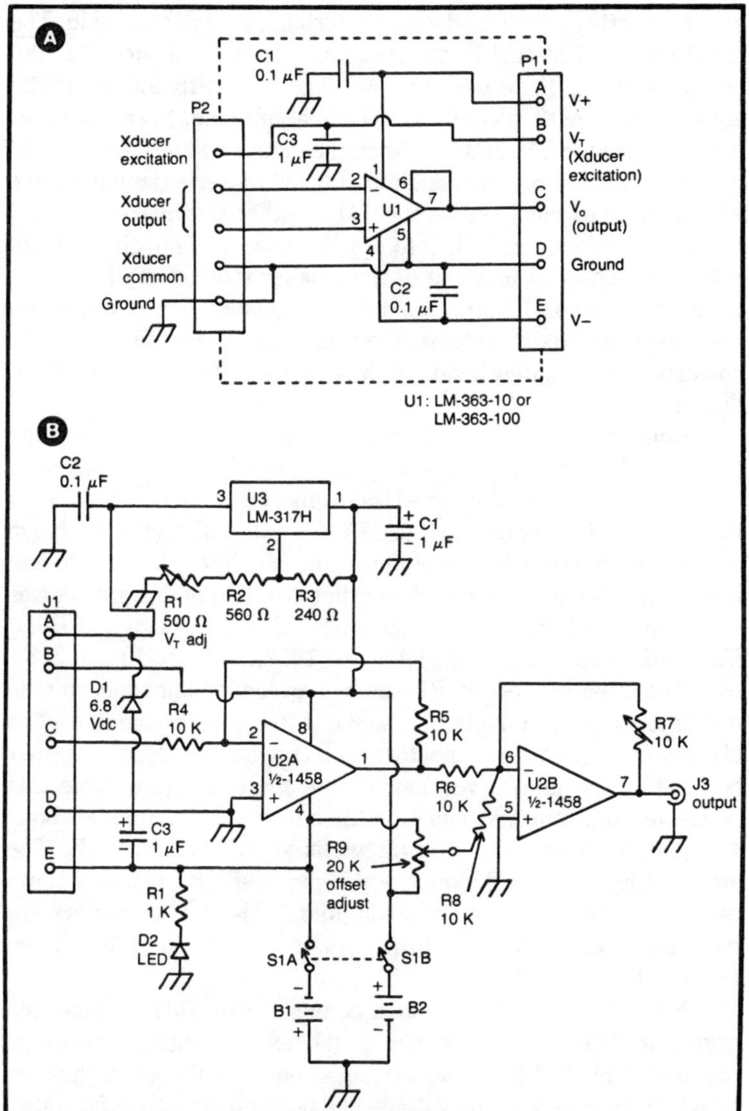

Fig. 9-4. A) Transducer amplifier, B) post-amplifier.

less (40 mg-force is a reasonable mean). Thus, we will reap benefits by increasing this voltage at the transducer end before transmitting it through the cable to the oscilloscope, strip-chart recorder, or computer A/D converter input. In Fig. 9-4A we have a single differential amplifier inside of a shielded box that is designed to attach directly to the transducer, or, through a very short cable. The amplifier is a National Semiconductor LM-363-x device ("x" denotes the gain). If we use the LM-363-10, then the output of the amplifier will be (3 mV)(10), or 30 mV at fullscale. If, on the other hand, we use the LM-363-100, then the output will be (3 mV)(100), or 300 mV at fullscale. Although the LM-363-x series amplifiers are a little more expensive than the 741 or other amplifiers, they are fixed gain and can be built into smaller spaces. This circuit (Fig. 9-4A) will easily fit into one of the smaller sizes of the little blue boxes by *Pomona Electronics,* or, in a smaller box (homemade, perhaps?) mounted directly on the transducer. It is also possible to construct the amplifier inside of the plug that attaches to the transducer!

Connector P2 must be the type which normally mates with the transducer. In the case of the amplifer which I built, the connector was a chassis-mounted mate to the transducer connector, so, with the correct cinch collar on the *ITT-Cannon* connector we could mount the *Pomona* box containing the amplifier directly to the transducer. A photograph of the earliest version of this project was shown in one of my earlier TAB books, *How To Design and Build Electronic Instrumentation* (TAB No. 1012), Fig. 22-10., p. 388.

The other connector, P1, can be any convenient connector. In my case, I used a 7-pin *Amphenol* 126-series connector. The chassis-mounted male connector was P1 on the transducer amplifier box, while the female version was on the end of the cable that connected the transducer preamplifier to the main amplifier box.

The main amplifier box circuit is shown in Fig. 9-4B. The amplifier here is a 1458, which is an 8-pin miniDIP which contains a pair of 741-family operational amplifiers. The two amplifiers are both unity gain inverting followers, so the total gain will be (−1)(−1), or +1 total.

Excitation for the transducer comes from a 100 mA adjustable regulator (U3) that can be set to the exact excitation potential required. This LM-317H device has an output voltage which is set by R1. In most cases, this voltage will be set to exactly 5.00 volts.

Diode D1 is not used to provide excitation, as might be expected on first glance. The purpose of this diode is to protect the transducer from damage in the event the LM-317H fails and places the V+ voltage output onto the transducer excitation voltage line. If the LM-317H dies, then the zener diode will break over and short the line to ground. This diode should be a 5-watt type, and a small fuse at the input of the LM-317H might be appropriate (say, ¼ ampere).

If the scale factor of this system is inadequate, then we can adjust the overall gain by changing the resistance of one or more resistors in the main amplifier. The overall gain is:

$$A_v = A_{v(U1)} \times A_{v(U2a)} \times A_{v(U2b)}$$

We can vary $A_{v(U1)}$ only in steps, i.e., either ×10 or ×100, but the other two are set by:

$$A_{v(U2a)} = R5/R4$$

or,

$$A_{v(U2b)} = R7/R6$$

I personally prefer to keep the gain of U2b constant as it is designed (0 to 1), and change $A_{v(U2a)}$ when needed.

The *offset adjust* control (R9) is used to cancel any output voltage which exists when the output should be zero. In the case of a gram-force transducer, we will remove any mass from the transducer and adjust the potentiometer to zero volts output. In a fluid pressure transducer, we would open the transducer to atmosphere (the zero reference for gauge pressure) and adjust the potentiometer for zero output.

It is also possible to use an offset null such as R6 in Fig. 9-3 for nulling the transducer, and then use R9 in Fig. 9-4B as a *position control* for the oscilloscope or strip-chart recorder.

WHEATSTONE-BRIDGE ELECTRONIC THERMOMETER

Although there will be several electronic thermometers in this book, the design in Figs. 9-5 and 9-6 are based on the Wheatstone bridge, which we have been discussing in the previous section. No treatment of Wheatstone-bridge based instruments would ever be complete without a discussion of temperature measurements. The designs in Figs. 9-5 and 9-6 are simplified somewhat, but will work well for the intended purposes.

A *thermistor* is a *thermal resistor,* i.e., a resistor which has a resistance value that changes with changes in temperature. The thermistor is not totally linear, but can be combined in networks

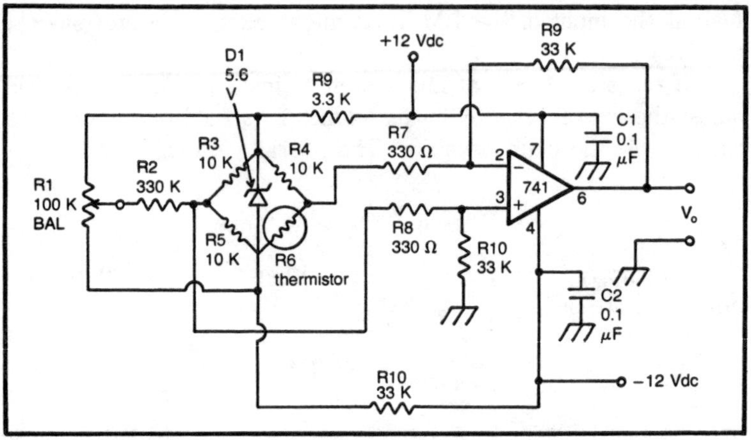

Fig. 9-5. Temperature amplifier (Wheatstone bridge/thermistor).

with other thermistors and resistors to linearize the transfer function (change of resistance per change of temperature). In fact, it is often the case that devices which we see as ordinary thermistors on the outside, are in reality a complex network of a half-dozen components inside of one shell.

Thermistors can have either *negative* or *positive temperature coefficients.* A negative temperature coefficient device has a resistance that *decreases* with increases in temperature, while a positive temperature coefficient device has a resistance that increases with increases in temperature. In either case, however, the unit of the transfer function is change-of-resistance-per-change-of-temperature (most usually $\Delta R/\Delta°$ C).

In Fig. 9-5 we are using a single thermistor in a Wheatstone-bridge circuit. There are three fixed resistors in the bridge, R3, R4, and R5. Resistors R3 and R4 are equal to each other, and should be precision types. The resistance of R5 is also shown as 10 K, but that is based on the assumption that the resistance of the thermistor will be 10 K at some specified reference temperate such as 0° C or some other easily measured point. In some cases, we can shift the temperature scale up or down a bit by making R5 variable, which also allows for variations in thermistor characteristics.

The most usual balance method is the potentiometer R1. This resistor is used to inject a current into one of the bridge nodes (via R2) that will null differences in resistances and other factors which would tend to cause an offset in the operational amplifier.

Excitation for the bridge is from a 5.6-volt zener diode, although one could just as easily have used any other voltage within the limits of the thermistor. The trick is to keep the power dissipation of the resistors in the bridge, especially the thermistor, below the point of *self-heating,* i.e., the power at which the temperature of the thermistor rises above ambient. This power dissipation is usually quite low, and will be specified in the thermistor spec sheet.

The output of the Wheatstone bridge is a differential signal, so we can use a simple differentially-connected 741 operational amplifier as the output stage. The design shown here is not one which I personally favor, however, because the input resistors (R7 and R8) are not at least five times the looking back resistance of the bridge. Of course, in a pinch, we could use only five times, so a 50 K resistor would be more appropriate in that position. Of course, that would mean a corresponding increase in R9 and R10, and probably would make a change of operational amplifier prudent. This circuit does work, however, and I have seen it in operation.

The temperature meter shown in Fig. 9-5 is not terribly linear over a large range, and may tend to drift with, of all things, temperature, because of the zener diode used for excitation. It is useful for many applications, however, so cannot be ignored.

Sometimes we are not much interested in the absolute temperature of a system or device, but the difference in temperature between two points. A differential temperature meter measures such "deltas." Examples of applications of differential temperature measurements is in medical electronics where certain types of bio-feedback are being used, and in environmental controls of homes and buildings where certain of the controls depend upon temperature differences rather than exact temperatures (e.g., the selection of heating source for hot water—solar or fossil fuel).

Figure 9-6 shows a simplified differential thermometer based on the circuit of Fig. 9-5. In this case, however, we have replaced one fixed resistor with another thermistor. For best results, the two thermistors should be matched as to temperature/resistance characteristic. We could also use a balance control (similar to R1 in Fig. 9-5) to null the bridge at a time when it is known that both thermistors are at the same temperature. This job can be done by placing them into the same environment for a period long enough for them to reach equilibrium.

ELECTROPHORESIS COLUMN METERING

During a lengthy period when I was earning my engineering degree I worked in the university's medical center, mainly in the

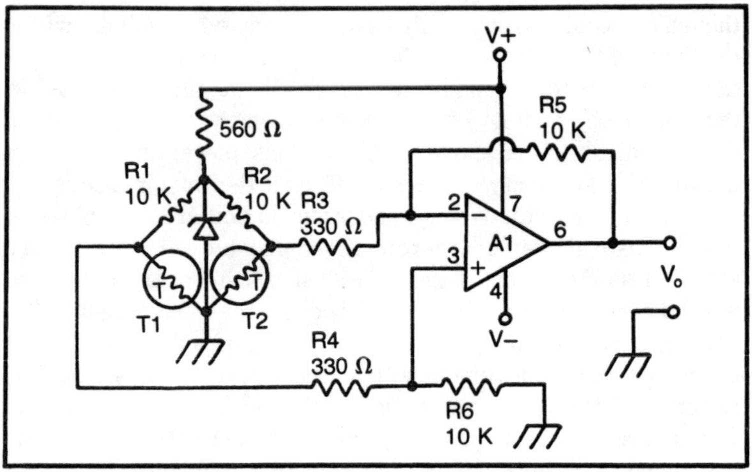

Fig. 9-6. Two-element thermistor bridge amplifier.

hospital. A friend of mine worked in a similar slot in the medical school for awhile and we occasionally had a little interchange of ideas. One day he came to me with a problem that seemed to baffle him concerning a device which his boss was building called an *electrophoresis column.* I don't even pretend to know a lot about electrophoresis, despite a Master's Degree in Medical Engineering, but I know that when protein materials are placed in an electrical field they migrate to various points in the field according to their characteristics. Thus, if a scientist wants to separate certain proteins out of some icky fluid ripped off from some human or animal subject, the fluid can be placed inside of a hollow glass tube such as shown in Fig. 9-7. A voltage field applied across the ends of the tube will create a potential that will cause the various proteins to migrate to various levels in the vertical column according to their characteristics. The scientist can then siphon off (or something) those which he or she wants and discard the rest. The hollow tube was about 1-inch i.d., and was about a 1-yard high. A platinum-wire electrode was attached through the bottom, and another at the top. A voltage field was then applied between the two electrodes and the electrophoresis column did its magic.

The key to successful 'phoresing, according to the scientist who wanted the thing instrumented, was keeping a constant current flowing in the column, which had an impedance of around 80 kohms. Since the current was 10 mA, then voltage had to be 800 Vdc. We had high-cost 0 to 100 mA *constant current source* (CCS) power supplies which were well-metered, but lacked the ability to meter more than one column at a time. My friend's boss needed to monitor a half-dozen or so.

The current was an absolutely essential measurement, and the voltage was deemed "nice to have." Also, the scientist was enamoured of the then emerging digital electronics technology so wanted digital readouts (sigh). The solution is shown simplified in Fig. 9-7. Deleted is the switch that permitted the voltage reading to be made using only one meter on all columns (deemed sufficient since they are all in parallel!).

The voltmeter is M1. In this case, we used the low-cost (then just coming down in price) 0 - 1999 millivolt digital panel meters and a precision voltage divider network. By using a 100 megohm, 1 percent instrumentation resistor (many such are on the market and we were able to "liberate" one from the engineering school student lab supply room), and a 100 kohm, 1 percent resistor, we would have our precision 1000:1 voltage divider. The meter would then read 0 - 1999 volts. As an alternative, shown in Fig. 9-7, we have substituted the 100 kohm precision resistor for a 91 kohm (need not be precision but should be a low tempco type) and a calibrating potentiometer. That way, we can calibrate the meter and also give the scientist something to diddle with (all "bench scientists" like to

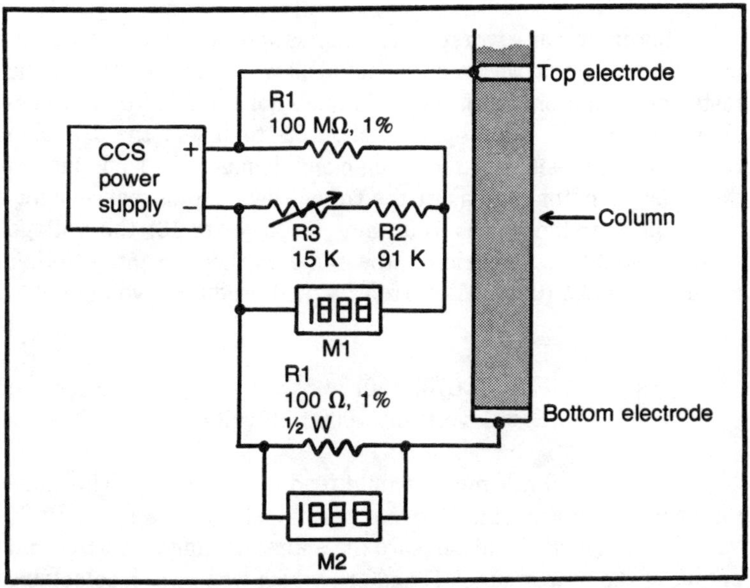

Fig. 9-7. Electrophoresis column metering.

diddle with their equipment, and are the bane of their technician's existence!).

The current measurement was the one which really stumped my friend, and was the reason for his visit to my lab. He simply could not find any reasonably priced digital current meters on the market. His original intent was to use a digital current meter with a shunt (if necessary), but that proved fruitless when a search of the catalogs of the school's suppliers yielded no current meters (*sigh,* ain't life in academic bureaucracies cruel!). The solution was elegantly simple, and my friend batted himself on the head when I showed it to him. The impedance of the column was around 80,000, and might go as high as 100 kohms or as low as 15 kohms. A 100 ohm precision resistor, therefore, would never be a large percentage of the total resistance of the column. In the worst case, the error introduced by adding a 100-ohm resistor in series with the column would be:

$$\text{Error} = \frac{(100\ \text{ohm})(100\%)}{(15{,}000\ \text{ohms})}$$

$$\text{Error} = \frac{10{,}000\%}{15{,}000}$$

$$\text{Error} = 0.67\%$$

This error was a worst case, and was well within the measurement tolerance that the scientist could allow. So the solution to the instrumentation problem, which is the whole point of this discussion (rather than electrophoresis!), was to insert the 100-ohm resistance in series with the column and then use the same kind of digital panel meter to measure the voltage drop across that resistor. Since the resistance was 100 ohms, a power-of-10, the voltage reading would be numerically the same as the current. In other words, a 10 mA (0.010 A) current, would produce a voltage of:

$$E = IR$$
$$E = (0.010)(100)$$
$$E = 1\ \text{volt, or, } 1000\ \text{millivolts}$$

The 0 - 1999 mV meter would read 0 - 19.99 mA. The same meter was also available in two other models. One was a 0 19.99 volt model (which would measure the entire maximum short-circuit output of the power supply!) and one which was switch selectable between 0 - 1999 mV and 0 - 19.99 volts.

There is no reason why we could not have used a 10-ohm resistor instead of 100 ohms. The 10-ohm resistor would have introduced less error (in fact, the reason why the 100-ohm was selected was because a 0.01 percent tolerance 100-ohm resistor was liberatable from the engineering school!). In the case of a 10-ohm resistor, the voltage drop would have been:

$$E = IR$$
$$E = (0.010)(10)$$
$$E = (0.1 \text{ volts}) = 100 \text{ mV}$$

In that case, we would have a 0 - 1999 mV meter reading 100 mV when 10 mA flowed, and 1000 mV when the maximum short-circuit current flowed. Most DPMs allow us to switch the decimal point around at will by grounding the correct terminal on the rear of the instrument, so we could have positioned the "dot" between the two zeros:"10.0" would denote 10 mA. Lest this imply more accuracy than was really present, however, we might want to use a 1-ohm resistor (when such can be obtained in precision types) and would then have a scale factor of 1 mV/mA, so 10 mV would indicate 10 mA.

Some DPMs allow the user to eradicate the "millivolts" or "mV" designation on the panel, thereby limiting user confusion when it measures something else indirectly, as in Fig. 9-7. Others permit us to insert photo negative appliques' that will indicate what is being measured. In the case of my friend, he simply eradicated the voltage designation and had the university sign shop make up engraved labels as needed.

A related, but opposite technique is used in modern digital multimeters, incidentally. Those instruments often use a constant current source across the ohmmeter terminals, and then use the digital millivoltmeter that forms the heart of the instrument to "read ohms." After all, when a 10 μA constant current passes through a 68-kohm resistor, a voltage of 680 mV is generated and can be read as ohms on the panel meter!

BASIC COLORIMETRY

In this section we will introduce you to basic colorimetry, a 19th-century technique (well, early 20th-century!) which still finds application in electronic instruments designed today! This method is presented not so much because you have some overpowering desire or need to know all there is to know about colorimetry and the related topic of densitometry, but to illustrate some basic instrumentation methods. The method illustrated here is comparison between a known sample and an unknown sample.

Colorimetry forms the basis of a lot of instruments, some automated some manual, found in medical and chemical laboratories. Figure 9-8 shows a basic colorimeter. Notice that it is a Wheatstone bridge (*that* again!) in which R1 and R2 form two of the branches, and a pair of photoresistors, PC1 and PC2, form the other two branches. Potentiometer R1 is a balance (BAL) control that is used to null out imbalances in the bridge when the bridge is being calibrated. A voltmeter, or a microammeter oftentimes a zero-center galvanometer, is used to indicate the output condition. Of course, we could always use a differential amplifier and further process the signal electronically. In actual fact, most instruments will do exactly that!

The two photocells are photoresistors, i.e., devices which exhibit a resistance that is proportional to the amount of light falling on them. If the same amount of light falls on both photocells, and the bridge is balanced, then the output of the bridge is zero. But, if one photocell receives more light than the other, then the bridge will become unbalanced and an output voltage results.

Normally, the bridge will be balanced with the sample being measured removed from the instrument. That way, the same amount of light from the collimated incandescent light will fall on both photocells, and R3 adjusted for zero output. The sample is then inserted into the instrument, and that will cut down the amount of light falling on PC2, thus unbalanced the bridge. This unbalance is read on the meter as proportional to some property of the same related to its transmissivity. In some cases, the property will reveal

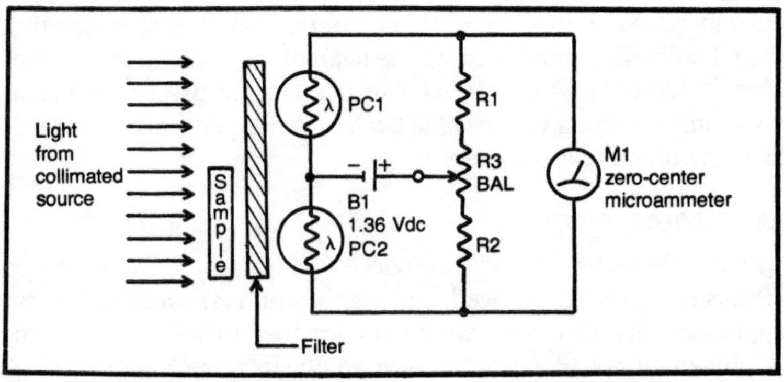

Fig. 9-8. Colorimetry circuit.

itself at only one color of light, so a filter is placed either at the collimated light source or between the light source and the photocells (as shown).

There are two ways this instrument can be used. In one method, the meter will be calibrated in the units being measured so its deflection after initially being nulled with the sample removed will tell the tale. In the other method, a mechanical dial connected to the BAL potentiometer contains the units being measured. When the sample is inserted, the instrument is unbalanced and then the operator renulls the bridge with R3. The amount of change required to renull after the sample unbalanced the bridge will be proportional to the property being measured. Of course, in electronic instruments today all of that renulling could be done with a computer and a *digital-to-analog converter* (DAC). The example shown is simplified. Although don't look down your nose at this instrument because it is simplified; this circuit represents very nearly the circuitry of a blood oxygen meter that was routinely used in medical labs until a short time ago, and is still used in many hospitals for very precision measurements which do not depend upon aging oxygen electrodes that the mass-screening units use.

Figure 9-9 shows one more example of a colorimeter. This instrument is the block diagram (simplified) of a CO_2 monitor. The original purpose of the instrument was to measure the CO_2 in the exhaled air of patients undergoing general anesthesia in a hospital. That measurement is considered of intense interest by some anesthesiologists. Basically, the circuit is the same as before, except that the collimated light source is replaced with an infrared source. The reason for this is that the infrared is absorbed by CO_2, but not by oxygen. If we compare the infrared absorption of exhaled patient air with the absorption of infrared in room air (which contains only trace amounts of CO_2), then we would be able to measure, or at least monitor the trend of, patient CO_2 exhalation. Photocell PC1 is exposed to the infrared source (which was an ordinary coffee pot heating element in one very good instrument) through an open glass tube through which room air was pumped. The other photocell, PC2, was connected to the exhalation line on the respirator, so sampled the patients exhaled air; a small pump kept the sample coming. Thus, the bridge would be unbalanced an amount proportional to the CO_2 content of the patient's exhaled air.

Calibration was simple but required a special gas mixture. The sample tube was first made to suck room air, and the bridge was balanced to zero output (V_o). This indicated 0-percent CO_2. Next,

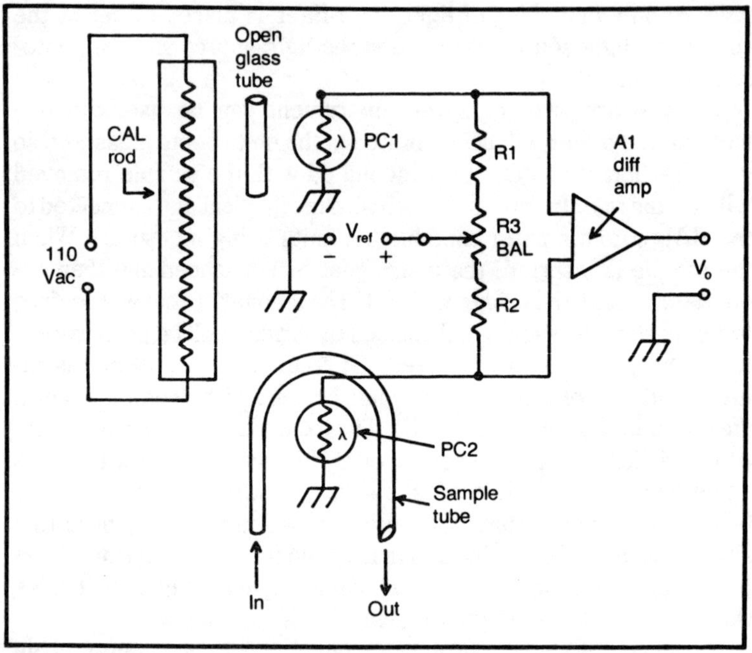

Fig. 9-9. Practical colorimetry example.

the tube was connected to a cylinder of calibration gas which was 95-percent pure oxygen and 5-percent pure CO_2. Then the gain of the differential amplifier (A1) was adjusted to give the voltage that represented 5-percent CO_2. In the practical case, the "span" (which is what life-sciences people tend to call gain controls) was adjusted to read "5.00" on the LCD panel meter.

The circuit shown in Fig. 9-9 is grossly simplified. Part of the added complexity is that infrared is not uniformly and linearly absorbed in CO_2. The transfer function is quite nonlinear, and that presents a bit of a problem. In the original meter which I worked on and repaired too many times, the linearization function was done by the archaic method of diode-op-amp "breakpoint" generators that "straightened out" the curve—but not very well. In today's microcomputer controlled models, however, we could easily input the raw data, and then use a look-up table in software to find the correct value to output to the meter.

A possible method for performing the same CO_2 (or similar) measurements is shown in Fig. 9-10. Here we have replaced the cal-rod with an infrared light-emitting diode (LED) and the photocells with infrared-sensitive phototransistors. The voltage across emitter resistor R3 would be proportional to the amount of infrared or light that fell on the transistor; a pair of these transistors will be usable in a differential amplifier to make our colorimeter. There are devices available that will be classified as optical switches or optoisolators that could be used in these cases. Such devices have a cutout slit that will permit the admission of a sample tube (which need be only very small). Those devices were developed for other purposes, but would work nicely for this measurement also.

RATIOMETRIC MEASUREMENTS

A densitometer is basically a light source and a photoreceptor that is configured to measure the amount of light transmitted between the two. This simple circuit is the basis of the colorimeter, which is basically little more than a differential densitometer. The receptor of our densitometer can be any form of device which will measure light intensity. In some cases we will use a photoresistive cell, CdS, etc., while in others we will use a phototransistor. There are also other receptors, up to and including photomultiplier tubes.

There is a little problem with most optical densitometers, however. We find that the output of the photoreceptor is proportional to the light falling on it. Under ideal circumstances, we will find that the only output variations are due to variations in the *sample* path. But in real, practical circuits, we find that a different

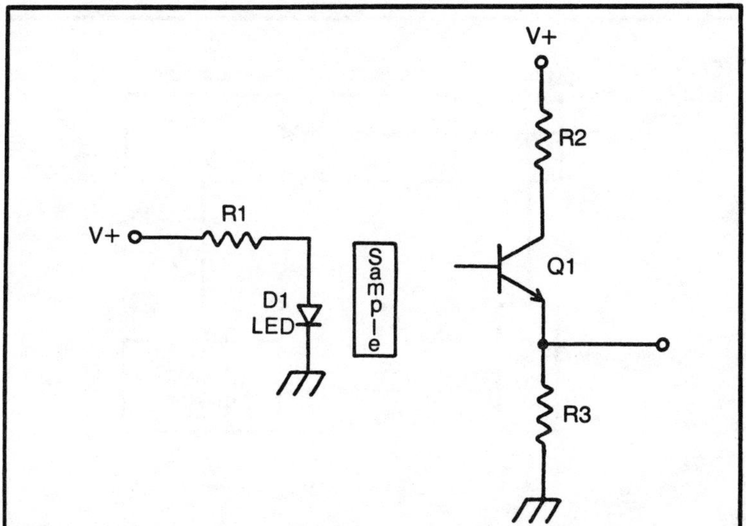

Fig. 9-10. Using LED/phototransistor.

situation exists. We find, for example, that the output of the light source varies. The problem might be variations in the power source, or, there might be other sources of variation in the light (or infrared) source. The power supply variation problem is partially solved by use of a regulated dc power supply for the light source. Another solution which also works nicely for other sources of variation, is shown in Fig. 9-11. This is an example of a ratiometric densitometer/colorimeter.

There are two densitometers (such as those which we used earlier), and each outputs a voltage that is proportional to the applied light intensity, $V1$ and $V2$. The light source is passed through a beam-splitter which causes the beam to exit in two orthogonal (i.e., right angle) fractions. The beam that is applied to densitometer no. 2 is used as a reference, while that applied to the densitometer no. 1 is passed through the sample first. The two output voltages are applied to an analog voltage divider to form an output voltage of:

$$V_{ot} = \frac{KV_2}{V_1}$$

Where:

V_{ot} is the output potential

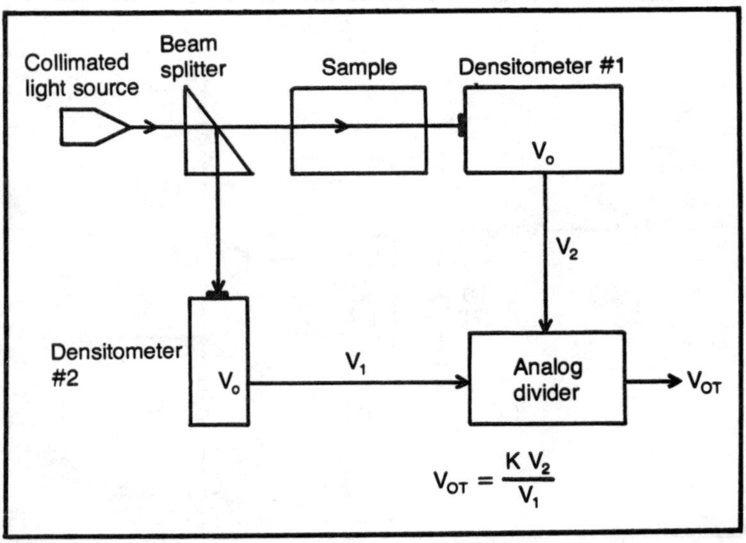

Fig. 9-11. Ratiometry circuit.

V_1 is the reference path output voltage
V_2 is the sample path output voltage
K is a constant (usually 10 or 1/10)

The idea of this circuits is to apply an input to the divider that is proportional to the applied light source, but not the sample. Since the output voltage contains a factor of this signal, errors due to variation in the light source are cancelled out.

Note what happens here. If the light intensity goes up, then both V1 and V2 will also rise. But V2 is divided by V1, so the increase is also divided out. Similarly, a reduced light output causes the voltage to reduce.

We can also do this same job in a microcomputer instrument. The divider would, of course, be done as software rather than hardware, but the concept is the same. The densitometers would have to either be digital-output types, or, be connected to analog-to-digital converters (A/D).

TEMPERATURE MEASUREMENTS

A measurement job which continuously pops up is the measurement of temperature. We have already seen how temperature can be measured electronically using a thermistor and operational amplifier, but there are also IC versions. One of the earliest IC temperature sensors was the LM-3911 device, with others following in short order. The *Precision Monolithics, Inc.* REF-05 device is basically a 5-volt voltage reference regulator, but has an output terminal that produces a voltage which is proportional to the absolute temperature in degrees Kelvin (*note:* ° K are the same as ° C, which is the common Celsius or centigrade temperature scale, but have a different reference point as "zero." In the Celsius system, the zero-degree point is the freezing point of water. In the Kelvin system the zero-degree point is so-called absolute zero. There is a 273 degree difference between these two scales. To convert from ° K to ° C, just add +273 degrees to the Kelvin temperature. In other words, 273° K = 0° C).

Solid-state IC temperature sensors are based on the properties of the PN diode junction. The usual sensor is one or more transistors inside of the IC package. We know from elementary transistor theory that the voltage drop across a PN junction in a diode or transistor is given from:

$$V_g = (KT/q)(\text{Ln } I_f/I_s)$$

Where:

V_g is the voltage drop across the PN junction

K is Boltzman's constant

q is the electronic charge

Ln indicate the use of natural logarithms

I_f is the forward current

I_s is the reverse saturation current

The reverse saturation current is relatively constant, while the K and q terms *are* also constant. If we can conspire to feed the PN junction from a constant current source, then the only variable in this equation is the temperature T, which is in degrees Kelvin.

There have been published a lot of circuits that use diodes and transistors (which are often diode-connected by shorting together the base and collector terminals) as the temperature sensing element. You can see this phenomenon with a simple dc ohmmeter and a rectifier diode. Connect the ohmmeter such that it will measure the forward resistance (something which some electronic ohmmeters will not do!) of the rectifier diode. Next, heat the diode package with the tip of a soldering iron, or, even with your hand. You will note that the resistance of the diode will vary with temperature!

There are several different integrated circuit temperature sensors on the market. In this book we will consider just two. One is the *Analog Devices, Inc.* AD-590 device, while the other is the *National Semiconductors, Inc.* LM-335 device.

The *Analog Devices, Inc.* AD-590 device is basically a current-source temperature sensor with a transfer function of approximately *one-microampere per degree Kelvin* (1 μA/° K). The AD-590 is a two-terminal device, so is extremely easy to apply. Figure 9-12 shows the simplest calibratible circuit using the AD-590. Since the AD-590 produces a current of 1 μA/° K, the output voltage will be 1-millivolt per degree Kelvin when the resistance is 1000 ohms. Of course, we can make a rough thermometer by simply using a single precision 1000-ohms resistor instead of the two resistors shown here. But that circuit would not be calibratable, and you would have to put up with errors. That error will not be too much of a problem in some applications, so the designer will be able to simplify the circuit a little bit by removing the potentiometer.

The potentiometer R2 is used to adjust the temperature/voltage output. Note that the potentiometer and fixed resistor will vary the total resistance back and forth across 1000 ohms, so that errors can be compensated for. For most users, it is probably best to use ice-water in which ice and water exist together and are in equilibrium. Under this condition, the temperature is 273° K, so the output voltage should be adjusted to 273 millivolts (0.273 V).

For many other readers, especially those with lower accuracy requirements, can use room temperature. Measure the room temperature in degrees Celsius, and then convert to degrees Kelvin by adding 273. Then adjust the output voltage for the current room temperature. *Note:* the sensor should have been on for about 15-minutes to reach equilibrium with the environment. For example, let's assume that the room temperature is 25° C, so the Kelvin temperature is (25 + 273)° K or, 298° K. In this case, the output voltage will be 298 mV. One factor in adjusting the temperature is the accuracy of the measurement of room temperature . . . not all home thermometers are terribly accurate.

If your only thermometer is a Fahrenheit model, then you can convert to Celsius using the standard formula

$$^\circ C = \frac{^\circ F - 32}{1.8}$$

or, for degrees Kelvin:

$$^\circ K = \frac{^\circ F - 32}{1.8} + 273$$

Example:

The room temperature is 70° F. What is this temperature expressed in ° K?

$$^\circ K = \frac{70 - 32}{1.8} + 273$$

$$^\circ K = \frac{38}{1.8} + 273$$

$$^\circ K = 21.1 + 273 = 294.1$$

In this case, we would adjust the output voltage for 294.1 millivolts. This reading may be a little difficult to obtain with many hobbyist-type voltmeters, but 294 millivolts is close enough.

One problem with the circuit of Fig. 9-12 is that the calibration can be done at only one point. Since no transducer is perfectly linear, this can result in errors on one end of the scale or another.

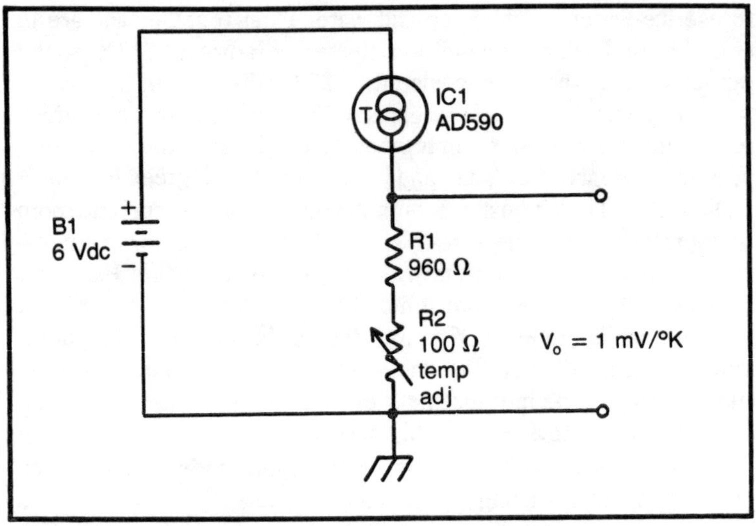

Fig. 9-12. Simple AD-590 thermometer.

The circuit in Fig. 9-13 allows two-point calibration, and thereby straightens out the curve a little bit. It also offers the use of an operational amplifier, which will reduce the output impedance considerably.

The operational amplifier is connected in the inverting follower configuration. One input of the circuit is driven by a 10.000-volt reference potential (V_{ref}), which must be developed from a

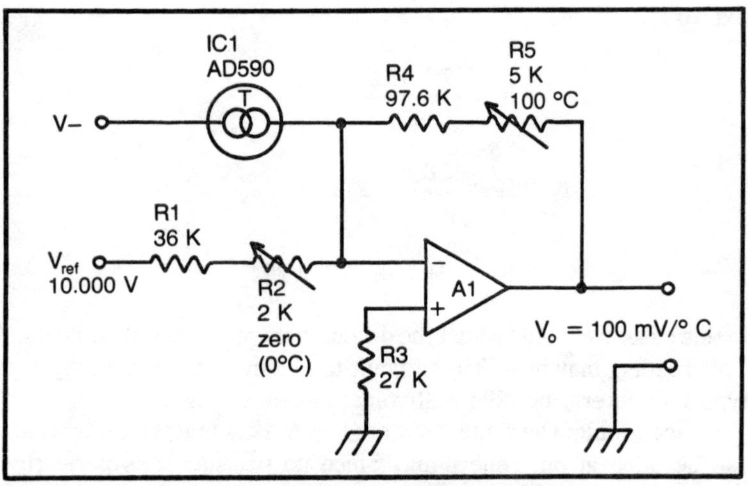

Fig. 9-13. AD-590 thermometer with two calibration points.

reference voltage source. The precision of this circuit is partially dependent upon the precision of this voltage. It *must* also be stable, which means one of the reference sources given earlier in this book are useful. We can also use an AD-581 *Analog Devices, Inc.* IC reference source. This circuit produces the zero-point calibration.

The other input of the operational amplifier is the current of *IC*1, an AD-590. Recall this current is 1 μA/° K. The effect of the V_{ref} input is to offset the voltage such that V_o will be zero when the Kelvin temperature is 273° K. This means that the output will be porportional to the Celsius temperature because 273° K = ° C.

The gain of the operational amplifier is set by the feedback resistors. Since one of these resistors is a potentiometer, we can adjust the output voltage at one other point as well as zero. The recommended temperature is 100° C, which is the temperature of boiling water at sea level.

The other temperature sensor which we will consider is the *National Semiconductor, Inc.* LM-335 device. This IC temperature sensor is available in two different packages, as shown in Fig. 9-14. The plastic TO-92 package is shown in Fig. 9-14A, while the metal case is shown in Fig. 9-14B (this is the same as the TO-46 transistor case, which is similar to the case of the AD-590). The TO-92 case device is designated LM-335Z, while the TO-46 version is LM-335H.

There are also two other versions of the device. The LM-335 is useful over the range −10° C to +100° C. If a wider range is needed, then the LM-135 will cover −55° C to +150° C, and the LM-235 operates over −40° C to +125° C.

The most elementary circuit for the LM-335 device is shown in

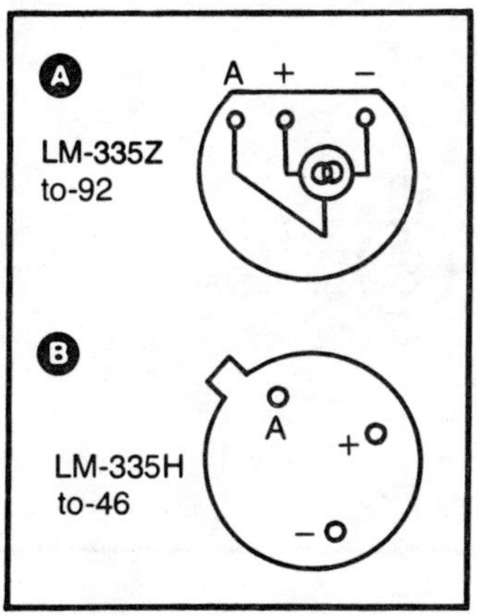

Fig. 9-14. LM-335 thermometer element packages.

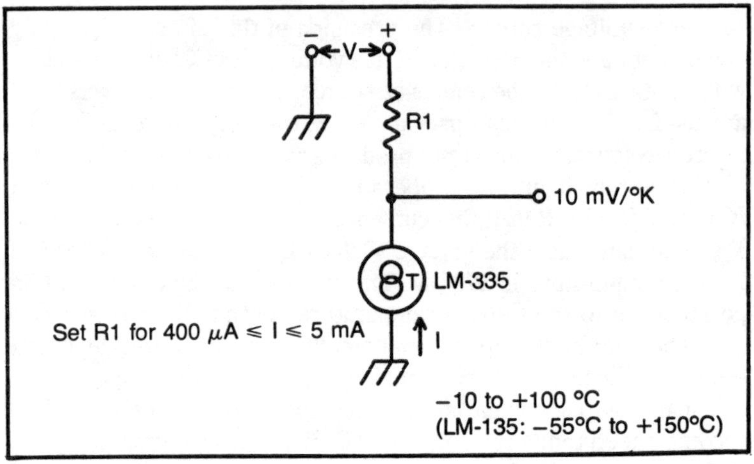

Fig. 9-15. LM-335 circuit.

Fig. 9-15. This circuit is a simple series circuit in which resistor R1 is selected according to Ohm's law to produce a current of 400 microamperes to 5 milliamperes. The output voltage will be 10 mV/° K, so there will be an offset for Celsius temperatures that must be compensated later in the system. This particular circuit will be useful for monitoring applications where the precision is not needed, and the "critical threshold" above or below which catastrophe happens is relatively wide. If you need for precision, then you will want to use the circuit of Fig. 9-16.

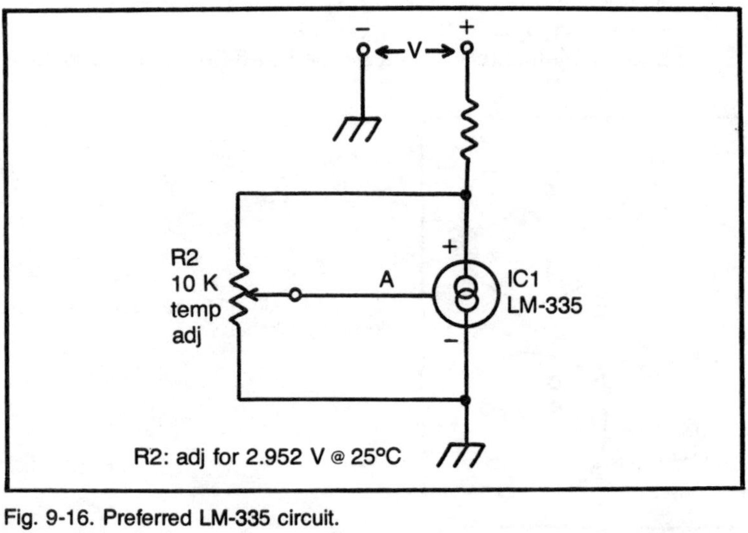

Fig. 9-16. Preferred LM-335 circuit.

Figure 9-16 shows a calibratable version of the previous circuit. Unlike some of the other IC temperature sensors on the market, the LM-335 is a three-terminal device in which the third terminal is for adjustment of the output voltage. Figure 9-16 shows the use of a 10-kohm linear potentiometer to adjust the output. The recommended procedure is to adjust the output voltage for 2.982 volts at a temperature of 25° C.

As is also true in many other instrument designs, the potentiometer is preferably a multi-turn "trimpot" type. This type of component will allow you to have greater resolution in the adjustment of the circuit.

Few readers will really want to measure temperature in terms of degrees Kelvin, but many will want to measure temperature in degrees Celsius. The ° C is the same *size* as the ° K, but the reference point is different. Since most of our electronic temperature sensors produce output scale factors in terms of mV/° K, we will find an offset for the Celsius scale. If the scale factor is 1 mV/° K, then the output offset for the Celsius thermometer will be 273 mV, and if 10 mV/° K the offset will be 2730 mV, or 2.73 volts. We can overcome this using a ° K–° converter circuit, such as shown in Fig. 9-17.

The circuit of Fig. 9-17 is merely a unity gain inverting follower with an offset potential null circuit. The output of this circuit will be of the opposite polarity as the input signal. So, if the input scale factor is 10 mV/° K, the output will be −10 mV/° C. This

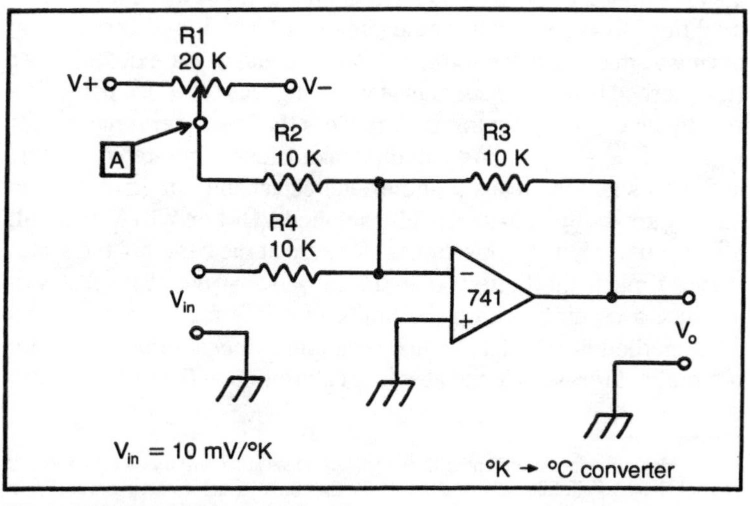

Fig. 9-17. Kelvin-Celsius converter.

circuit uses gain constants selected for the 10 mV/° K input signal. If you want to use it with 1 mV transducers, then replace R3 with a 100-kohm resistor in order to increase the gain to ×10. The adjustment procedure for this amplifier is as follows:

1. Turn on the circuit and allow to sit for 15-minutes while the temperature transducer equilibrates to the environment. Transducer output is connected to V_{in} of Fig. 9-17.
2. Measure the room temperature in degrees Celsius*.
3. Convert to degrees Kelvin by adding 273 to the reading e.g., 26° C = 299° K.
4. Adjust potentiometer R1 in Fig. 9-17 for approximately 2.99 volts at point "A."
5. Measure V_o with a 3½-digit (or better) digital voltmeter, or, an analog voltmeter capable of discriminating 10 mV.
6. Adjust R1 for the output reading that corresponds to the temperature measured in step-2. In this case, it would be 260 mV.

Note: V_{in} = 10 mV/° K. For 1 mV transducers change step-3 to produce a voltage of 0.299 volts, or, insert a second gain of −1 stage between the transducer and the input of Fig. 9-17. This would make the transducer effectively a 10 mV/° K unit.

For those readers who lack the voltmeter necessary to adjust this machine, there is another procedure . . . and I might add, it is also useful for those who do have good quality voltmeters. In this modified procedure, we would create a 0° C ice-water bath by having a small jug of ice water in which ice and water exist together (the method is not without fault, for finding true 0° C "ice point" is a real technical trick, but for most readers the ice-water temperature will be close enough). We can then take a less accurate voltmeter, and then short the leads together and adjust the *zero* knob to read exactly zero. If you have an old-fashioned VOM or VTVM that will allow you to adjust the instrument for zero at the center of the scale, then so much the better. Use the most sensitive scale that will produce a reading, within the limits of safety for the instrument. This method is called an "equal deflection" measurement. We are not really interested in the absolute calibration of the instrument as

* Make this measurement with an ordinary glass and mercury thermometer. If you only have a Fahrenheit thermometer, use the conversion formula given earlier in this chapter.

we are the knowledge of what it reads when a known voltage is applied. Since the output of the circuit in Fig. 9-17 will be 0.00 volts when the ice water and thermal probe equilibrate, we want to know what the voltmeter will read with zero volts input—that is the purpose of shorting the leads together. After that, the circuit can be attached to the voltmeter, and R1 adjusted to produce the same reading as the shorted leads produced.

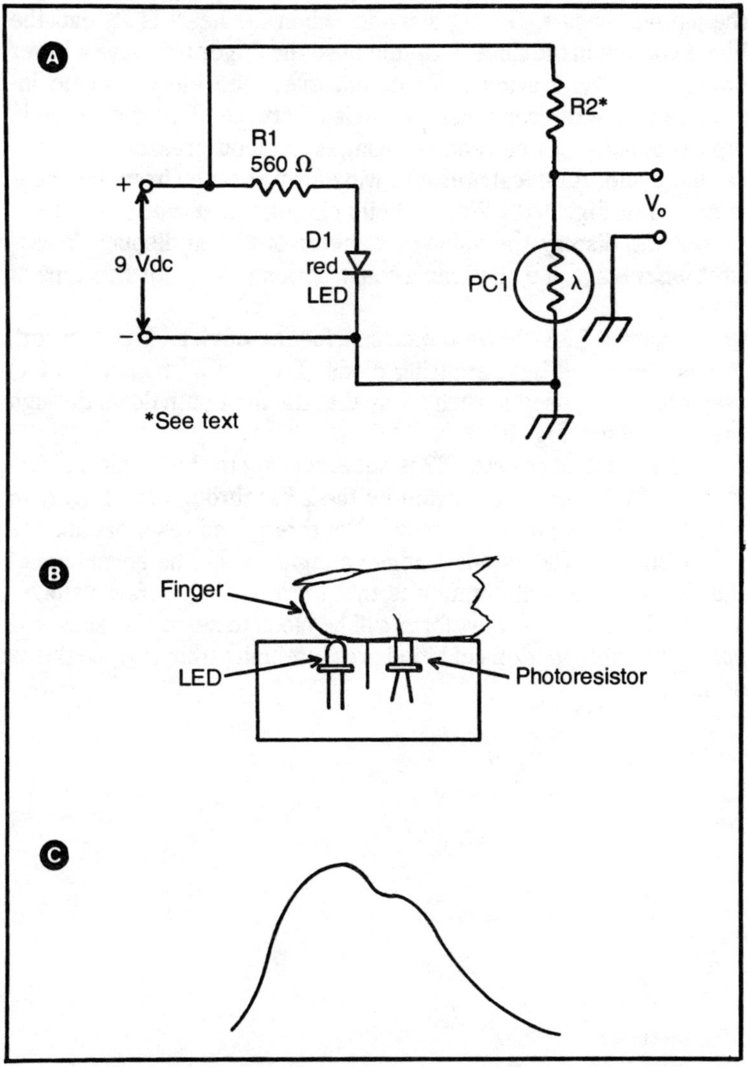

Fig. 9-18. Photoplethysmograph A) circuit, B) construction, C) waveform.

PHOTOPLETHYSMOGRAPH

Figure 9-18 shows a simple device that will allow the user to keep track of heart rate. This sensor is used extensively in medical electronics, and also in those low-cost (and some not so low cost!) exercise heart-rate monitors. In those instruments, a sensor is slipped over the thumb or index finger and registers your pulse. The basis for the operation of this instrument is that blood flow is pulsatile, i.e., it flows in pressure pulses caused by the beating of the patient's heart. During diastole, when the heart is at rest, the blood volume in the finger is minimal so the finger will have a lower optical density. During systole, however, the blood volume increases so the optical density will also increase. This difference in otpical density can be read as changes in blood pressure over the cardiac cycle. A typical pressure waveform as taken from the finger is shown in Fig. 9-18. We can build circuitry that will count these pulses and display the value on a meter or digital display. *Radio-electronics* magazine once ran a construction project for this type of device.

Figure 9-18A shows the circuit for the device. The "sensor" consists of a red light-emitting diode (D1) and a photocell (PC1) assembly juxtaposed in such a way that the light path flows through the finger (see Fig. 9-18B).

The value of resistor R2 is set according to the nominal resistance of PC1 when illuminated by the LED through the finger. In many cases, this circuit will consist of three fixed resistors and the photocell in a Wheatstone-bridge configuration. The advantage of the Wheatstone configuration is that there is little offset voltage. The minimum on the waveform will be close to zero; this sensor is *not* calibratable in terms of blood pressure units (mm Hg), so this is no loss.

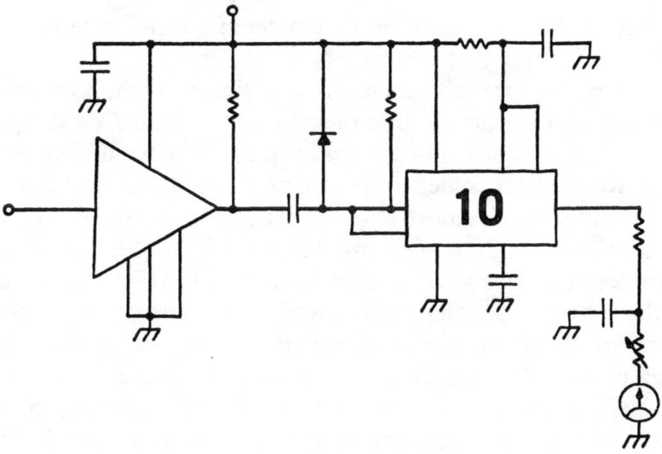

A Timer Primer

An electronic timer is a circuit or device that will measure some specific period of time, or, will provide some indication that it is active for a specific period of time. Various electronic circuits have been, and still are, used as timers but in the context of this book we will consider RC timers. Resistor-capacitor networks can be designed to provide a timer indication, and are in relatively widespread usage.

The RC network can be used with a variety of devices, including bipolar transistors, unijunction transistors, field effect transistors (of both types), digital IC devices and certain special purpose timer IC devices. The circuits used with ordinary transistors, unijunction transistors and field effect transistors unfortunately suffer from more problems than do the IC versions. One of the principal differences is that the transistor type of timers suffer from duration instability as the supply voltage changes. In addition, it is normally true with RC timers that the duration will change with temperature changes. In part, this is due to the temperature coefficient of the resistors and capacitors used in the RC network. But those factors are easily controlled by using low temperature coefficient components. The transistor type of timer also suffers additionally from temperature problems because the characteristics of the transistor can change just enough to make the duration shift. This problem is especially acute in unijunction transistor timers. That leaves the digital IC and special-function IC timers for us to consider. In both cases, these timers seem a little more immune from the problems that normally beset RC timers.

Before dealing with various forms of timer circuit, however, it would pay us to review a little elementary capacitor/resistor network theory, at least on the descriptive level. That would tell us just how an RC network comes to be used as a timer in the first place.

Consider the demonstration circuit shown in Fig. 10-1A, and the associated timing waveforms shown in Fig. 10-1B. Recall that the capacitor serves as a "storage tank" for electrical charge, and that the voltage appearing across the capacitor is proportional to the amount of electrical charge stored (the energy in a capacitor is stored in the insulating dielectric between the two plates). In the circuit of Fig. 10-1A, the capacitor *C* is fed electrical charge from a battery or other potential source (*V*) through resistance *R*. The resistor serves to limit the rate at which charge can be transferred from *V* to *C*. Initially, the switch is set in position "A," so any charge stored in the capacitor will be drained off; current from one plate of the capacitor will flow through the switch and resistor *R* to balance the current on the opposite capacitor plate. To begin our discussion, keep the charge curve of Fig. 10-1B in mind as we describe what happens when the switch is moved to position "B." Initially, the capacitor voltage is zero, so the entire source voltage *V* will appear across the resistor (in this case, $V_r = V$). At first, the capacitor charge increases very rapidly, but it soon begins to slow down until it reaches a point of equilibrium with the source. Although in theory, it is impossible for the capacitor to ever reach exactly the source voltage, for all practical purposes it will reach 99.999 percent of the total in a short while. The rate of increase of the voltage is said to be exponential because it obeys the following equation:

$$V_c = V\,(1 - e^{-t/RC})$$

Where:

V_c is the voltage across capacitor C
V is the source voltage
e denotes the use of natural logarithm system
t is the duration of which V is applied to the circuit
R is the resistance in ohms
C is the capacitance in farads

Note: The product "RC" is called the *time constant* of the RC network, of which more later.

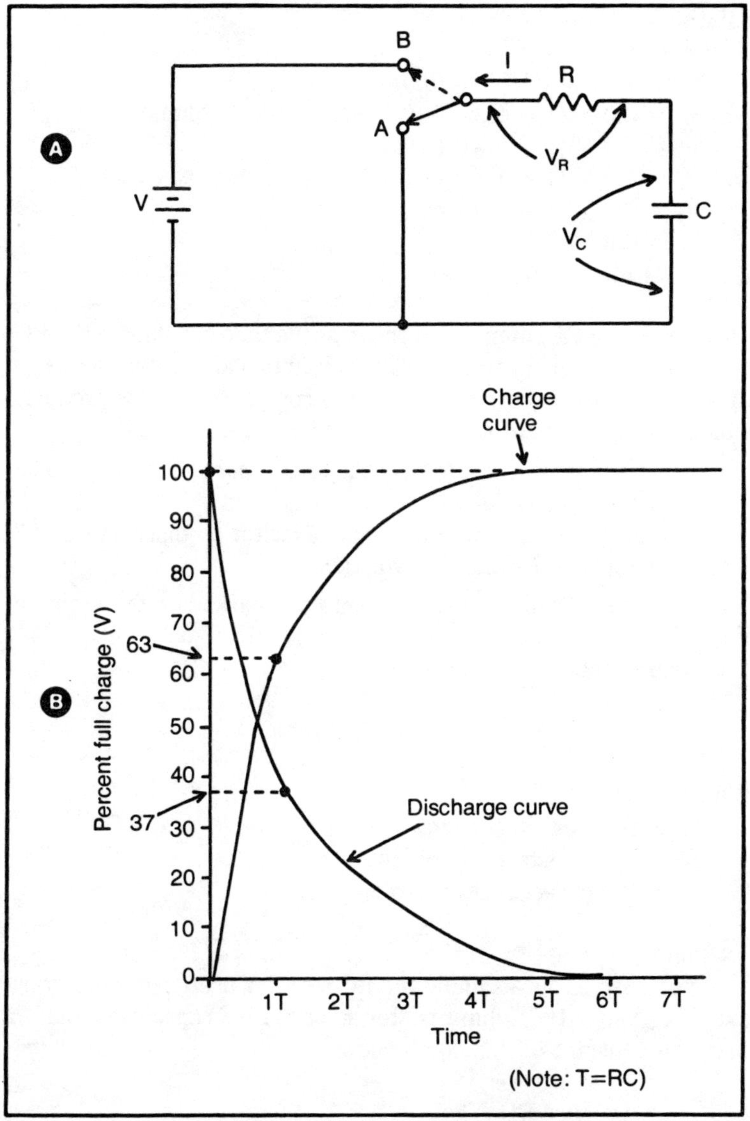

Fig. 10-1. A) RC timer network, discharge state, B) charging waveform.

Example:

An RC network consists of a 0.1 μF capacitor and a 100 kohm resistor. Find the voltage that appears across the capacitor 13 milliseconds (0.013 seconds) after the switch in Fig. 10-1A is toggled from position "A" to position "B"; V = 20 volts.

Solution

$V_c = V(1 - e^{-t/RC})$
$V_c = V(1 - EXP(-t/RC))$
$V_c = (20\ V)(1 - EXP(-0.013/(10^{-7}F)(10^5\ ohms))$
$V_c = (20\ V)(1 - EXP(-0.013/10^{-2}))$
$V_c = (20\ V)(1 - EXP(-1.3))$
$V_c = (20\ V)(1 - 0.2725)$
$V_c = (20\ V)(0.7275)$
$V_c = 14.55$ volts

The factor RC, mentioned above as the time constant of the RC network, is very important to timer circuits and usually figures in the equation that sets the values of the components. This product, which has the units of seconds, is defined simply as either:

1. The time required for a capacitor to charge from zero to 63 percent of its fully charged voltage, or,
2. The time for a fully charged capacitor to discharge to 37 percent of its charged voltage.

Both the 37 and 63 percent points are marked on the graph in Fig. 10-1B.

The time constant equation is:

$$t = RC$$

Where:

t is the time in seconds
R is the resistance in ohms
C is the capacitance in farads

Example:

Let's use as our example the RC network used in the previous example, i.e., a 100 kohm resistor and a 0.1 μF capacitor. What is the time constant of this combination?

$$t = RC$$
$$t = (10^5\ \text{ohms})(10^{-7}\ \text{farads})$$
$$t = 10^{-2}\ \text{seconds}$$
$$t = 0.01\ \text{seconds} = 10\ \text{milliseconds}$$

Normally, a capacitor in an RC network is considered fully charged, or fully discharged, after a period of five time constants (5t), a fact that is reflected in the curves of Fig. 10-1B.

When the capacitor discharges, the rate of change is very rapid at first, but then slows down. Again, the curve is exponential but in this case it is an exponential decay. The governing equation is:

$$V_c = Ve^{-t/RC}$$

or,

$$V_c = V(EXP -t/RC)$$

Example:

Assuming a 20 Volt power supply, calculate the voltage of the RC network 8 milliseconds (0.008 seconds) into the discharge cycle, if the capacitor had been allowed to be fully charged initially.

Solution:

$V_c = (V)(EXP -t/RC)$
$V_c = (20\ V)(EXP\ (-0.008\ s/(10^5\ ohms)\ (10^{-7}F))$
$V_c = (20\ V(EXP(-0.008\ s/10^{-2}s)$
$V_c = (20\ V)(EXP -0.008/0.01)$
$V_c = (20\ V)(EXP\ 0.8)$
$V_c = (20\ V)(0.45)$
$V_c = 9$ volts

The accuracy of the RC time constant directly affects the accuracy of the timer circuit. With precision or hand-selected components that are low drift, we can obtain accuracies of 1 percent or better. Carbon composition resistors and "ordinary" capacitors have a nasty habit of having severe temperature problems, so one should opt for precision, low-tempco resistors and capacitors of either silver mica, polystyrene or polycarbonate. Also please note that precision capacitors are a lot harder to obtain than precision resistors. One is advised to either hand select capacitors on a digital capacitance meter or General Radio bridge, or, make the resistor adjustable to compensate for problems of the capacitor tolerance.

Now that we thoroughly understand capacitors (!?!), we are ready to apply that knowledge to the problem of timers.

There are several different schemes for making timers. Rather than give you a series of projects, I would rather show you three different approaches and then let you design your own timers. Figure 10-2 shows a circuit that is sometimes called a *half-monostable multivibrator* (for reasons that escape me, for it behaves

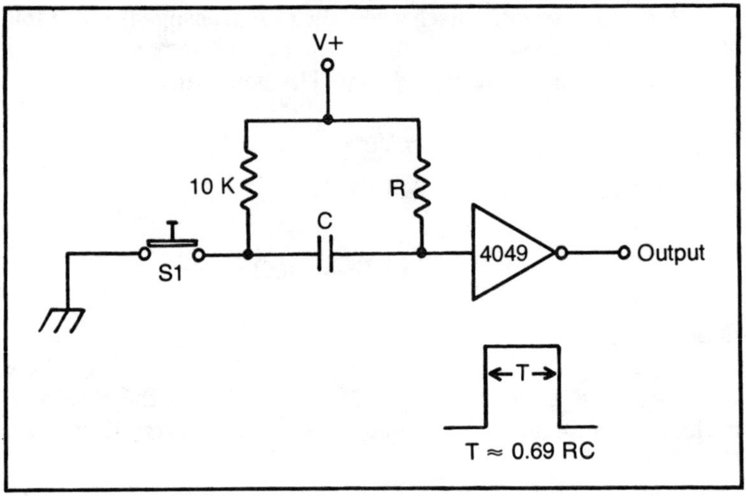

Fig. 10-2. Simple CMOS monostable multivibrator.

like a full monostable multivibrator!). Any monostable multivibrator has but one stable state (hence its name!). It will remain dormant in the stable state until triggered by some external event, at which point it will toggle to the unstable state. The monostable multivibrator will remain in the unstable state for a fixed duration, usually designated as *T*, and then automatically revert back to its stable state. In the circuit of Fig. 10-2, this is shown by the pulse waveform. It is LOW initially, but following trigger it will snap HIGH for a period of time (t) and then drop LOW again to await another trigger event. The monostable multivibrator is sometimes called a *one-shot* circuit because it produces one output pulse for each input pulse.

The circuit of Fig. 10-2 is a one-shot, and it produces a pulse of approximate duration $t = 0.69RC$, where t is in seconds, R is in ohms and C is in farads. When switch $S1$ is open, capacitor C is essentially discharged because both plates are at the same potential. This condition places a HIGH on the input of CMOS inverter 4049. When S1 is closed, however, one end of the capacitor is grounded, so the capacitor can begin charging. Initially, the voltage across the capacitor is zero, so the inverter thinks its input is at ground potential (LOW). Thus, the LOW input will produce a HIGH output. The output of the 4049 inverter will remain HIGH until the voltage across the capacitor increases to the point where the inverter thinks it sees a HIGH input. Normally, this is the midpoint between V− and V+. In most digital electronic systems, especially where TTL devices are mixed with CMOS, the V− is 0-volts and V+ is +5 volts. The trigger point will be ½ (5 V − 0 V), or 2.5 volts. We can thus derive the T = 0.69 RC formula:

$$V_c = V\,(1 - e^{-t/RC})$$

Since a CMOS device is being used, V_c = ½V, so:

$$\frac{V}{2} = V\,(1 - e^{-t/RC})$$

$$\frac{V}{2\,V} = 1 - e^{-t/RC}$$

$$0.5 = 1 - e^{-t/RC}$$

$$-0.5 = -e^{-t/RC}$$

$$\mathrm{Ln}(\,0.5) = \mathrm{Ln}\,(e^{-t/RC})$$

$$-\,0.69 = -t/RC$$

$$0.69\ RC = t$$

The so-called half-monostable multivibrator circuit of Fig. 10-2 is often used for the reset circuit on microcomputers. When the operator wants to reset the computer, he or she will simply press the button and a pulse will appear on the reset line. If we delete the 10 kohm resistor and the switch, and simply ground the capacitor, then we will also have a power-on reset circuit. Such a circuit is used in a microcomputer to initialize the computer when power is applied. At the instant the dc power supply is turned on, the capacitor charge will be zero so V_c will be zero also. The output of the inverter will snap HIGH and will remain HIGH for a period of time set by the RC time constant. In order to make sure that everything happens according to Hoyle, the typical microcomputer power-on reset circuit will use a 1 megohm resistor and a 1 μF capacitor, yielding a pulse duration of:

$$T = 0.69\ RC$$

$$T = (0.69)(10^{6}\ \text{ohms})(10^{-6}\ \text{farads})$$

$$T = (0.69-)(10^{0}\ \text{seconds})$$

$$T = 0.69\ \text{seconds}$$

Although 690 milliseconds seems like an excessive time in computer work, it allows all circuits to become energized and prevents lack of synchrony.

A second approach to the timer problem is shown in Fig. 10-3. This circuit is also a monostable multivibrator, but is a full monostable. The circuit of Fig. 10-3 is based upon the 4013 Type-D CMOS flip-flop. Let's review the rules of a typical Type-D flip-flop of the type represented by the 4013 device. In a Type-D, the data (HIGH or LOW) applied to the D-input will be transferred to the Q-output when the clock line is HIGH. When the clock drops LOW again, the Q-output will retain the data at its last valid state. Thus, a Type-D flip-flop will "remember" a single bit of binary data. If *set* and *clear* lines are provided, as they are on a Type-D flip-flop, then they will operate in the following manner completely distinct from the clock circuit:

1. A HIGH applied to the *set* input will force the output to Q = HIGH;
2. A HIGH applied to the *clear* (CLR) input will force the output to Q = LOW

It is the second property that is of interest to us in this discussion, namely the operation of the clear line.

The clock (CLK) line of the 4013 is used in Fig. 10-3 as the timer trigger input. Since the D-input is wired permanently HIGH, and trigger pulse received on the CLK line will force the Q-output HIGH. At that time (see Fig. 10-4) the HIGH on the Q-output will

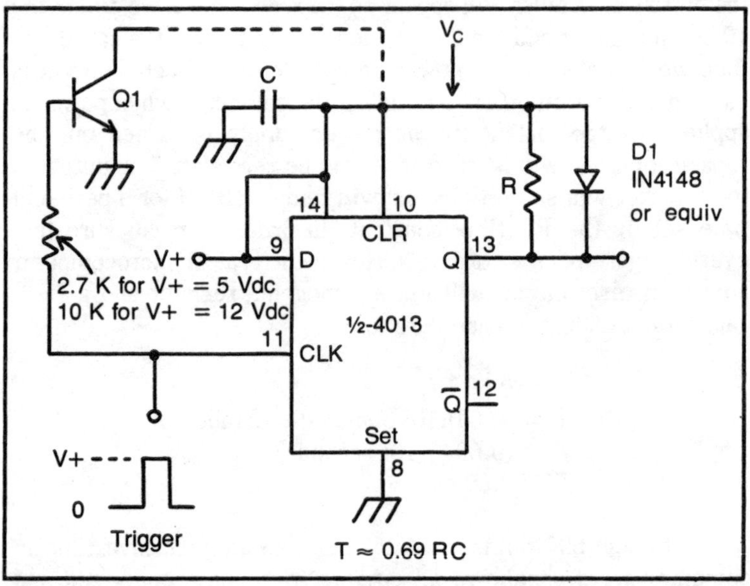

Fig. 10-3. 4013 CMOS monostable multivibrator.

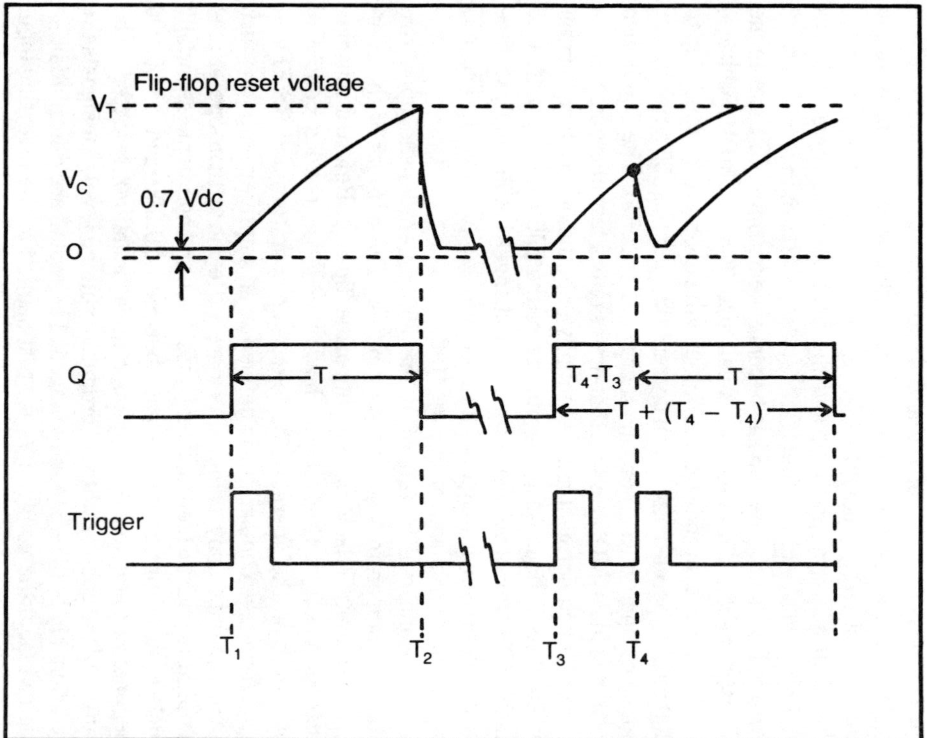

Fig. 10-4. Waveform.

cause the capacitor C to charge through resistor R. When the voltage across the capacitor reaches V_t, the voltage required to reset or "clear" the flip-flop, the Q-output will snap LOW again.

The capacitor will charge through resistor R when Q is HIGH, and when Q is LOW it will discharge through diode D1. If this diode were not present, the capacitor would be forced to discharge through resistor R, and that would yield a refractory period (when no new triggers are recognized) equal to the output duration t. But the diode permits rapid discharge of C once Q is LOW.

The transistor shown in this circuit is used to retrigger the input. Normally, this circuit will not respond to new trigger commands until after the period t plus the short refractory period of D1 is expired. But with transistor Q1 shunted across C, we can dump the charge across C at any time and start afresh. The total time will be t plus the already expired portion of t that was spent prior to the subsequent trigger pulse.

Still another approach, and the final one which we will consider, is the ubiquitous 555 IC timer device shown in block diagram form in Fig. 10-5. The 555 has become the foremost IC timer in the industry because it is simple, small size (8-pin miniDIP package) and low power. In addition, its output will either sink or source up to 200 milliamperes of electrical current (making the 555 a real power house compared with certain other timer circuits). The 555 is also available in a CMOS version that draws only a few microamperes of current when dormant and essentially only the output load current when active. Because of the nature of the 555 timer circuit, it is essentially free of problems caused by changes in the supply voltage (a problem which plagues UJT circuits, for example!). The 555 block diagram is shown in Fig. 10-5.

The internal workings of the 555 timer depend upon two voltage comparators and an RS flip-flop. The outputs of the comparators control the R and S inputs of the flip-flop. Before discussing the operation of this circuit, however, let's spend a couple of moments on the operation of voltage comparators and RS FFs.

A voltage comparator is defined as an amplifier with too much gain. Indeed, many comparators are little more than operational amplifiers with no feedback resistors. This arrangement essentially makes an amplifier with a gain between 20,000 and 1,000,000 depending upon the type of operational amplifier selected. The output of the comparator will produce a voltage level that signals the relationship between input voltages. The inputs are differential, so we can label the voltages V^- for that applied to the inverting input

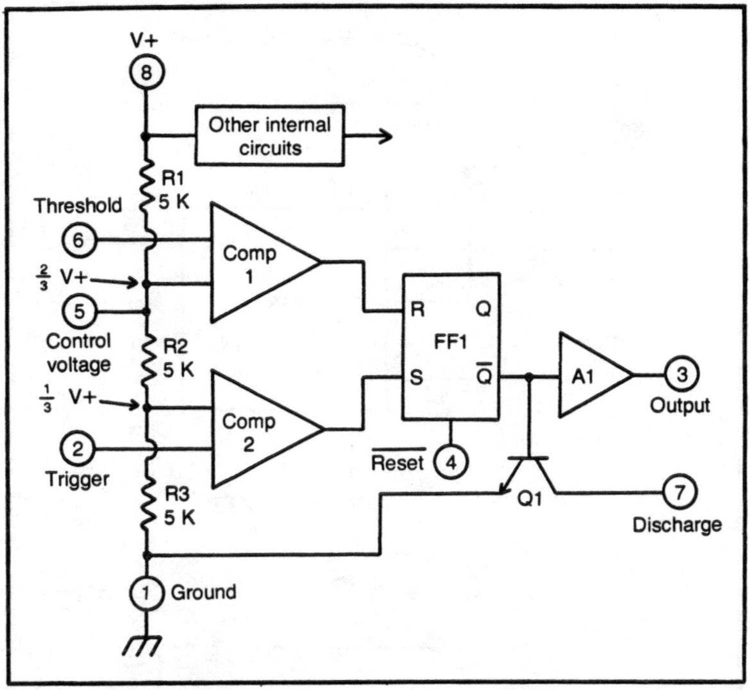

Fig. 10-5. Type 555 IC timer, internal circuit block diagram.

and V^+ for that applied to the noninverting input. If the comparator has bipolar power supplies, then the output will be as follows:

a. If V^- greater than V^+, then the output will be HIGH-negative;
b. If V^- less than V^+, then the output will be HIGH-negative;
c. If V^- equals V^+, then the output will be zero.

In the case of a monopolar power supply, as is found in the 555, there can be no HIGH-negative state so (a) above would also yield a zero output.

An RS flip-flop is also called a bistable multivibrator because it has two stable states. If an active signal is applied to the *set* (S) input, then the outputs go to the condition in which Q = HIGH and NOT−Q = LOW; similarly, an active signal can be applied to *reset* (R) to force the outputs to the condition of Q = LOW and NOT−Q = HIGH. The term "active" input is used instead of HIGH or LOW because both NAND and NOR gates can be used to construct RS

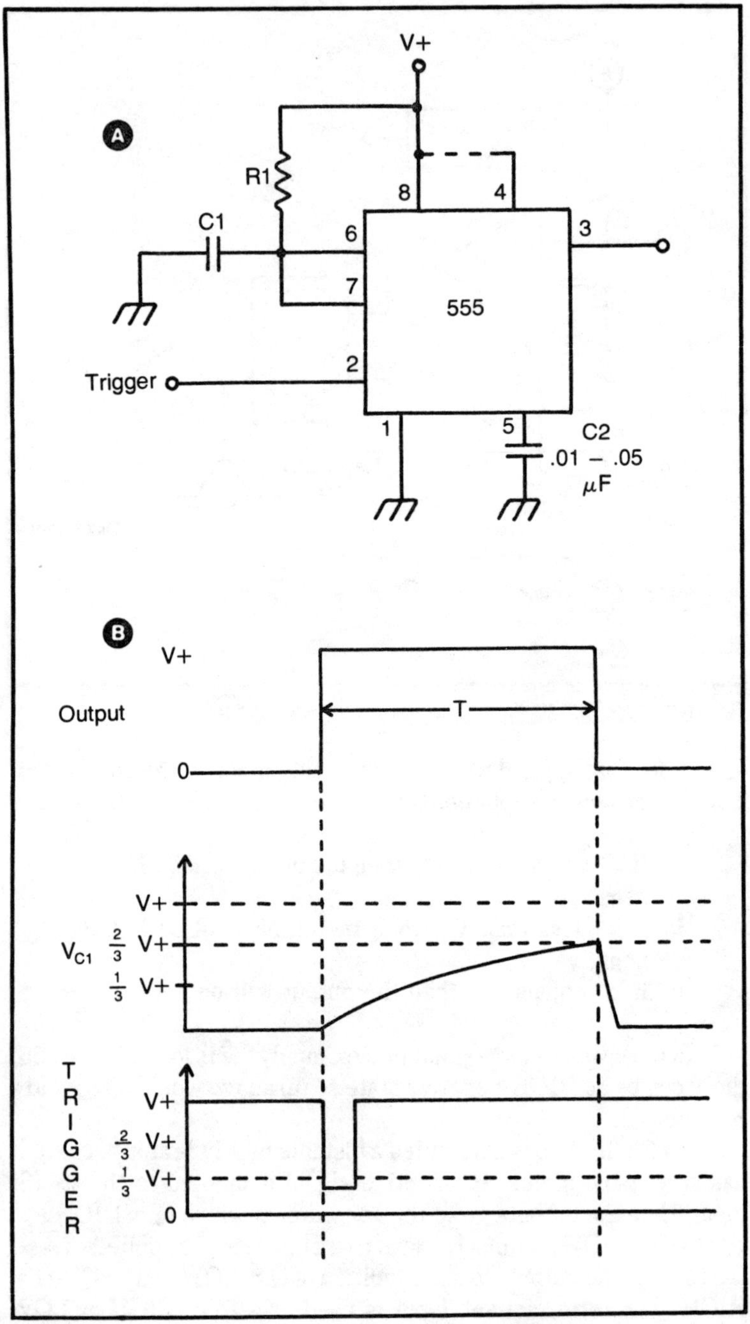

Fig. 10-6. A) 555 monostable multivibrator, B) timing diagram.

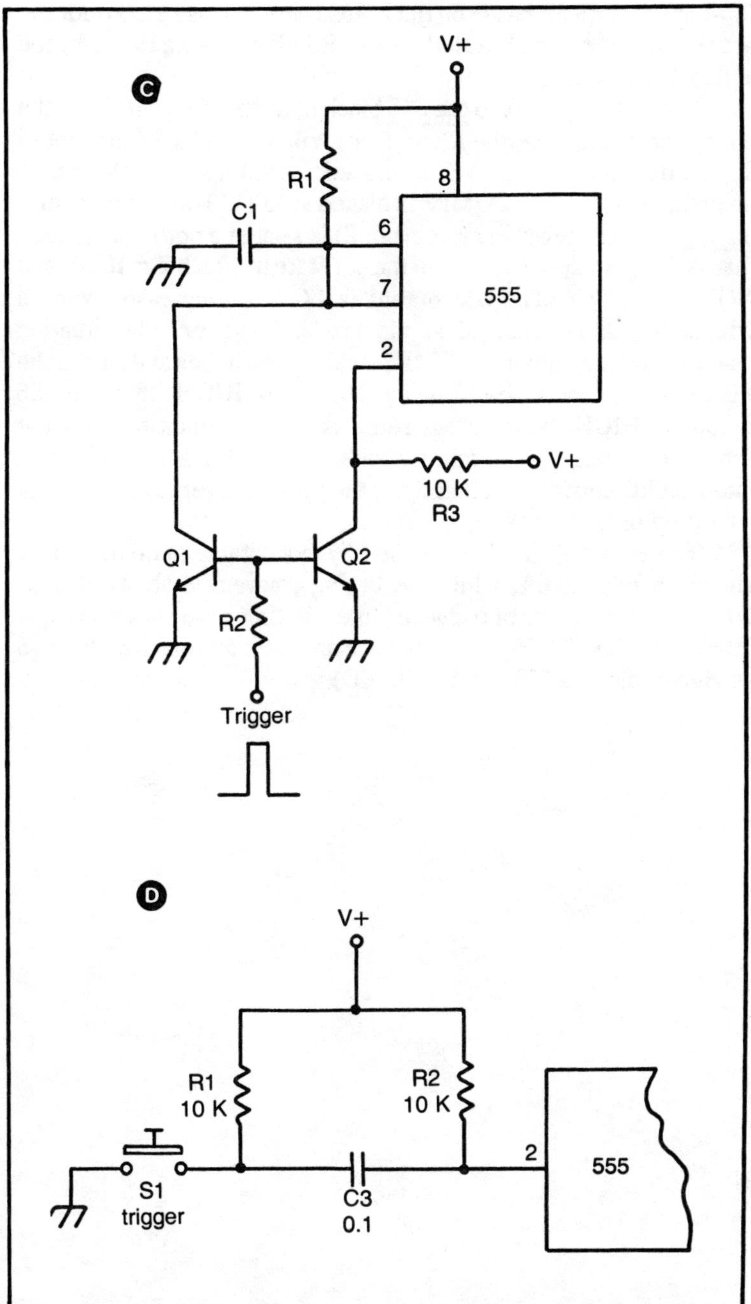

Fig. 10-6. C) retriggerable version of the circuit, D) manual triggering.

flip-flops. An RS FF based on the NAND gate will use a LOW for the active input state, while a NOR-based RS FF will use a HIGH for the active input signal.

Now let's get back to the 555 block diagram of Fig. 10-5. In the normal dormant state, the $\overline{\text{RESET}}$ line will be tied HIGH in order to keep it disabled. The COMP1 is biased to a voltage of ⅔-(V⁺) on its inverting input, while COMP2 is biased at ⅓-(V⁺) on its noninverting input. The trigger input (pin no. 2) is kept at a potential greater than ⅓-(V⁺), so the FF is kept in the set state in which Q = HIGH and NOT-Q = LOW. The 555 output is LOW in this case. When a trigger signal is received at pin no. 2, however, the situation changes and the output of COMP2 suddenly switches and allows the FF to go into the mode in which NOT-Q is HIGH, thus the 555 output is HIGH. This output remains in this condition until the threshold voltage is brought up to ⅔-(V⁺) by the action of the an external RC network. At that time, the FF will revert to the original condition of Q = LOW.

The external circuitry for the 555 monostable multivibrator is shown in Fig. 10-6A, while the timing diagram is shown in Fig. 10-6B. Two methods of triggering the 555 timer are shown in Figs. 10-6C and 10-6D. Here we have a method for making the 555 retriggerable (10-6C), or not (10-6D).

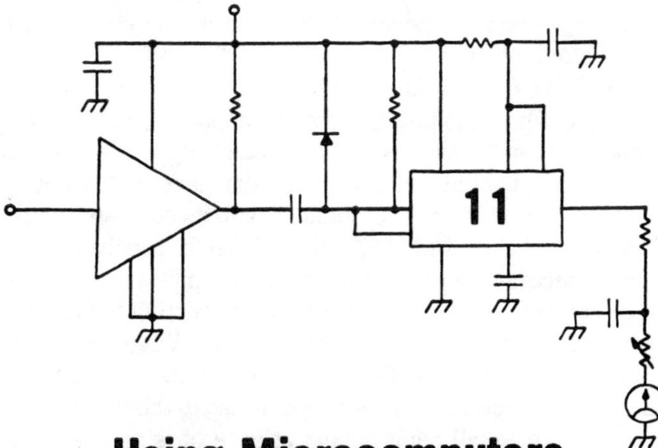

Using Microcomputers in Your Project Designs

The microcomputer has taken over the imagination of the public, and has become a primary tool in project design. In this chapter we will discuss some of the methods used to interface computers and the "outside world." Other circuits, some of them analog circuits, will also be used with the computer in many cases. The treatment of this topic is necessarily limited to a few of the topics which are of greatest interest. For readers who desire a deeper understanding, let me not modestly recommend my own TAB book, *Microprocessor Interfacing* (TAB book No. 1396). Also, if you have a lot of analog-to-digital or digital-to-analog conversion problems, let me also recommend my book *Microcomputer Interfacing: A/D-D/A* (TAB book No. 1271).

MAKING CUSTOM COMPUTER KEYBOARDS

Computer keyboards tend to be universal, and that makes it less than useful for certain applications that require custom keyboards. Examples are retail "point of sale" terminals, and scientific instruments. There are a couple of options open to us, however. First, we could use a standard ASCII or hexadecimal keyboard with custom keycaps. Some manufacturers offer their keyboards with removable clear-plastic keycaps that can be equipped with typed labels under the clear cover. In other cases, when a manufacturer is going to buy a sufficiently large quantity the keyboard company will *make* custom keycaps. This option is only open to large projects, however, because of the cost of making printed custom keycaps.

Another possibility is to make a circuit such as Fig. 11-1A. Here we see what amounts to a microcomputer input port that is connected to a bank of switches. The 74LS244 device is basically an IC containing eight noninverting buffers that have tri-state outputs. The output of each buffer is not dumped onto its corresponding data bus line until the correct *chip-select* signal is received. On the 74LS244 device, there are two active-LOW chip-select lines, each of which controls four of the internal noninverting buffer stages; pins 1 and 19 are the two chip-select terminals. When either 1 or 19 is brought LOW, then the four buffers controlled by that pin are connected to the output terminals, hence also to the data bus. Since we want to activate all eight buffers at once (it is assumed that an eight-bit input port is needed), both chip-select terminals are connected together.

The signal that turns the input port on and off is called $\overline{\text{IN}}$. This signal is developed from the control signals of the microcomputer. In the Z80 device, for example, the $\overline{\text{IORQ}}$ and $\overline{\text{WR}}$ signals will go LOW. In addition, there will be an eight-bit address on the lower byte of the address bus (A0 - A7) that determines which port is being called. A circuit is provided (discussed elsewhere in this book under the title "Microcomputer Interfacing") that will decode this address and provide a LOW output only when the correct port is called. Thus, we will have three signals which must *all* be LOW when the port is to turn on. A single three-input NOR gate will serve to generate the $\overline{\text{IN}}$ signal.

Each input line of the 74LS244 is connected to a grounded switch and a 3.3 kohm pull-up resistor that is connected also to + 5 volt dc. When the switch is open, the input will be HIGH and when the switch is closed the input will be LOW. Thus, we can define when the switch is closed or open by checking that bit.

In most microcomputers, the "keyboard" will be scanned periodically to see which input bits are HIGH and which are LOW. In this case, a switch is LOW when it is active. Thus, when switch S7 is closed (as if the operator was selecting the function represented by S7), then its corresponding bit will be LOW. Although only rarely will it be necessary to activate two switches at a time, so we can easily prioritize the switch by the programming protocol.

There are several different protocols that can be used to recognize the keyboard. We could, for example, input the keyboard data to the accumulator and then follow a masking protocol that uses

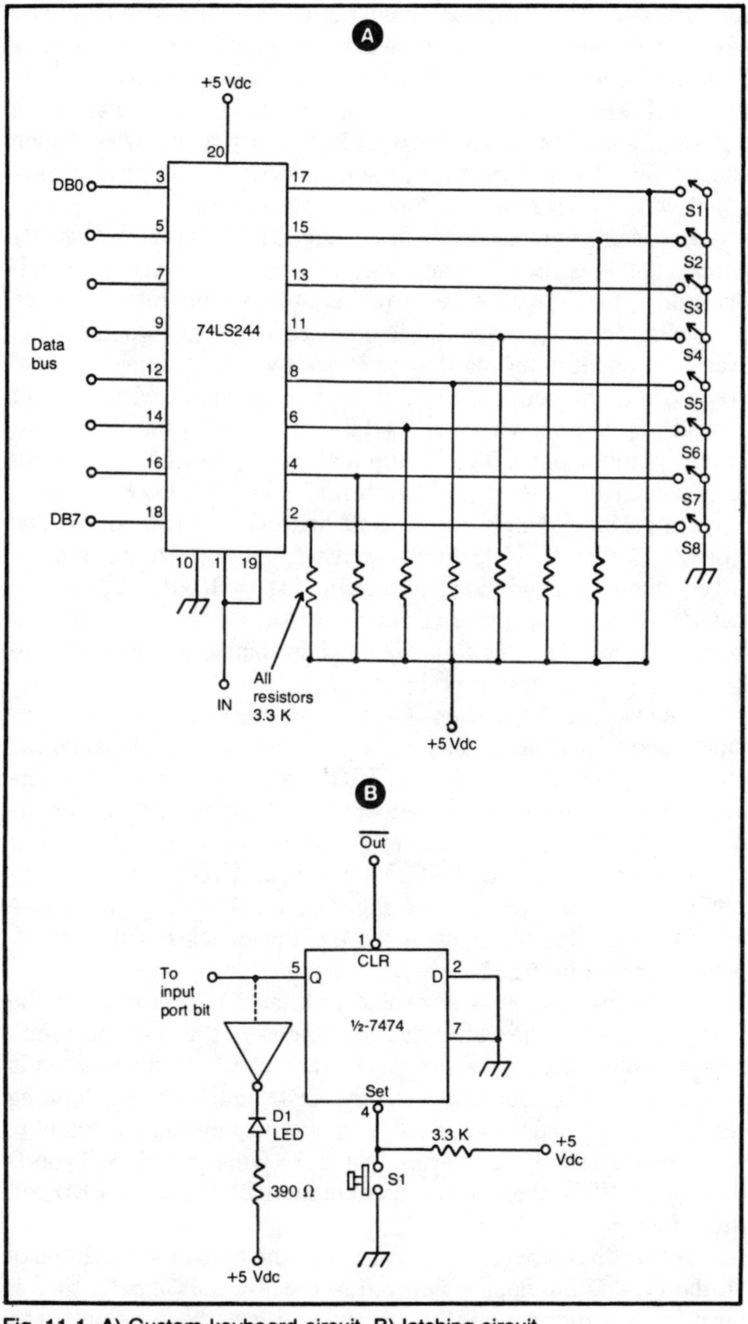

Fig. 11-1. A) Custom keyboard circuit, B) latching circuit.

the logical-AND instruction to null out all but the desired bit. The selection is then made (according to the need) by a ***branch on zero*** or ***branch on nonzero*** instruction. We can also use a shift instruction to find the LOW bits. Some microcomputers have a shift instruction that causes bit-7 of the accumulator into the carry bit. We can then use a *JUMP on carry clear* or *JUMP on carry set* instruction combined with a branch on zero/nonzero instructions.

In cases where the keyboard of Fig. 11-1A is scanned rapidly enough (at least 80 Hz scan rate, or one scan every 12.5 milliseconds), then it will be easy to ensure that any operator action on the switch is recognized. But in other cases, we will not want to scan the computer keyboard so rapidly and will want some means to "remember" the keystroke operation. We can do this with the 7474 Type-D flip-flop shown in Fig. 11-1B. Let's recall some of the rules of the operation of the Type-D flip-flop, as regards the *clear* and *set* terminals. When the *clear* (CLR) terminal is made LOW, then the outputs go immediately to the state with Q = 0 (i.e. LOW) and Not-Q = 1 (i.e. HIGH). Similarly, when the *set* terminal is made LOW, the outputs go to the condition of Q = HIGH and Not-Q = HIGH. We can use these terminals of the 7474 to operate our keyboard bits. The *clock* and *D-input* terminals are grounded to keep noise from causing problems.

The keyboard switches of Fig. 11-1B are connected to the *set* inputs and to ground. Again, a pull-up resistor is used. When the switch is open, the *set* input is HIGH, so is inactive. When the operator presses the switch, however, the *set* input will momentarily drop LOW, thereby causing the Q-output to go HIGH. The Q-output of the Type-D flip-flop will remain HIGH until the computer "reads" the input. This flip-flop, incidentally will be connected to the 74LS244 inputs in place of the switch-resistor combinations shown in Fig. 11-1A.

If the application is amenable to a lamp indicator to let the world know when the switch has been pressed, then use the circuit shown in the dotted lines in Fig. 11-1B. This LED driver circuit is standard practice, and connects the LED and a current-limiting resistor as the load between the V^+ power supply and the output of an open-collector TTL inverter. When the Q-output of the Type-D flip-flop is HIGH, then the inverter output is LOW, so the LED will turn on.

After the computer scans the input, and reads the condition of all the type-D flip-flops connected to the 74LS244 inputs, then it must issue a dummy $\overline{\text{OUT}}$ command to reset the flip-flops. The

$\overline{\text{OUT}}$ signal is connected to the *CLR* inputs of the flip-flop, and will force the Q-output LOW.

MICROCOMPUTER INTERFACING

One of the essential skills required of almost anyone engaged in electronics either vocationally or avocationally is microcomputer interfacing. This skill is, fortunately, easily acquired and the principles are easily understood. The interfacer must be familiar with the microcomputer control signals. Although there are various different control signal schemes, they are all basically similar. In this discussion, we will use the 6502 microprocessor chip as an example. That microprocessor is a memory-mapped device that assigns I/O ports to memory locations.

The 6502 microprocessor uses two main control signals, the R/$\overline{\text{W}}$ line and the Phase-2 clock line in addition to the sixteen-bit address bus to control memory read and write operations (hence also input—i.e., read—and output—i.e., write—operations). The action of these signals are listed below:

Phase-2 Clock. This signal is the complement of the Phase-1 clock signal, and follows the Phase-1 clock. Hence, when Phase-1 is HIGH, then Phase-2 is LOW; and *vice versa.* Both Phase-1 and Phase-2 have the same frequency and are derived from the *Main Clock*.

R/$\overline{W}$. This signal is called read-write and is HIGH for a read (input) operation and LOW for a writer (output) operation.

Address Bus. The sixteen bits of the address bus define the specific address of the memory location being called. Thus, a total of 65,536 (i.e., 64 K) locations can be uniquely specified.

The timing signal for the read cycle of the 6502 microprocessor is shown in Fig. 11-2A, while the write cycle timing is shown in Fig. 11-2B. The read cycle is the input operation, while the write is an output operation (remember, the 6502 is memory-mapped). The Phase-1 and Phase-2 clock signals are shown in Fig. 11-2A. Note that they are complementary. The read and write signals become active during the Phase-2 portion of the clock signals. Since this is a read operation, we look for both Phase-2 clock and R/$\overline{\text{W}}$ to be HIGH at the same time. Note that the address becomes valid during the Phase-1 segment of the clock cycle and remains valid during the entire duration of the Phase-2 segment. Thus, we can use an address decoder with an active-HIGH output (i.e., SELECT) and a three-input NAND gate to indicate when the correct signals are present. The data from the memory location or input port being

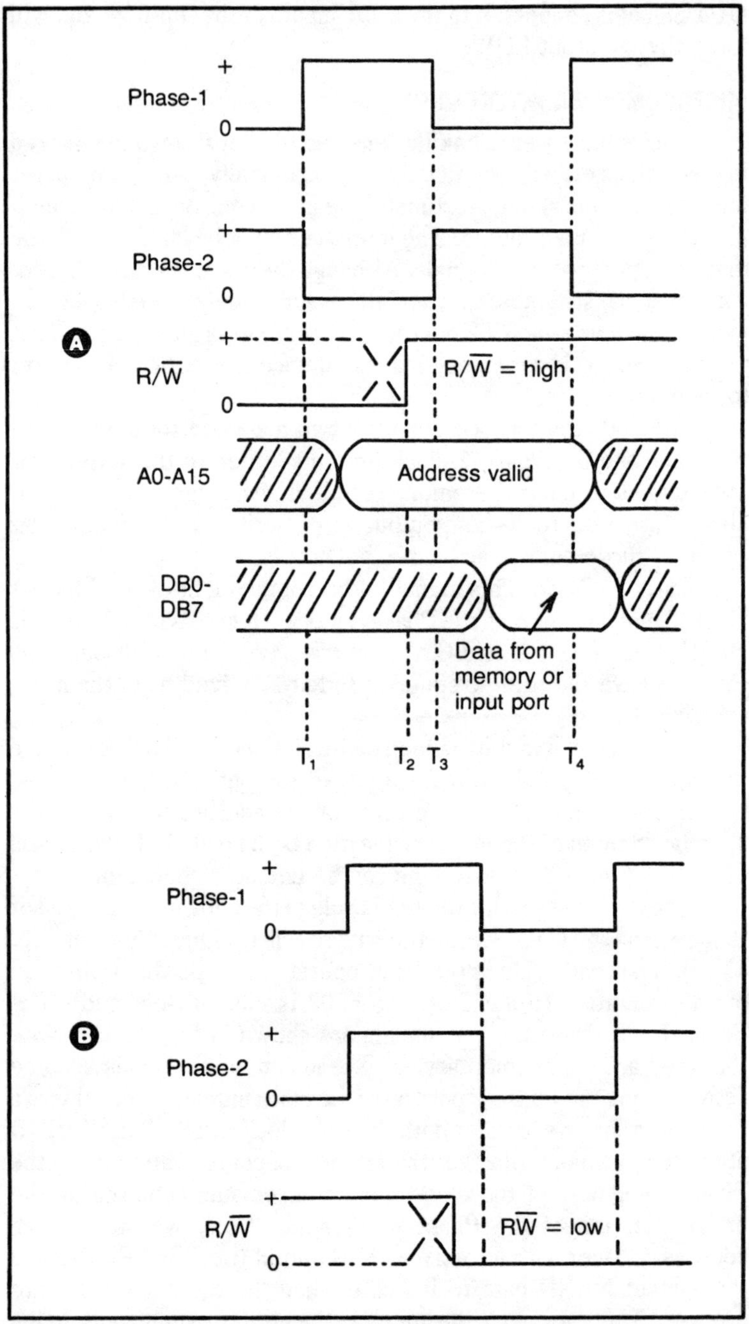

Fig. 11-2. A) 6502 timing signal for read cycle, B) for write cycle.

addressed during this read operation is transferred to the data bus(DB0 - DB7) during the last 300 nanoseconds of the Phase-2 clock segment.

Figure 11-2B shows the operation of the control signals during the write (i.e. output) operation. For the sake of simplicity we have eliminated the address bus and data bus signals from this drawing, and home in on only the Phase-2 clock and $R/\overline{W}$ signals. Recall that the read/write signal will be LOW during the write operation. Thus, sometime late in Phase-1 the $R/\overline{W}$ signal will drop LOW. When the Phase-2 clock also drops LOW we will know that a write operation is taking place. All that we need is an active-LOW address decoder signal ($\overline{SELECT}$) to be compatible with this signal. A three-input NOR gate will perform the decoding job in that case.

ADDRESS DECODERS

The SELECT and $\overline{SELECT}$ signals are generated by decoding the address bus. The design goal is to produce a signal that will become active (either HIGH or LOW as set by the need) if and only if the correct address is present on the sixteen-bit address bus (A0 - A15). There are several strategies that can be followed in this endeavor, but we will discuss only a few of the more common. The circuit shown in Fig. 11-3 can be used whenever we have a single (or a low number of I/O ports in a memory-mapped system, or, alternatively, like the Z80 the microprocessor will permit distinct I/O addressing situations. We can assign the I/O port to a place in the upper 32 K of memory. This technique is especially useful where the computer does not use the entire complement of memory (frequently the case in small computers). In Fig. 11-3 we use only three bits, i.e., A13, A14, and A15. Thus, the address of this location is E000 hex (i.e., 57.344_{10}), or 1110000000000000^2. Of course, any combination of addresses in which the three high-order bits are HIGH will trigger the circuit, so one must ensure via programming and system design that no other address in this pattern will be called by the microprocessor. This chore is not too difficult in small systems where no other memory locations above

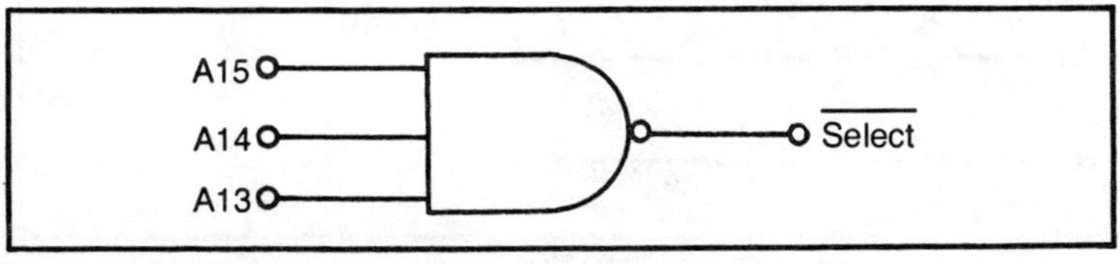

Fig. 11-3. Three-input gate address decoder.

32 K are needed. Many instruments and control systems based on microprocessors will use considerably less memory, so this approach can be quite practical. In one study performed by a microprocessor chip manufacturer, it was found that 80 percent of all control system applications programs used less than 2 K of memory, while 98 percent used less than 8 K of memory; little is lost in such systems by applying the I/O port assignments to the upper 32K.

Recall the rules of operation for the NAND gate: The output will go HIGH if any one of the inputs is made LOW; *all* (three in this case) inputs must be HIGH before the output will be LOW. Since the three inputs of the NAND gate shown in Fig. 11-3 are connected to address bits A13-A15, we must have these bits HIGH (as discussed above) for the output to be LOW. The circuit in Fig. 11-3 is, therefore, an active-LOW address decoder so will be produce a $\overline{\text{SELECT}}$ signal output.

The circuit of Fig. 11-3 is limited to three bits. Where we want a larger number of bits to be considered in the address, we have to use a larger gate. The circuit in Fig. 11-4 is an eight-bit address decoder, and is probably the most standard circuit in the business (a lot of companies use it). The 7430 NAND gate, a TTL device, will provide the ability to decode up to eight bits of the address bus. On

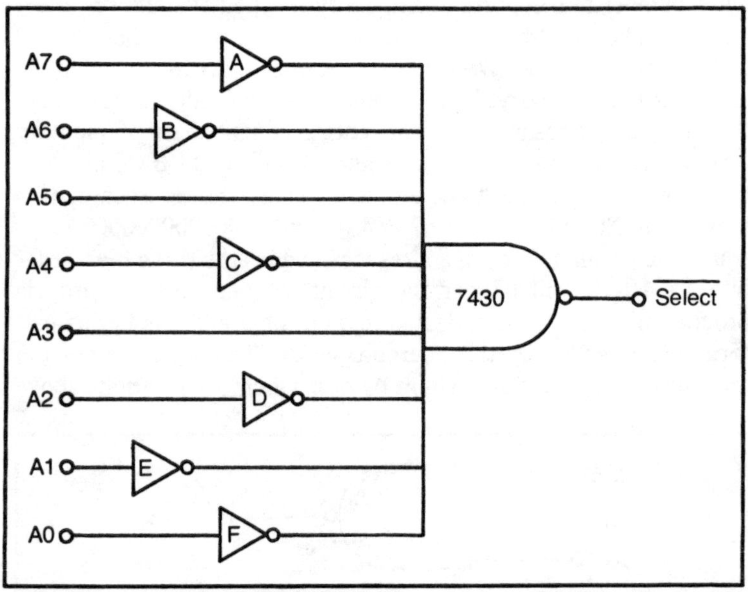

Fig. 11-4. 7430 address decoder example.

the Z80-based machine, this circuit will provide all of the decoding needed for 256 input/output ports. The Z80 uses the lower eight bits of the sixteen-bit address bus to provide the I/O address (it is not a memory-mapped machine). In the 6502 and similar machines, this circuit can be used in a limited fashion, or, combined with another eight-bit decoder to wring out all sixteen bits (of this, more later).

The 7430 is an eight-bit NAND gate, so all eight inputs must be HIGH in order for the output $\overline{\text{SELECT}}$ to be LOW. Of course, the only natural address which will do this trick is FF hex (i.e., 11111111_2). But, if we place inverters in some of the input lines, then we will be able to make the circuit more flexible. In the specific example shown, we are using the lower eight bits of the address bus, and inverters are placed in A0, A1, A2, A4, A6, and A7. This means that those lines will present a HIGH (logical-1) to the 7430 input if the address bus line is LOW (logical-0). Thus, the single address that will yield 11111111_2 at the 7430 inputs is 11010111 (D7 hex).

Of course, there is no reason why we have to use this specific address. We can switch the inverters around to any bit which we like. In fact, many microcomputer plug-in card manufacturers will provide from three to six (there are lots of six-inverter "hex" chips on the market) inverters on the circuit card with jumpers. The user will select which, if any, inputs are to be inverted by correct placement of the jumpers. In fancier systems, there may be on-board DIP-switches that will allow you to set the address that will be selected by that card. There is also no reason why sequential address bus bits must be used for the address decoder. If we wanted to use some unusual combination in which elements of both upper and lower bytes of the address bus are used, then there is not one reason why it could not be done.

The output of the circuit in Fig. 11-3 is active-LOW, so the 7430 output will drop LOW only when the correct address is placed on the address bus.

The eight-bit address decoder can be used for almost all microprocessor I/O applications, and many memory applications. In some cases, the memory will be broken into blocks that require a single eight-bit decoder plus one or more two to four-bit decoders in order to select the particular memory location. In such a system, some of the higher order bits of the address lines permitted to the system will form a "block decode" scheme (see *Microprocessor Interfacing* by J. J. Carr, TAB Book No. 1396). Figure 11-5 shows a system for combining active-LOW outputs to provide higher order decoding. In this case, we have selected two eight-bit decoders to form a single sixteen-bit SELECT signal. We have two eight-bit decoders of the sort shown in Fig. 11-4, so the outputs will be active-LOW ($\overline{\text{SELECT}}$). One decoder examines A0 - A7, while the other examines A8 - A15. These $\overline{\text{SELECT}}$ outputs ($\overline{\text{SELECTLO}}$ and $\overline{\text{SELECTHI}}$) are connected to the input of a two-input NOR gate. Recall the rules of operation for the NOR gate: if any input is HIGH, then the output is LOW, but if all inputs are LOW then the output is HIGH. Since the outputs of the eight-bit decoders are active-LOW, and they are connected to the NOR gate inputs, the NOR output will be HIGH if and only if the correct sixteen-bit address is present on A0 through A15. Thus, the output of the NOR gate in Fig. 11-5 is a SELECT signal.

DEVICE SELECT PULSES

The SELECT/$\overline{\text{SELECT}}$ signals are combined with the read/write and other control signals (depending upon the chip used) to form system IN and OUT (or $\overline{\text{IN}}$/$\overline{\text{OUT}}$) signals. The IN/$\overline{\text{IN}}$ signals become active for an input operation, while the OUT/$\overline{\text{OUT}}$ signals are active for output operations. In this section, we will discuss methods for generating the IN and OUT system signals. Figures 11-6 and 11-7 show two methods for generating these signals.

Recall from our earlier discussion that the 6502 microprocessor chip uses the Phase-2 clock and R/$\overline{\text{W}}$ signals in conjunction with the address decoder SELECT or $\overline{\text{SELECT}}$ signals to generate input or output (read or write) operations. In Fig. 11-6 we use these signals along with a pair of three-input NAND gates and an inverter to make the $\overline{\text{IN}}$ and $\overline{\text{OUT}}$ active-LOW signals. From our earlier discussions you should remember that the two control signal lines

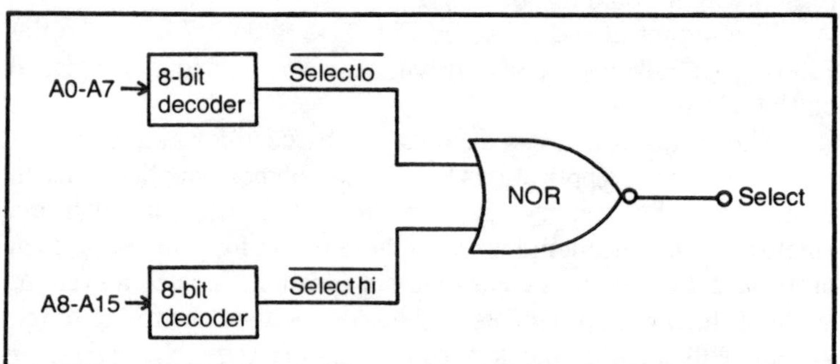

Fig. 11-5. Using multiple 8-bit address decoders for 16-bit or higher decoding.

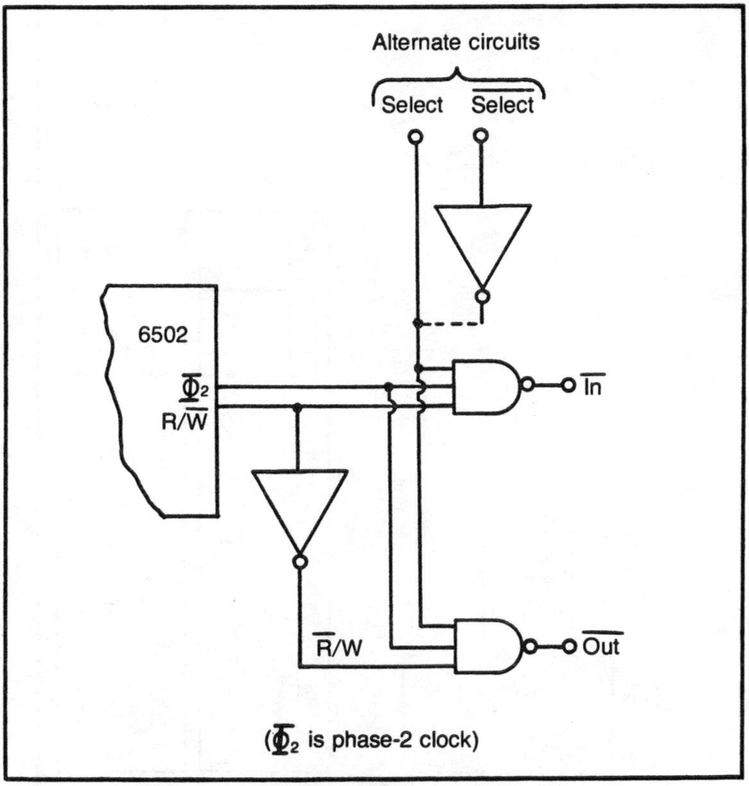

Fig. 11-6. Generation of 6502 $\overline{\text{IN}}$ and $\overline{\text{OUT}}$ system signals.

(Phase-2 and R/W) will be HIGH for an input operation (which is a read to the chip). Thus, we can use the NAND gate inputs connected directly to the control signals for this operation. The third input of the NAND gate will be connected to either a SELECT signal or an inverted $\overline{\text{SELECT}}$ signal (see dotted line connection in Fig. 11-6). Thus, when all three criteria are met, the three inputs of the NAND gate will be HIGH, so the output ($\overline{\text{IN}}$) will drop LOW.

The output operation follows exactly the same criteria, except for the $R/\overline{W}$ line being LOW during write (output) operations. Thus, we have to invert $R/\overline{W}$ to form $\overline{R}/W$ in order to drive the second gate. In the event that the output operation to the selected address is called, the output of the second gate ($\overline{\text{OUT}}$) will drop LOW.

The normal fan-out of TTL chips is ten, which means that the gates will drive up to ten standard 1.6 mA TTL inputs. We can use this circuit without additional help if and only if not more than ten devices are turned on (the usual case). If more are to be driven, then

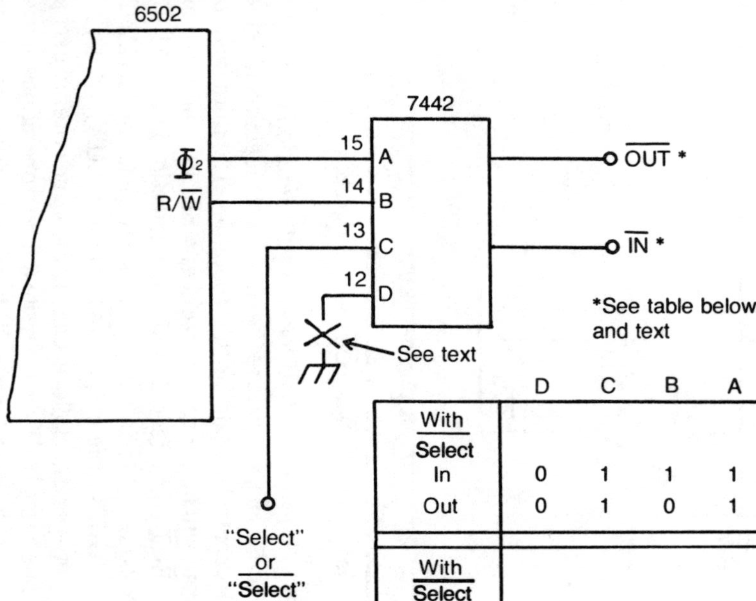

	D	C	B	A	Decimal Digit	Decimal Pin
With Select						
In	0	1	1	1	7	9
Out	0	1	0	1	5	6
With $\overline{Select}$						
In	0	0	1	1	3	4
Out	0	0	0	1	1	2

Fig. 11-7. Alternate method.

we must turn to a high-current or high-power buffer/driver IC.

In some systems, the designer has made a true system IN/OUT signal (or signals), so will not use the address decoder. The individual I/O ports or memory blanks will use the address decoder (of its own) and either IN/$\overline{\text{IN}}$ or OUT/$\overline{\text{OUT}}$ (as the case may be) to turn on the circuits. I suspect at you will find most small computers use the system signals, rather than producing all signals on the same board.

A second method is shown in Fig. 11-7. This circuit is also one of the more popular types on the market, and uses an older TTL IC device—the 7442. The 7442 TTL device is a *BCD-to-1-of-10-decoder.* More simply put, the 7442 has ten active-LOW outputs to represent the decimal digits from 0 to 9. It also has four input lines that are weighted according to the binary coded decimal (BCD) system of representing decimal digits with a four-bit binary word. This system uses the following four-bit words:

Decimal	*Binary/BCD*
0	0000
1	0001
2	0010
3	0011
4	0100
5	0101
6	0110
7	0111
8	1000
9	1001

Thus, when the binary word 0111 is applied to the BCD inputs of the 7442 device, the output that represents decimal "7" (i.e., pin no. 9) drops LOW.

In Fig. 11-7, we apply the two control signals, i.e. Phase-2 clock and R/$\overline{\text{W}}$ to the *A* (weight = 1) and *B* (Weight = 2) inputs of the 7442 device. The *C* input (weight = 4) is connected to the SELECT or $\overline{\text{SELECT}}$ lines. The D input (Weight = 8) can be either grounded (made permanently LOW) or it can be used to enable/disable the chip. For this particular circuit, the D-input will be LOW for enable and HIGH for disable.

The table shown in Fig. 11-7 shows the various codes for the 7442 inputs for $\overline{\text{IN}}$ and $\overline{\text{OUT}}$ signals depending upon whether the

SELECT or $\overline{\text{SELECT}}$ form of address decoder signal is selected by the designer. In the event the SELECT signal is chosen, the D-input will be HIGH when the correct address appears on the address bus. The $\overline{\text{IN}}$ signal will be generated by the word 0111 (decimal 7), so pin no. 9 of the 7442 becomes the $\overline{\text{IN}}$ terminal. Note the meaning of the binary word 0111 in this context. It means that the D-input is LOW, and that SELECT, Phase-2 and R/$\overline{\text{W}}$ are all HIGH (exactly the conditions required for a read operation in which the address decoder is active-HIGH). Other binary words are used for $\overline{\text{SELECT}}$ systems.

The D-input can be used in some cases for the address decoder. We can make a simple address decoder using the D- and C-inputs in a manner similar to the NAND gate decoder shown earlier in this section. We can, for example, connect the D-input to the A15 address bus bit, and the C-input to the A14 address bus bit. Thus, we will only get an $\overline{\text{IN}}$ or $\overline{\text{OUT}}$ signal when both A14 and A15 are HIGH or LOW, depending upon the protocol set up by the designer.

The 7442 device is limited to ten outputs, so will only respond to four bit binary words up to 1001_2 (i.e., 9_{10}). The limit for a four-bit word, however, is sixteen digits (called 0 - 9, A-F in hexadecimal notation). Thus, if we can use all four bits to their limit, we can generate more combinations:

Decimal	*Hexadecimal*	*Binary (Four-Bit)*
0	0	0000
1	1	0001
2	2	0010
3	3	0011
4	4	0100
5	5	0101
6	6	0110
7	7	0111
8	8	1000
9	9	1001
10	A	1010
11	B	1011
12	C	1100
13	D	1101
14	E	1110
15	F	1111

The TTL 7415x devices are useful for this type of decoder. The 74154 is particularly useful because it will perform as a sixteen digit version of the 7442, but also because it allows the user to select whether the outputs will be active-HIGH or active-LOW (instead of active-LOW only as in the 7442). The 74154 device is sometimes used as an address controller.

TYPICAL INPUT/OUTPUT PORTS

There are quite a number of devices in the manufacturer's catalogs that can be used for input and output ports in microcomputers. Some of them are designed especially for this service, and while quite expensive are often quite capable devices. Others are merely CMOS or TTL chips that can be pressed into service. We have, for example the 8216 and 8226 devices made by Intel in the former category and the almost totally equivalent 74125 and 74126 devices from the standard TTL lines.

A simple output port used on many microcomputers is shown in Fig. 11-8. This circuit is designed around the standard TTL 74100 device. The 74100 is a dual quad-latch. A "latch" is basically a

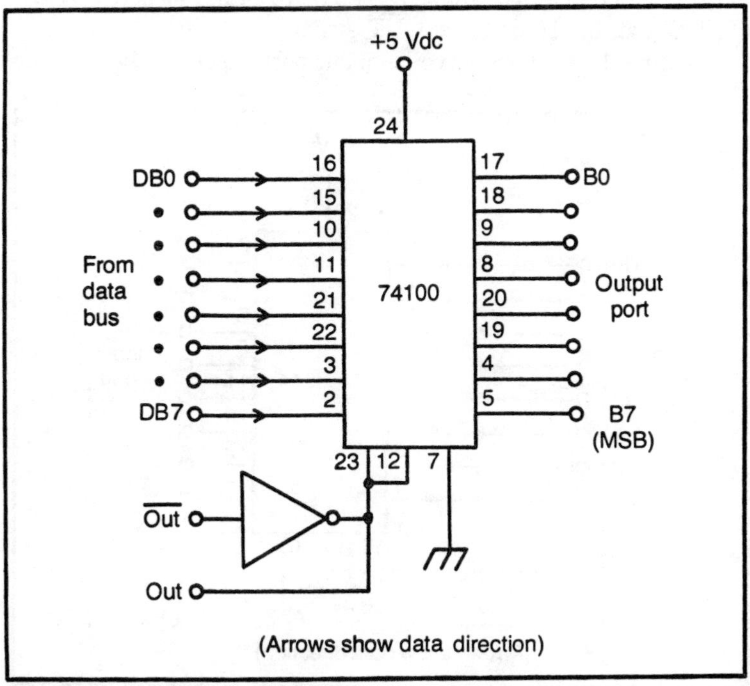

Fig. 11-8. Example of an output port.

Type-D flip-flop without a NOT-Q output. Thus, a latch will have a clock line, an input (called "D" in FF terminology) and a Q-output. When the clock line goes HIGH, the data appearing on the input is tranferred to the output line. When the clock drops LOW again, the data transferred to the output *remains* on the output, thus is said to be "latched." In a quad latch circuit (of which several exist in TTL and CMOS IC lines), the inputs and outputs of four latches are independent, while all four clock lines are connected together to form a "strobe" line. The 74100 has *two* quad latches in the same package. The clock lines are arranged into two terminals that each handle one entire quad latch. Thus, when we connect these terminals (pins 12 and 23) together, we gain the capability of strobing eight input lines from the computer data bus (DB0 - DB7) to the respective output lines at one time. The 74100, therefore, makes an almost ideal "output port" for a microcomputer.

The eight inputs of the 74100 are connected to DB0 through DB7 of the microcomputer data bus, while the eight corresponding output lines become the output port and will be routed to an appropriate connector. The strobe lines are connected together to form a master strobe line that will be turned on by either an active-HIGH OUT signal, or, an inverted $\overline{\text{OUT}}$ signal.

Figure 11-9 shows a typical input port based on the tri-state

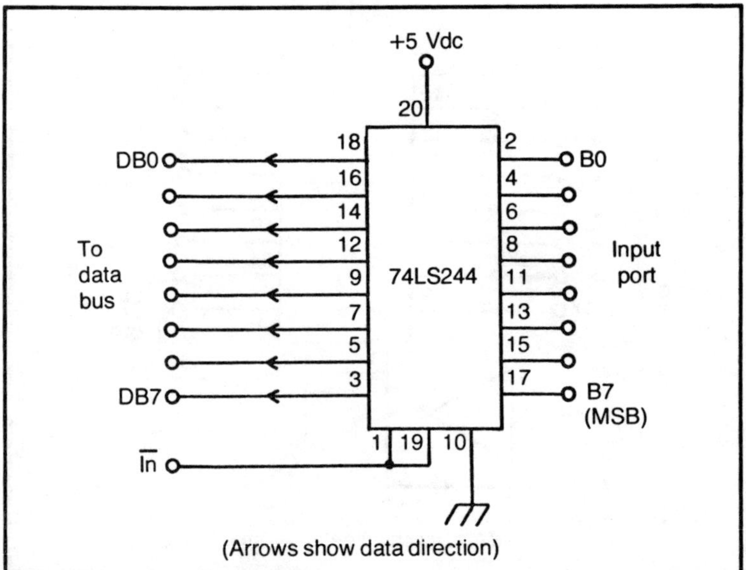

Fig. 11-9. Example of an input port.

output TTL 74LS244 device. This circuit was discussed in detail earlier, so will be just mentioned in passing here. The 74LS244 is normally the chip of choice unless one wants to use one of the more complex special purpose I/O chips made by the microprocessor manufacturers. The 74LS244 and the 74100 are both low in cost and easily available in hobbyist and industrial electronics supply houses.

DRIVING HIGH-VOLTAGE OR HIGH-CURRENT LOADS WITH A DIGITAL PULSE

There are many cases where we will want to drive a high voltage or high current load from a computer or other digital circuit. All of these computers operate from a relatively low-voltage (+5 volts is most common with one or two computers using ± 12 volts), and are totally incompatible with high voltage loads. The typical output current drive capacity of computer microprocessor chips is 1.8 to 3.6 milliamperes, which is totally unacceptable. Even output ports based on chips such as the 74100 (a common selection) will drive only 15 to 20 milliamperes per bit. Obviously, we will need a means for driving much higher current loads.

Figure 11-10 shows several circuits which can be used to drive these "high" loads. The version in Fig. 11-10A is used for driving a moderate size relay that can, in turn, be used for controlling 115 volt ac circuits. The 2N3053 device can handle most small relays. The base of the transistor is driven from the output of a TTL inverter. If a device with a fan-out of 10 is selected (typical), then 18 milliamperes of drive current are available. An open-collector TTL inverter is used.

The relay coil is connected in parallel with a 1N4007 rectifier diode. The purpose of this diode is to protect the circuit from high voltage spikes generated when the relay is de-energized. This phenomenon is called "inductive spike," and is capable of killing the transistor and other semiconductor devices in the circuit. The rectifier is normally reverse biased by the dc power supply. But when the inductive spike is generated, it will have a polarity opposite the power supply, so the diode will be forward biased under that condition. The spike will be clipped to 0.7 volts, rather than hundreds of volts.

If the transistor can be connected directly to the output of a computer, then we can delete the inverter. The 2.7 kohm resistor can be connected to one bit of a computer output port. When that bit goes HIGH, then the transistor will be forward biased so the

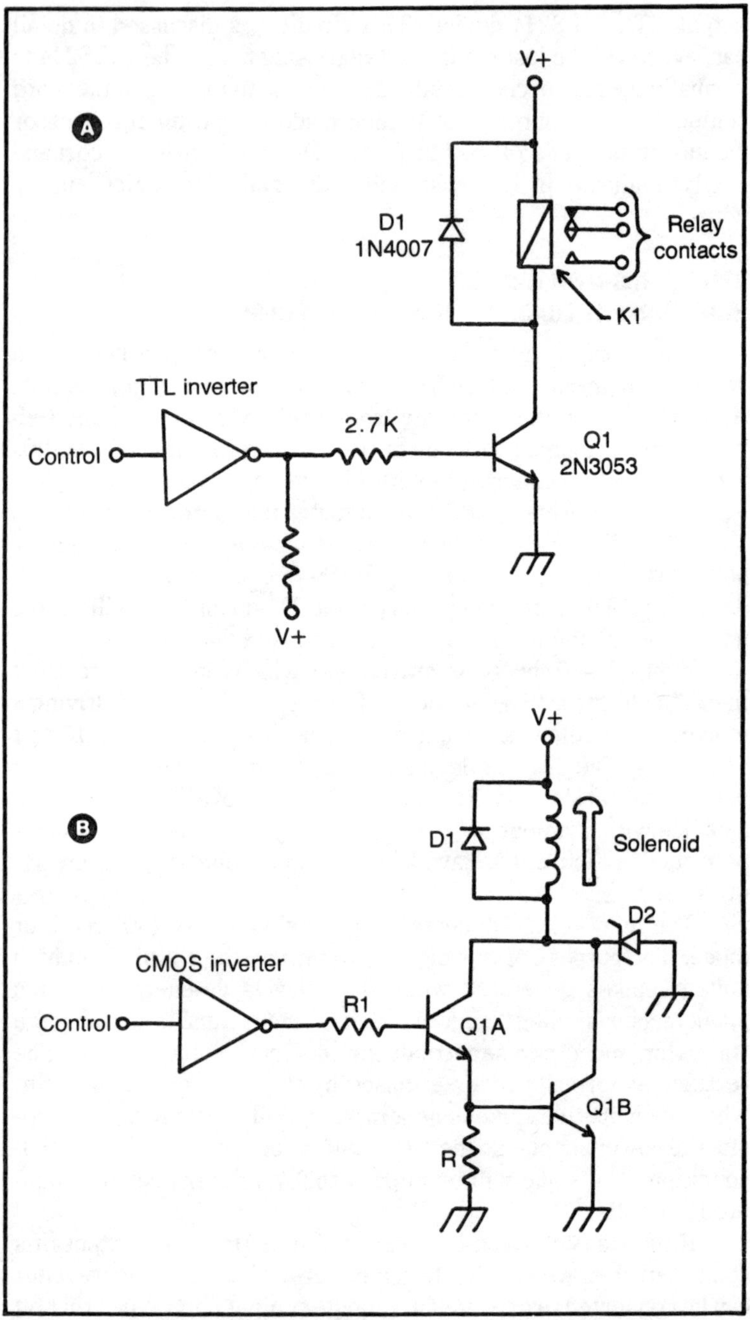

Fig. 11-10. A) Relay driver, B) high current relay or solenoid driver.

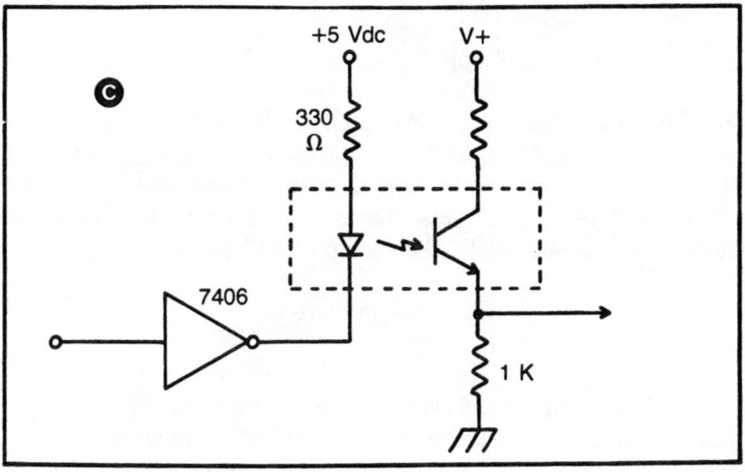

Fig. 11-10. C) optical isolation technique.

transistor will turn on. If the bit goes LOW, then the transistor is unbiased so the transistor is turned off. Thus, a HIGH output from the computer will turn on the relay, and a LOW will turn off the relay.

There is, however, a fundamental limitation to this circuit. The current gain of the transistor is the ratio of the collector current to the base current, and is expressed as the "*beta*" or H_{fe}:

$$H_{fe} = I_c / I_b$$

The gain (H_{fe}) is a function of the transistor selected, and will be typically 80 to 120 for most common "grab bag" transistors (higher gains are also available, but you will have to select them from the charts). The collector current, I_c, is a function of the relay and V+, and is given by:

$$I_c = V^+/R_k$$

Where:

I_c is the relay current in amperes
V^+ is the power supply voltage
R_k is the relay coil resistance in ohms

Thus, if we have a 28 Vdc relay with a 200-ohm coil resistance, then the current will be:

$$I_c = (28\ \text{Vdc})/(200)$$
$$I_c = 0.14\ \text{amperes, or, 140 milliamperes}$$

We must make a couple of determinations before these circuits can be operated successfully. For example, if the output drive current of the computer is limited, then we will have to select a transistor with a higher beta. Thus, if we have a 1.8 milliampere output (i.e., fan-out of 1), then the *beta* will be

$$H_{fe} = 140\ \text{mA}/1.8\ \text{mA}$$
$$H_{fe} = 78$$

The value calculated above is a minimum value. We must also account for transistor aging, which could potentially reduce beta gain, and for the fact that transistor gain figures are typically not what is given in the simplistic version of the spec-sheet. In most cases, there will be "typical," "maximum," and "minimum" values for H_{fe} described in the specifications sheet. The usual practice is to select a transistor with 20 percent reserve (multiply by 1.2, thus making the gain needed $1.2 \times 78 = 94$), and then use that for the minimum value.

The beta problem can be overcome in many cases by using a so-called ***superbeta*** circuit, also known as the Darlington amplifier or Darlington pair. An example is shown in Fig. 11-10B. The Darlington pair consists of two transistors, shown here as Q1A and Q1B. The collectors of the two transistors are connected together, while Q1A is used as an emitter follower to drive the base of Q1B. The gain of the circuit is the ***product*** of the individual *betas* of Q1A and Q1B. In other words:

$$H_{fe} = H_{fe(A)} \times H_{fe(B)}$$

Where:

H_{fe} is the total *beta* gain
$H_{fe(A)}$ is the *beta* of transistor Q1A
$H_{fe(B)}$ is the *beta* of transistor Q1B

Thus, if the two transistors have the same *beta* gain, then the total *beta* gain is the square of the individual *beta* gains. Suppose, for example, the *beta* gains are both 100 (a moderate and typical figure). The total *beta* is

$$H_{fe} = (100)(100)$$

$$H_{fe} = 10{,}000$$

A *beta* of 10,000 certainly qualifies as a "superbeta" transistor.

Most semiconductor manufacturers now make Darlington amplifier transistors. Typically housed in a TO-3 power transistor case, these *superbeta* power transistors are capable of collector currents to 30 amperes with gains typically 5000 to 25,000. Assuming that we have an output current available of 3.2 milliamperes (i.e., a fan-out of two), then the total collector current which we can drive with a *superbeta* of 10,000 is:

$$I_{c(max)} = (H_{fe})(I_b)$$
$$I_{c(max)} = (10{,}000)(0.0032 \text{ amperes})$$
$$I_{c(max)} = 32 \text{ amperes}$$

In the example shown in Fig. 11-10B, the Darlington transistor is shown driving a solenoid-and-plunger. Most of these circuits require high currents, far beyond the capability of Fig. 11-10A. Thus, we will need a high-current superbeta transistor in order to drive the circuit. Also, I might add, the diode shunting the high-current coil of the solenoid must also be scaled up.

Diode D2 is sometimes used as an added limit on the inductive spike. The zener potential of D2 is selected to be approximately 150 percent of V+. If the voltage at the collector is negative, then it will be clipped to 0.7 volts, and if positive it will be clipped to V_z of D2.

There are also other methods for handling loads that are incompatible with the microcomputer. We could, for example, use an optoisolator for this purpose. An optoisolator is a device (IC) which has a light emitting diode (LED) positioned such that it will illuminate the base of a phototransistor when it is turned on. Thus, turning on the LED will also turn on the phototransistor. An example is shown in Fig. 11-10C. Here we have an optoisolator connected such that the LED is the collector load of an open-collector TTL inverter. When the input of the inverter is HIGH, then the output is LOW, so the LED is turned on. But, if the input of the 7406 is LOW, then its output is HIGH and the LED is turned off. If the output of the transistor side is connected to a device such as a silicon controlled rectifier, then the pulse applied to the input of the inverter will in turn turn on the SCR.

DIGITAL-TO-ANALOG CONVERTER (DAC) FOR MICROCOMPUTERS

A Digital-to-Analog Converter (DAC) is a device that will

convert a binary word to a proportional analog voltage. These devices are sometimes used to produce some sort of display output from a computer to an oscilloscope, strip-chart recorder, X-Y recorder, servomotor or other device. DACs were once relatively costly devices, but now we can obtain integrated circuit types at a very low cost. The popular DAC-08, introduced by *Precision Monolithics, Inc.*, is now available from several sources, including *Advanced Micro Devices, Inc.* and *National Semiconductor.* The *DAC-0808* device used in Fig. 11-11 is from *National Semiconductor,* and is available widely in mail-order stores and from those parts stores that deal with amateurs and hobbyists. The one which I bought to test the circuit of Fig. 11-11 (on my Timex T/S-1000) was from a *Jim-Pak* display at a local distributor.

The circuit in Fig. 11-11 uses the DAC-0808 (or any variant of the DAC-08 device) in the configuration called ***unipolar straight binary***. In other words, the output potential will be of one polarity only and will be 0-volts when the applied binary word is 00000000_2, and full-scale positive (i.e., +10-volts in this case) when the applied binary word is 11111111_2.

The output of any DAC is the product of the reference voltage (or current), and a fraction made up of the applied binary word in the numerator and the full-scale binary word in the denominator (or, their decimal equivalents). For an 8-bit DAC, therefore, the output voltage will be:

$$V_o = V_{ref} \times \frac{nnnnnnnn}{11111111}$$

Since 11111111 is 256_{10}, we can rewrite this expression in the form:

$$V_o = \frac{V_{ref} N_{10}}{256}$$

Where:

V_o is the output potential in volts
V_{ref} is the reference potential in volts
N_{10} is the decimal equivalent of the applied binary word
256 is the decimal value of 11111111_2

In the case of a 10 volt reference supply, we could rewrite this equation in the form:

$$V_o = \frac{(10\ V)(N_{10})}{256}$$

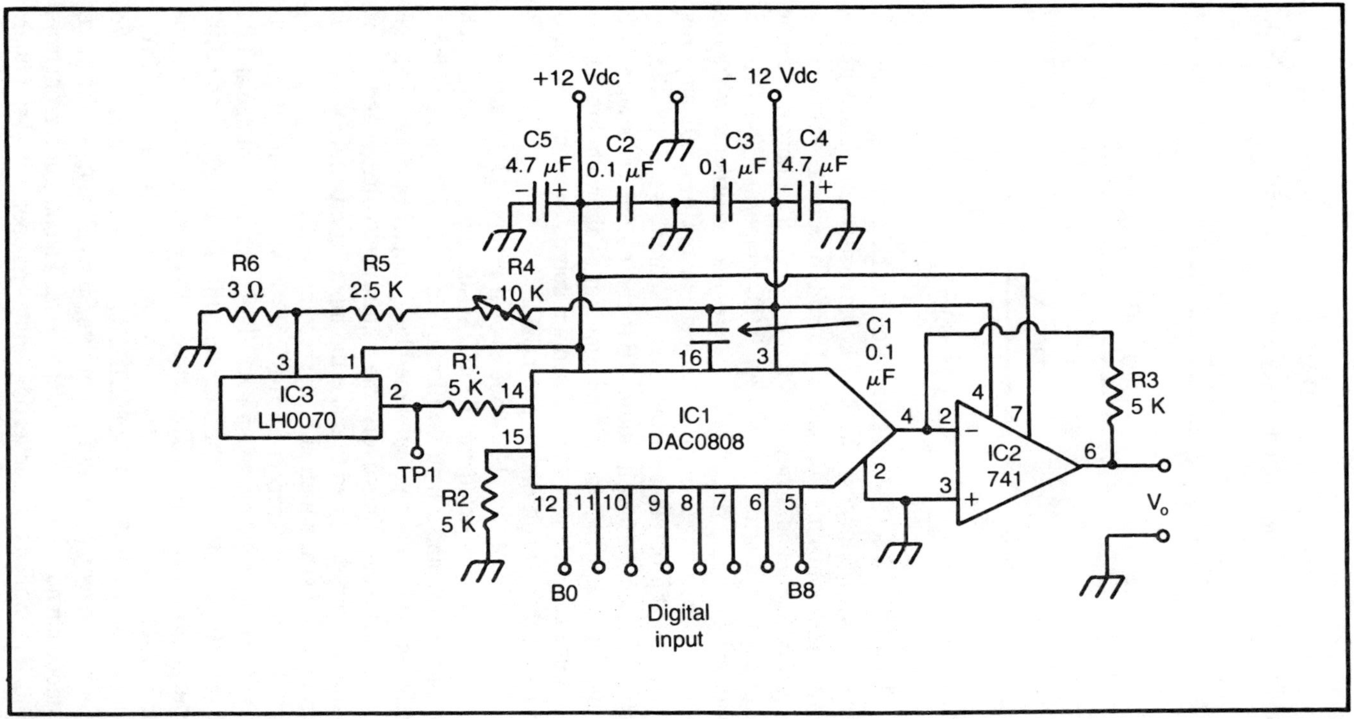

Fig. 11-11. Digital-to-analog converter.

Example

Find the output voltage of a DAC with a 10-volt reference (e.g., as in Fig. 11-11) if the applied binary word is 10101101_2 (i.e., 173_{10}).

$$V_o = \frac{(10\ V)(N_{10})}{256}$$

$$V_o = \frac{(10\ V)(173_{10})}{(256)}$$

$$V_o = \frac{(1730_{10}\ V)}{(256)}$$

$$V_o = 6.76\ \text{volts}$$

The DAC-08 device is a current-output DAC and uses a current reference source. Since we have a 5 kohm resistor in series with the I_{ref} terminal (pin no. 14), and a 10-volt reference voltage, the reference current is 2 mA (0.002 amperes). The output current maximum is the same value, i.e., 0.002 amperes. The DAC-08 device can be output directly to a load resistor in order to make the output into a voltage. Since an output current of 0.002 amperes exists, and the output resistance is 5 kohms, we can make the circuit produce a 10 volt output. But the output resistance is 5 kohms. By adding the operational amplifier as shown in Fig. 11-11, we reduce the output resistance from 5 kohms to about 100 ohms and retain the 10 volt output potential.

The inputs of the DAC-0808 shown in Fig. 11-11 are TTL-compatible. The basic design of the DAC-08 and DAC-0808 is such that the digital binary inputs will respond to a wide range of input levels, and this can be programmed by a special reference voltage pin. In this case, however, the reference voltage and other parameters are designed to make the inputs compatible with TTL sources (the usual situation in microcomputers).

The reference current for the DAC-0808 is applied to pin no. 14, and is developed by resistor R1. The reference voltage at TP1 will be 10.00 volts, so the current will be 10/5000, or 0.002 amperes (2 milliamperes). The actual voltage is adjusted by potentiometer R4, which controls the reference supply IC, an LH-0070.

The output of the DAC-0808 is a 0 to 2 mA current. There are actually two complementary outputs, pins 2 and 4. We have grounded pin no. 2, so it is not usable here. The output current from pin no. 4 is applied to the inverting input of an operational amplifier.

The output voltage developed by the 741 operational amplifier is given by:

$$V_o = I_o \times R3$$

Given the values of 5000 ohms for R3, and 2 mA for the maximum value of I_o,a the maximum output voltage will be:

$$V_o = (0.002 \text{ A})(5000 \text{ ohms})$$
$$V_o = 10.000 \text{ volts}$$

Thus, when the binary input is 00000000, the output will be 0.0 volts, and when the binary input is 11111111 the output voltage will be 10.00 volts.

The DAC-0808 will operate to power supply voltages of ± 15 Vdc, but in this case ± 12 Vdc power supplies are specified. The reason for this selection is that most microcomputers which offer other than +5 volt dc will offer ± 12 Vdc. If your computer offers ±15 Vdc, then go ahead and use it.

The capacitors on the power supply lines are necessary to keep the DAC from responding to noise. The 0.1 μF units must be placed as close as possible to the body of the DAC, while the 4.7 μF capacitors can be placed a little further away. Like the power supply 0.1 μF units, the frequency response capacitor (C1) must be as close as possible to the body of the DAC-0808.

In some cases, it will be better if a capacitor is placed across resistor R3. The output of the DAC will contain a substantial amount of noise because of the step-changes inherent in digital devices. The high frequency noise can be reduced by placing a capacitor across R3 of sufficient value to round off the steps of the output function. In most cases, it is wise to use only as much frequency response as is needed for the desired application. The value of the capacitor will be found from:

$$C_{\mu F} = \frac{10^6}{2\pi (5000)(F_{Hz})}$$

or, by collecting the constants:

$$C_{\mu F} = \frac{32}{F_{Hz}}$$

Where:

$C_{\mu F}$ is the capacitance in microfarads shunted across R3 in Fig. 11-11

F_{Hz} is the upper-end −3 dB point in the frequency response

Example

An ECG signal has a frequency response of 0.05 Hz to 100 Hz. Find the value of a capacitor to be shunted across R3 in Fig. 11-11 that will permit a digitized ECG signal to be reproduced properly.

$$C_{\mu F} = 32/F_{Hz}.$$
$$C_{\mu F} = 32/100 \text{ Hz}.$$
$$C_{\mu F} = 0.32 \text{ (use } 0.33\ \mu\text{F, a standard value).}$$

A/D CONVERTER FOR Z80-BASED MACHINES

In order to process analog signals in a digital computer it is necessary to use an analog-to-digital converter (A/D or ADC). The ADC will produce an *n*-bit binary word that is proportional to the applied analog input voltage (or current, in some cases). There are several different coding schemes, but in general, we will find unipolar and bipolar inputs, and both straight binary and offset binary output words (not to mention BCD and other combinations which find special applications). A "unipolar" input is one in which the permissable input voltages are all of one polarity. For example, 0 to +10 volts, 0 to +5 volts, 0 to +2.56 volts, 0 to +5 volts, 0 to −5 volts and so forth are "unipolar" input voltage ranges. A bipolar input voltage range is of the form $-V_{in}$ to $+V_{in}$, for example, −10 volts to +10 volts, −2.56 volts to +2.56 volts and so forth.

Straight binary is the type of coding in which the minimum voltage is represented by 00000000, while the maximum voltage is represented by 11111111 (or vice versa). For example, in a unipolar straight binary system in which the voltage range is 0 to +10 volts, we can represent 0 volts with 00000000 and +10 volts with 11111111.

In bipolar systems, a little dilemma exists. There are always an even number of possible states in a binary word of eight-bits length: 256. If we have a bipolar input voltage of, say, −10 volts to +10 volts we must do one of two things. Either the ranges must be uneven by the value of 1-LSB voltage, or, we must create equal ranges and no zero. If we opt for equal ranges, then there will be no unique state for the value 0-volts. In the case of the 8-bit ADC with a −10 to +10 volts range, each 1-LSB state is worth 80 millivolts. Thus, we have the possibility of zero being equally well represented by +80 mV and −80 mV, which leads to the idea of *plus-and-minus-zero*.

Figure 11-12 shows a simple, low-cost ADC that will work on almost any Z80-based microcomputer. In this particular case, the circuit was intended for the *Timex T/S-1000* computer. The Z80 microprocessor chip uses the following control signals for I/O port control:

$\overline{\text{IORQ}}$ This active-LOW input-output request signal indicates that the I/O operation is taking place. In other words, when the $\overline{\text{IORQ}}$ drops LOW and I/O operation is happening. The $\overline{\text{IORQ}}$ must be combined with $\overline{\text{RD}}$ for input operations, and $\overline{\text{WR}}$ for output operations.

$\overline{\text{RD}}$ *Read* signal. This active-LOW signal tells the world that a read (i.e., input) operation is taking place.

$\overline{\text{WR}}$ *Write*. This active-LOW signal indicates that a write (i.e., output) operation is taking place.

A0-A7 Lower-order eight bits of the address bus contains the address of the I/O port. Since the I/O address is 8-bit long, the Z80 can uniquely address up to 256 different I/O ports, numbered 000 to 255.

These control signals must be used in interfacing the ADC shown in Fig. 11-12. The combination for an input operation are: $\overline{\text{IORQ}}$, $\overline{\text{RD}}$ and the correct address on A0 - A7. For an output operation, the $\overline{\text{IORQ}}$, $\overline{\text{WR}}$ and the correct address on A0 - A7 are required.

The integrated circuit ADC used in Fig. 11-12 is the National Semiconductor ADC0801-4 device. This low-cost device is easily available from mailorder sources, and is also available from local outlets in the *Jim-Pak* components displays.

The ADC0801-4 device is specially designed to work with microprocessors. In this case, we are connecting the ADC as if it were an input port. The device contains its own internal clock circuit, which is RC-timed by way of R1 and C1. The clock frequency is given by:

$$F_{clk} = 1/1.1R_1C_1$$

Where:

F_{clk} is the clock frequency in hertz (Hz)
R_1 is the value of R1 in ohms, usually 10,000 ohms
C_1 is the capacitance of C1 in farads

With the values shown in Fig. 11-12, the clock frequency is:

$$F_{clk} = \frac{1}{(1.1)(R_1)(C_1)}$$

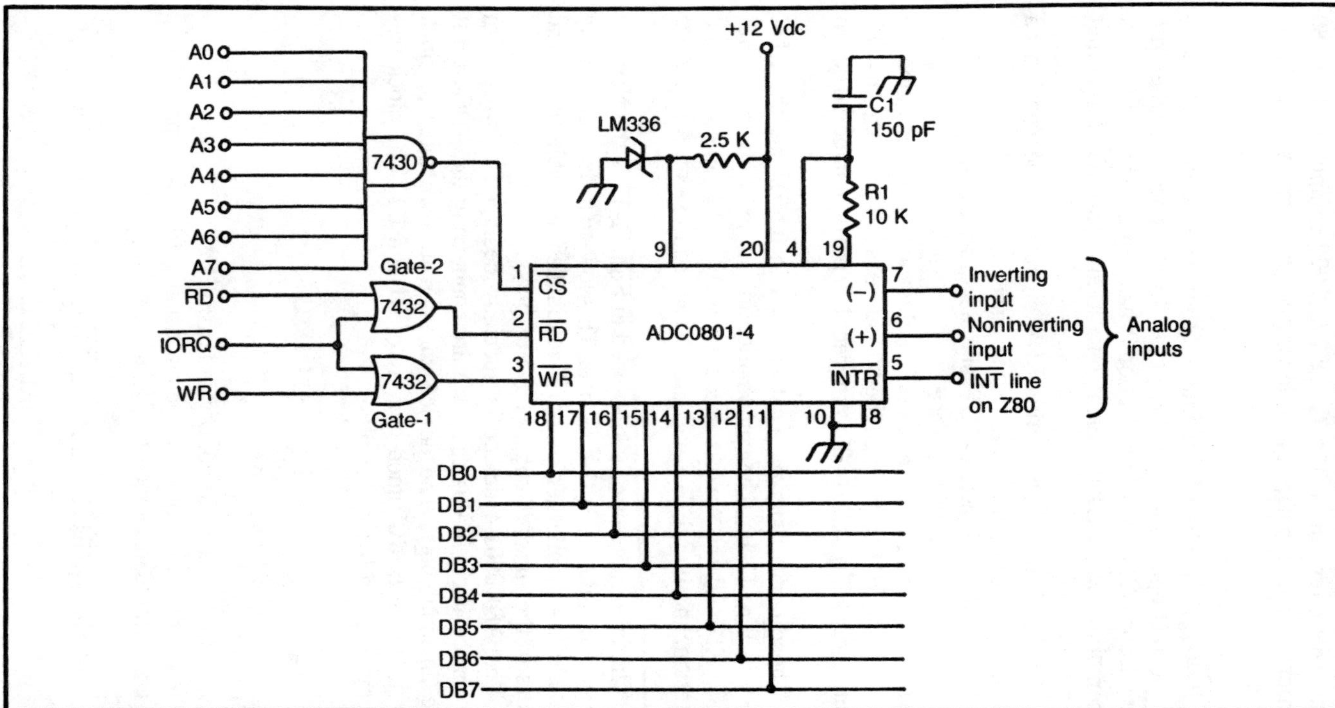

Fig. 11-12. ADC0801-4 A/D converter.

$$F_{clk} = \frac{1}{(1.1)(10^4)(1.5 \times 10^{-10}\ F)}$$

$$F_{clk} = 1/1.65 \times 10^{-6} = 600{,}000\ Hz = 600\ kHz$$

The clock period is 1/F, or, in this case, 1.65 microseconds. The ADC is a successive approximation A/D converter, so will make a full-scale conversion in $n + 1$ clock pulses, where n is the bit-length. Since this is a 8-bit device, the conversion will occur in 9 clock pulses, or, in time units:

$$\frac{1.65\ \mu S}{\text{pulse}} \times 9 \text{ pulses} = 14.85 \text{ microseconds}$$

Thus, we can count on a conversion to take place every 14.85 microseconds.

The input voltage range is 0 to +2.5 Vdc, so each 1-LSB transition is worth 2.5 Vdc/256, or 0.0098 volts (9.8 millivolts). The input is unipolar, and the output is straight binary.

Three control pins are provided on the ADC0801-4 device: $\overline{CS}$, $\overline{RD}$ and $\overline{WR}$. The $\overline{CS}$ is an active-LOW chip-select, while $\overline{RD}$ is a read, and $\overline{WR}$ is a write. The conversion sequence is initiated by making $\overline{CS}$ and $\overline{WR}$ simultaneously LOW. The $\overline{RD}$ control is also used as an *output enable*. When $\overline{RD}$ is HIGH, the output lines are tri-stated, so they will float at high impedance across the data bus. In this condition, the ADC0801-4 is all but transparent to the Z80 device, and will not load the bus. When $\overline{RD}$ is LOW, however, the output register of the ADC0801-4 is connected to the output pins, so will dump its contents onto the data bus.

A pair of OR gates are used to decode the Z80 control signals $\overline{IORQ}$, $\overline{WR}$ and $\overline{RD}$. Let's review the rules of operation for an OR gate:

1. If either input is HIGH, then the output is HIGH, and
2. If both inputs are LOW, then the output is LOW

Since the control inputs of the ADC0801-4 are active-LOW, we will want the outputs of the 7432 OR gates LOW when the condition that activates the $\overline{RD}$ and $\overline{WR}$ exists. This means that the two inputs of either 7432 must be LOW in order for the 7432 output to be LOW. One input of each 7432 is connected to the $\overline{IORQ}$ signal from the Z80. The remaining input of the 7432 will be connected to either $\overline{WR}$ or $\overline{RD}$. Thus, for an *output* operation, the $\overline{WR}$ and $\overline{IORQ}$ will drop

LOW, so the output of gate-1 will drop LOW for all output operations. Thus, the $\overline{WR}$ control input of the ADC0801-4 will go LOW for all output operations. Similarly, during input operations $\overline{RD}$ and $\overline{IORQ}$ goes LOW, so the output of gate-2 will also go LOW. In that case, the $\overline{RD}$ line of the ADC0801-4 will go LOW.

The chip select line ($\overline{CS}$) is stimulated by the address decoder (7430). The address decoder in this circuit is an eight-input NAND gate. The rules for the operation of such a gate are as follows:

1. If *any* one input is LOW, then the output is HIGH, and
2. All eight inputs must be HIGH for the output to be LOW

In order to select the ADC0801-4, all eight inputs of the 7430 device must be HIGH. The inputs are connected to bits A0 through A7 of the address bus, so the only address which will cause a LOW output on the 7430 is FF_{16}, or 11111111_2. This address is the equivalent of 256_{10}. If any other address is wanted, then place an inverter in series with the appropriate input address line. The idea is to make all inputs of the 7430 HIGH when the correct address is on the line. If we wanted 10110011 to be the address, then we could connect five of the address lines (A0, A1, A4, A5 and A7) directly to the 7430, while the other lines (A2, A3 and A6) must first be inverted before being applied to a 7430 input.

The machine language program which is needed to use this converter will accomplish the following:

1. Start a conversion, it will cause both $\overline{CS}$ and $\overline{WR}$ to go LOW simultaneously, and then be reset HIGH. This is done with the following type of sequence:

   perform a dummy output operation to locate FF(hex).

2. Perform some other task or execute NOP (no operation) commands while awaiting an interrupt request.
3. When an interrupt request is received, the Z80 must perform an input read from port FF(hex).

The reference voltage for this A/D converter is supplied by an LM-336 IC device operated from the +12 volt power supply. If a more stable reference is needed, then use one of the circuits described earlier.

EIGHT-CHANNEL MULTIPLEXED A/D CONVERTER

The A/D converter shown earlier in this book was a single-channel model. Only one analog input can be accommodated. In Fig. 11-13, however, we see a circuit that will accommodate up to eight different analog inputs in one A/D converter. These inputs are not all converted at one time, but are converted in sequence with each other. The eight inputs are selected with the three-bit address word, AD0, AD1 and AD2 according to the following table:

Channel	*AD0*	*AD1*	*AD2*
0	0	0	0
1	0	0	1
2	0	1	0
3	0	1	1
4	1	0	0
5	1	0	1
6	1	1	0
7	1	1	1

The clock signal that is applied to the CLK input of the ADC0808 must be TTL-compatible, which means that it should be between 0 volts and 0.8 volts when LOW, and between +2.4 volts and +5 volts when HIGH. In this particular circuit, the clock signal is generated by a Motorola MC4024P voltage controlled oscillator chip. The control voltage (pin no. 2) input is applied directly to the +5 volts dc power supply, so no control is possible. The operating frequency is set solely by the capacitor C1. The particular capacitor selected will yield a clock frequency in the 500 kHz range.

The +5.00 volt reference potential needed for the ADC0808 is developed in a *Precision Monolithics, Inc.* REF-02 device. This device is an IC reference voltage regulator, and will produce an output potential that is settable via potentiometer R1 to exactly +5.00 volts. You may wish to use some other method of obtaining the reference voltage (of which, several were given earlier in this book). If your accuracy requirement is low, then use a zener diode or an LM-309H IC three-terminal voltage regulator. You may also use any of the zener-diode-operational-amplifier circuits shown earlier, as well as some of the newer IC reference sources (which are not shown here).

Other than the channel selection, the control protocol for this

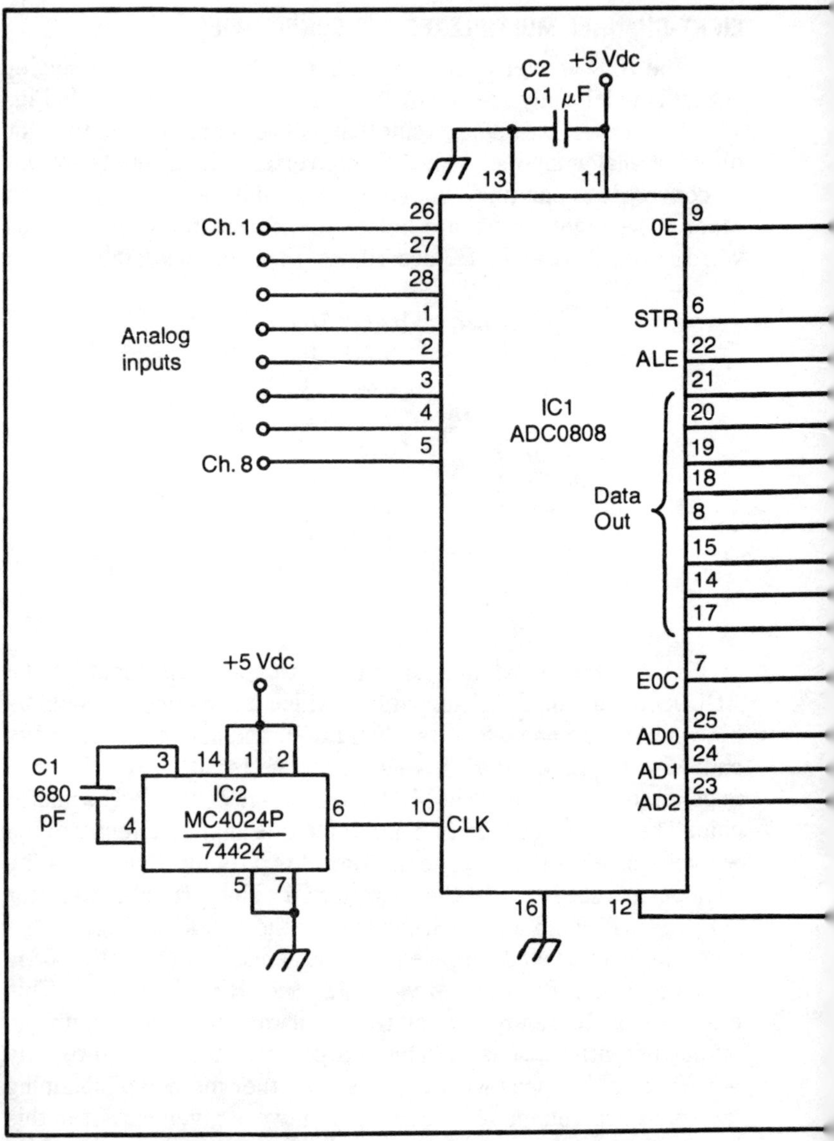

Fig. 11-13. ADC0808 A/D circuit.

device is similar to that described above for the other ADC circuit. We create our start signal by executing an output command. A difference in this circuit, however, is that we must specify the channel being selected. In the other circuit, any "dummy" output, i.e., an output instruction to any unused location, would be suffi-

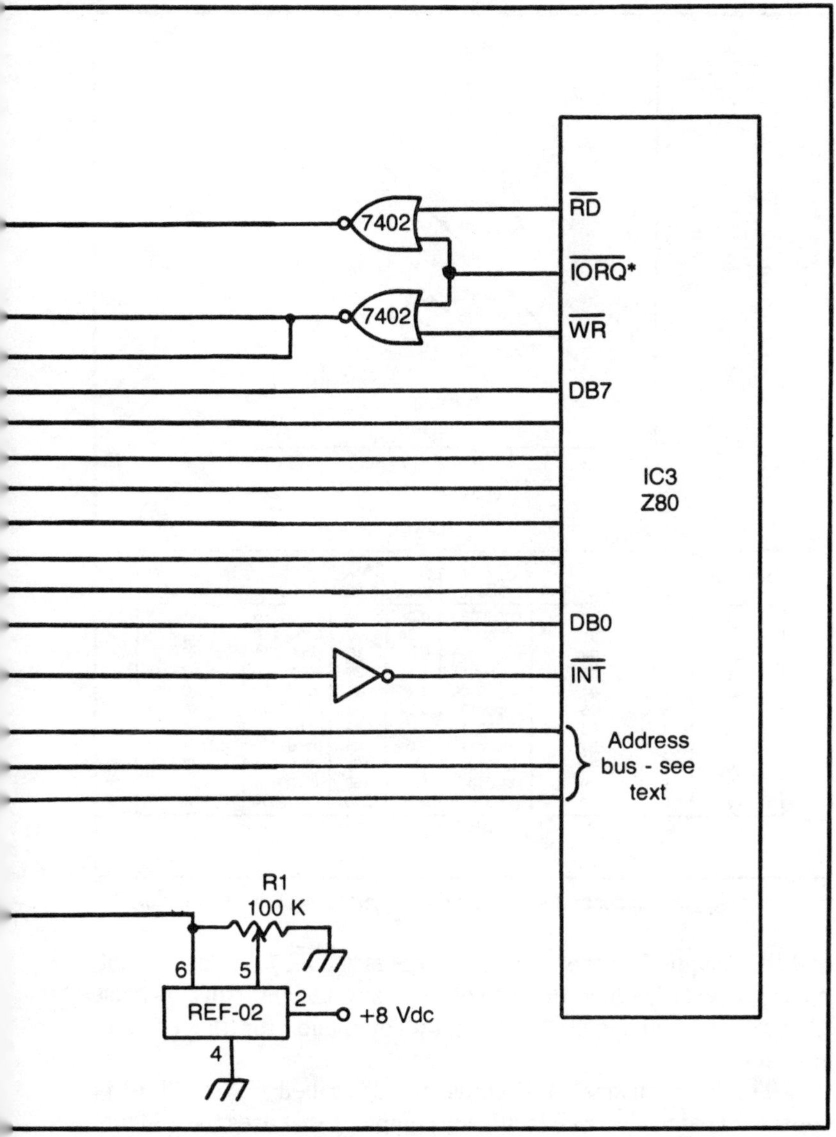

cient. In this circuit, though, we must output to the channel (0 - 7) that we wish to use. Thus, if we want to select channel 2, we would have to output to Channel-2 when generating the start pulse. The ADC0808 will signal the Z80 that it is completed its action by generating an *end of conversion* (EOC) pulse. This pulse is inverted

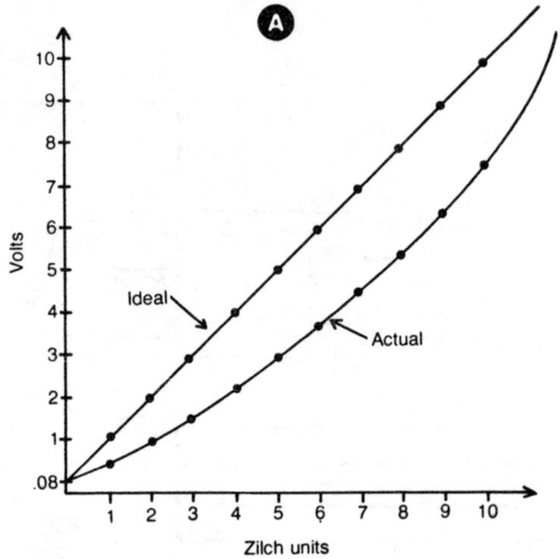

B

Zilch Units	Ideal			Actual			
	Volts	Hexadecimal	Binary	Volts	Scale	Binary	Hexadecimal
0	0	00	00000000	0.08	2	00000010	02
1	1	01	00000001	0.24	6	00000110	06
2	2	02	00000010	0.60	15	00001111	0F
3	3	03	00000011	1.12	28	00011100	1C
4	4	04	00000100	1.76	44	00101100	2C
5	5	05	00000101	2.52	63	00111111	3F
6	6	06	00000110	3.32	83	01010011	53
7	7	07	00000111	4.2	105	01101001	69
8	8	08	00001000	5.4	135	10000111	87
9	9	09	00001001	6.72	168	10100101	A5
10	10	0A	00001010	8.40	210	11010010	D2

Fig. 11-14. A) Transducer calibration curves, B) hypothetical calibration table.

and then applied to the interrupt request ($\overline{\text{INT}}$) line of the Z80 microprocessor. When the Z80 responds to the interrupt, it must generate a *read* signal, i.e., input command, for the channel selected.

The eight-channel A/D converter described in Fig. 11-13 is capable of being interfaced with any computer that uses the Z80, or, directly to a Z80 microprocessor chip. In the former case, the computer will have to make the bus available to the user. The Timex T/S-1000 is an example of a simple, low cost computer that has the bus available to the user on a rear-panel interface connector. All of the Z80 control signals, plus all bits of both the address and data busses, are available on that connector.

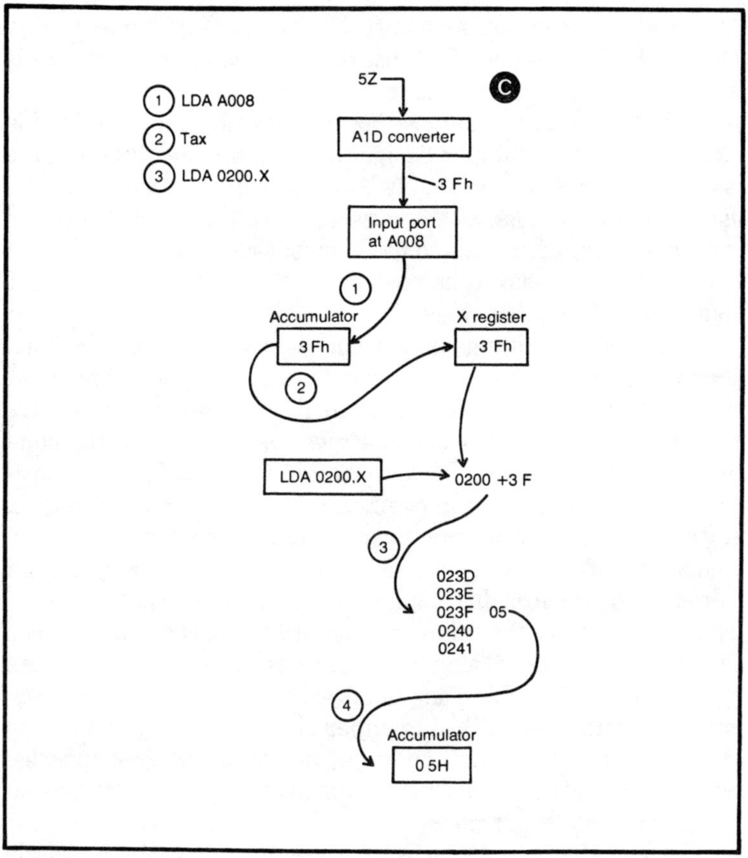

Fig. 11-14. C) programming example.

TRANSDUCER LINEARIZATION BY COMPUTER

A *transducer* is a device which produces a voltage or current output that is proportional to some non-electrical input parameter. Examples are force, pressure (which is force per unit area), tem - perature, displacement and so forth. Ideally, the transducer will be linear, i.e., there will be a straight line relationship between the input parameter and the output voltage. Let's create a situation with a hypothetical unit—the Zilch unit. We do this in order to let people know that the method works with all forms of units. The ideal curve in Fig. 11-14A is what we want to see. In this curve, the scale factor is 1-volt per Zilch (1-V/Z), and the ideal curve is a straight line with a slope of 1 and a Y-axis crossing at 0,0. But the real transducer is not so nice and linear, as shown by the "Actual" curve in Fig. 11-14A. In this case, we find that the curve is not straight, but is bowed. Thus, there is a degree of nonlinearity.

There have been many methods of curve linearization used in electronics instrumentation design. If we can assume that the curve is parabolic, then we can simply build an analog square-rooter and use the output voltage as the quasi-ideal voltage. But that trivial case is not very often open to us. In most cases, the curve will not have some nice easy relationship to the ideal but will have a complicated function instead.

An attempt at fitting odd-ball curves was the diode breakpoint generators used in some instruments. These devices broke the curve into a given number of small linear segments which are represented by the tangent (derivative) of the curve at the point represented. When we consider the circuit linear (at least over some narrow range), but in error, it is merely a case of change of slope. We may thus use an ordinary amplifier to convert the input signal to the ideal case. The diodes used in the breakpoint generator were biased to turn on that amplifier at the specified input voltage. If we wanted to break the curve into, say, 16 different segments, then a total of 16 operational amplifiers, 32 resistors and 32 diodes were needed. In addition, these amplifiers and diodes tended to be temperature sensitive so we had to either keep the temperature in the breakpoint generator constant or provide temperature compensation. Because of this temperature sensitivity, the diode breakpoint generator is no longer popular.

Still another method, which became practical about a decade and a half ago with the advent of special-function hybrids and/or integrated circuits, is to use the input voltage to generate the mirror image of the actual curve, and then sum the two voltages to form a linear curve.

The currently popular method for linearizing transducers is to use a computer. This is especially easy since the computer is low cost, and will probably be used with the transducer or other analog instrument whether or not the curve is linear.

Figure 11-14B shows the table of Zilch units, ideal conditions and the actual input signal. We are assuming an eight-bit microcomputer. Thus, under the scale factor column we will find 0-volts input represented by 0 (i.e., 00000000_2 or 00h), and +10 volts by 256 (11111111_2 or FFh). Let's consider a hypothetical case. Figure 120C shows the case where 5 Zilch units are applied to the A/D converter that serves the computer, and the nonlinear output produces a binary word of 3Fh. We see in the table of Fig. 11-14B that 5-Z is represented by 05h ideally, and 3Fh actually. In the computer we are using, the microprocessor is a 6502, so the input ports are memory-mapped. In this particular case, the input port is located at A008h, so the data is input to the accumulator register through input port (memory location) A008h. The data in the accumulator after this operation is the 3Fh that was developed by the A/D converter.

We use the data input (Fig. 11-14C) from the A/D converter is used to index the computer to look up the correct value (05h) in a look-up table. Since an eight-bit computer is used, the look-up table will consist of 256 bytes, or one entire page of computer memory (00h to FFh). The 6502 (and most other microprocessors) have an absolute indexed addressing mode. The actual address is the absolute address given by the second and third bytes of the code added to the contents of either the X or Y registers. For example, suppose the contents of the X-register are 06h when the following code is encountered:

LDA nnnn	AD	6502 op-code for absolute indexed LDA
nn	00h	low-order address byte
nn	02h	high-order address byte

The absolute address is 0200h, and the effective address is 0200h + 06h, or, 0206h. In the case of a transducer linearization technique, therefore, we can select the absolute address to be the lowest location in the look-up table, for example 0200h. All elements of the table are held in the address range 0200h to 02FFh. Thus, since 5-Z is represented in actual terms by 3Fh, we can store the data 05h at location 023Fh in the look-up table, and then using *absolute,X* addressing mode. The result will be a linearized version of the input data. Although there is a quantization error, the result is a lot nearer the ideal than previously was the case.

The microcomputer linearization method is all but immune to temperature changes, so despite any initial errors, the error will be consistent throughout the temperature range. This is almost as good as having absolute accuracy . . . but not totally.

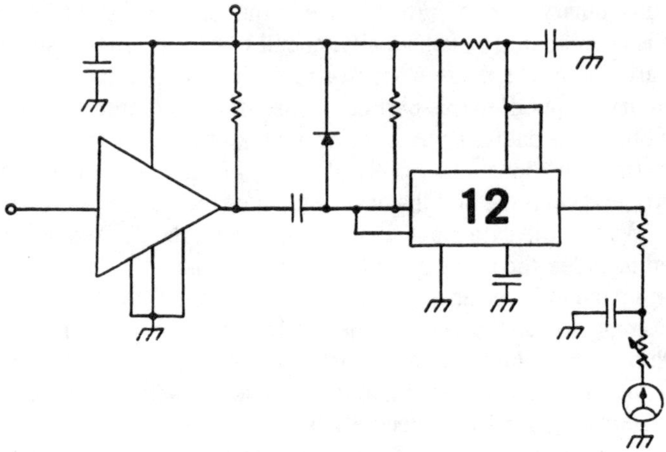

Miscellany

This is the fun chapter for me. In it, we will find a lot of different circuits that are used for a lot of different applications. There is little rhyme or reason for the organization of this chapter, it is somewhat a "stream of consciousness" treatment. These circuits have been culled from a lot of years of building electronic devices. I hope that you will find them useful.

OSCILLOSCOPE SWITCH

Some years ago the main oscilloscope for most users was a single-channel job that could only display one waveform at a time. Modern oscilloscopes, however, use two traces so will display two waveforms at a time provided that they are time-locked (super-fancy and very expensive models can display two non-time-locked waveforms!). The neat trick which these 'scopes use is not two electron beams, but a switching system which will apply first one waveform and then the other to the CRT deflection plates. Figure 12-1 shows an example of an electronic switch that will permit the display of two audio signals on the same oscilloscope. If your oscilloscope is already a two-channel job, then one of these circuits can be used to make each channel into a two-channel system, thus providing up to four channels of analog waveform display capability.

The basis for this circuit is the persistence of the CRT phosphor. The light emitting phosphor coating does not extinguish immediately after the electron beam flies past, but instead, will "per-

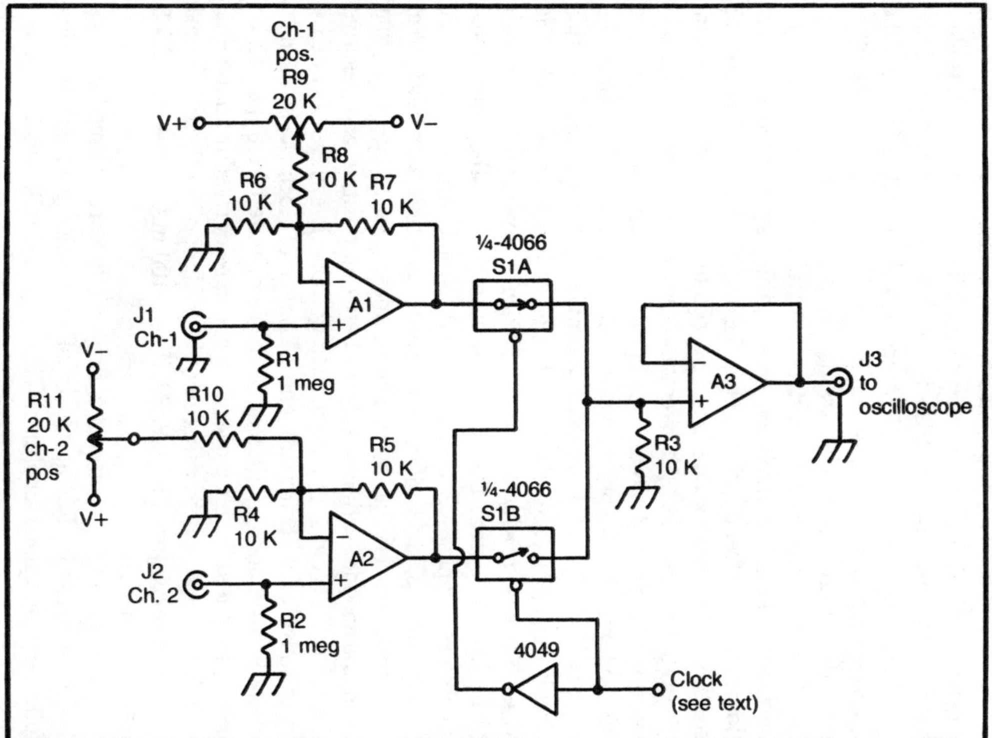

Fig. 12-1. Oscilloscope switch (2-channel).

sist" on the screen for a very brief instant. If you switch the signal being apply to the CRT deflection plates at a fast enough rate, then the persistence of the CRT will keep the waveform looking continuous, even though it is in fact chopped.

The switches in this case is a CMOS 4066 device, which contains four independent switches in one package. With a single 4066 package, therefore, we can build two of these circuits, thus making our oscilloscope into a four-channel instrument. The 4066 switch will be *turned on* when the control electrode (identified in Fig. 12-1 as a circle on the body of the switch) is LOW, and will be *turned off* when the control electrode is HIGH.

The control electrodes are fed such that one will be HIGH when the other is LOW, and vice versa. The inverter (4049) is used for this purpose. Switch S1B is fed from the input side of the inverter, while the control electrode of switch S1A is fed from the inverter output. Since the two ends of the inverter are complementary, the control electrodes are driven out of phase.

The clock frequency must be several times higher than the maximum frequency which you expect to display. For most applications, the clock frequency will be in the neighborhood of 100 kHz, or so. Since the oscilloscope is not synced to the chopper waveform generated by the clock, it is unimportant whether or not the frequency is accurately known. The clock must be a square wave, and can generate from any of a variety of circuits that fall under the rubric of *astable multivibrator*. The proviso is that the clock for this circuit must be compatible with the input levels expected by the 4049. The reason why we use this particular inverter is to permit it to be operated from the same power supply as the operational amplifiers. Typically, the op amp will operate from ±12 volt dc power supplies, and CMOS devices are compatible with this type of power source. The clock, therefore, must be a square wave that swings back and forth across zero between positive and negative limits. A TTL clock is not suitable because it swings between 0 volts and +3.2 volts (or so). The TTL type of clock can be used *if*, both the 4049 and the 4066 devices are operated from +5 volt dc power supplies. Such a power supply can be derived from the +12 volt dc power supply by using an LM-309H, or other three-terminal IC 5-volt regulator. The LM-309H is a 100 mA device in a TO-3 transistor package.

The output signals from the two switches are summed across a 10-kohm resistor, R3, at the input of a unity gain noninverting follower operational amplifier. This amplifier will buffer the electronic switch output. The frequency response of the circuit may well depend upon the operational amplifier selected. If a 741-family device is used, then the frequency response will be limited. The 741 is known as "unconditionally stable" because its frequency response is clubbed to death such that no unpredicted oscillations can occur.

Each channel of the electronic switch has its own input amplifier stage. The gain of this stage is two, so you must adjust the scale on the oscilloscope vertical input channel accordingly (at least in your mind). If you want to make each channel a gain of ten, so that at least the input attenuator markings are numerically correct, then adjust R6 and R4 according to the equation:

$$A_v = \frac{10\ K}{R} + 1$$

or, rearranged to give us *R*,

$$R = \frac{10\ K}{A_v - 1}$$

Substituting for A_v:

$$R = \frac{10\ K}{10 - 1}$$

$$R = 10\ K/9$$
$$R = 1.1\ \text{kohms}$$

Unless we had *individual* position controls, the two signals chopped by the electronic switches would simply superimpose on each other. With this circuit, though, there are a pair of position controls, one for each channel, R9 and R10. These potentiometers would normally be mounted on the front panel of the electronic switch in order to permit individual adjustment of the instrument.

In some cases, the action of the position control will be far too much, especially at lower signal ranges. We are assuming here that only the higher ranges of the vertical attenuator are used. A way to avoid this problem is to use lower value resistors for the potentiometer, and then series connect between the ends of the potentiometers and the V+/V− power supplies fixed resistors. Alternatively, we can increase the value of the resistor between the potentiometer wiper and the summing junction at the inverting input of the operational amplifier.

The input of each channel is fixed, and no attenuators are used. The input resistance is 1 megohm, and the configuration of the circuit in the noninverting follower. A pair of BNC jacks are normally used for the input connections, but you can use the same type of connector as is on your oscilloscope if you want to use the same probes and accessories.

60 Hz NOTCH FILTER

The ac power mains in the United States operate at a frequency of 60 Hz. Since these lines are seemingly everywhere, especially where electronic instruments are used, there is a large electrical and magnetic field with a 60 Hz frequency all over the environment. If you doubt this, then take an oscilloscope probe, and hold the tip. Observe what waveform appears on the screen of the CRT! It will be a distorted 60 Hz waveform. The same neat trick can be done with a sensitive ac voltmeter, and you will find that the signal has a substantial amplitude. One old (and probably only moderately accurate) rule of thumb tells us that the average signal-carrying wire will pick up at least 10 mV of 60 Hz signal for each foot of length. If the signal is low-level then we may find that it is all but obscured by the 60 Hz interference.

Unfortunately, the 60 Hz frequency used by the power lines, and which causes so much interference, is also right in the middle of the band where many electronic instruments work. Medical instruments typically have a frequency response of 0.05 to 100 Hz, and many other scientific instruments are also thus limited. Even instruments with wider bandwidths will include 60 Hz, so are susceptible to 60 Hz hum interference. What is needed, therefore, is a filter that will wipe out only 60 Hz and no other frequency. Such a filter is called a *notch filter.* Figure 12-2 shows a typical modest example.

The notch filter shown in Fig. 12-2 is an active filter based on the operational amplifier. The filter section of the circuit is a twin-T network consisting of R1 - R3 and C1 - C3. The component values used in this network must be matched. No filter is perfect, i.e., will take only the design frequency and none other, but is actually taking out the design frequency and several frequencies near the design frequency. The "Q" of the filter is defined at the ratio of the design ratio to the bandwidth:

$$Q = F_c/F_{bw}$$

Where:

Q is the quality factor, and nondimensional
F_c is the design center frequency, in hertz
F_{bw} is the bandwidth, in hertz

In the case of Fig. 12-2, the filter is designed to have a "Q" of 4, and a center frequency of 60 Hz, so the bandwidth is:

$$F_{bw} = F_c/Q$$
$$F_{bw} = 60/4$$
$$F_{bw} = 15 \text{ Hz.}$$

If the filter is symmetrical, then the bandwidth will be distributed approximately equally around the center design frequency, so the −3 dB points will be ±7.5 Hz above and below 60 Hz. The "notch," therefore, will fall off rapidly from 52.5 and 67.5 with a deep notch of about −40 dB at the center frequency.

The components used in this circuit, especially in the twin-T network must not only be matched, but be also accurate. There are many circuits where the component values are not terribly important, so long as the components are equally matched—this is *not* one of them. It is important to use the exact values, or performance will be deteriorated. The resistors can easily be bought as precision types, while the capacitors are a little more difficult. It is necessary

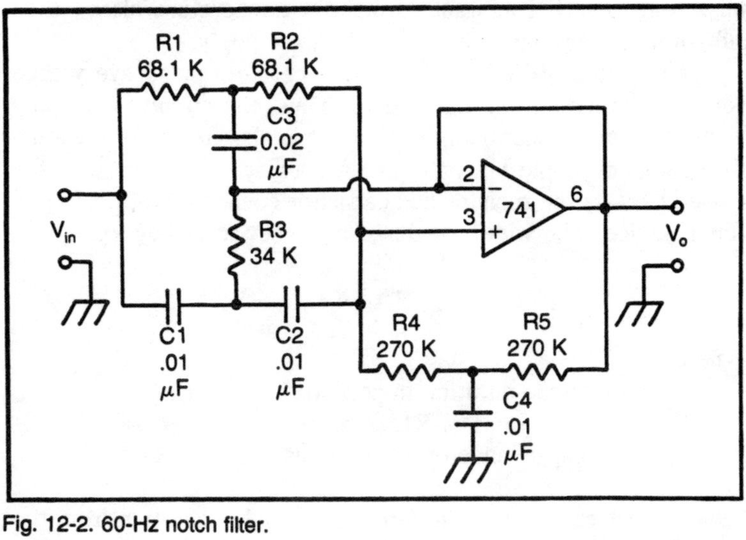

Fig. 12-2. 60-Hz notch filter.

to measure the capacitances in order to find exactly 0.01 μF and 0.02 μF units. You can, incidentally, use a pair of 0.01 μF units, identical to C1 and C2, connected in parallel for 0.02 μF. Precision capacitors are available, but they are hard to locate in most cases.

LIGHT/TEMPERATURE-OPERATED SWITCHES

There are times when we need some warning when a certain condition exists. For example, if the temperature either exceeds or drops below some certain limit, or, if light appears and/or disappears. The latter, incidentally, is often used for intruder alarms. A light beam will be directed at a photocell on the opposite side of a passageway. If an intruder enters, then the light beam is interrupted and the alarm sounds. The project shown in Fig. 12-3A is the sensor circuit for such an alarm.

The sensor transducer can be either a photoresistor (as shown in Fig. 12-3A), or, alternatively, a thermistor. A photoresistor is also known as a photocell, and is capable of varying its resistance with changes in light intensity. In most cases, the greater the light intensity, the less the resistance. You will have to check the specifications for the particular cell that you choose for this project, and set R2 accordingly. A thermistor, on the other hand, will vary its resistance with changes in temperature. Both negative and positive temperature coefficients are available (thermistors were described earlier in this book). In this project, however, we will use only photoresistors. If you want to make a temperature alarm, then substitute a thermistor, and vary R2 accordingly.

The action of Fig. 12-3A is to generate a pulse every time someone interrupts the light beam. Thus, the circuit is not only useful for burglar alarm applications, it is also useful for counting the number of people who enter or leave a space. The duration of the pulse is set by the resistor and capacitor connected to the 555 IC timer device. The width, or duration, is set according to:

$$t_{sec} = 1.1 R_1 C_1$$

Where:

t_{sec} is the pulse duration in seconds
R_1 is the resistance of R1, in ohms
C_1 is the capacitance of C1 in farads

In most cases, we will determine a pulse duration from the specific application, or, will pick one completely arbitrarily. The

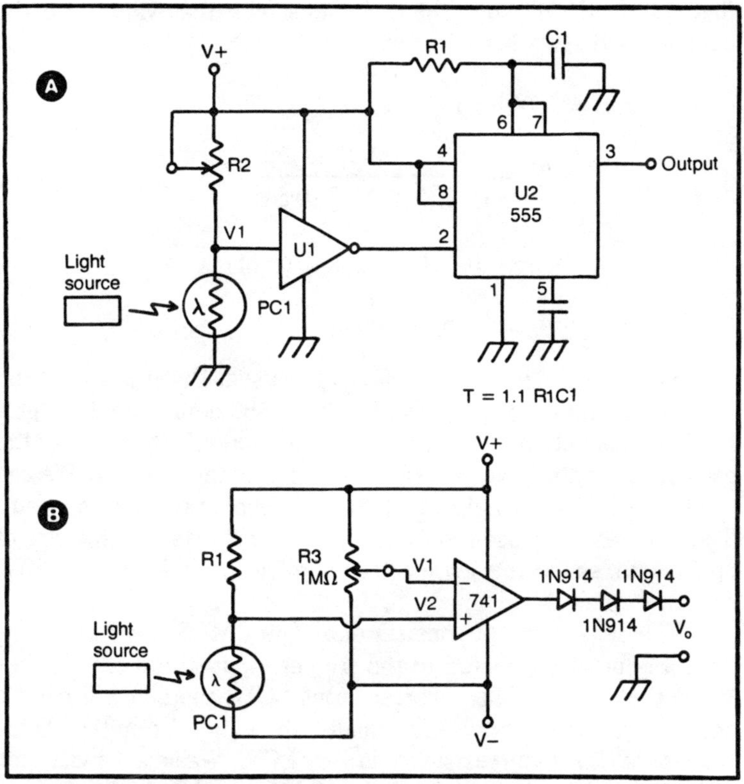

Fig. 12-3. A) Light activated trigger, B) alternate circuit.

usual procedure, then is to select R1 and C1 values that will meet the time specification. Since there are far more easily available "standard" resistor values than capacitor values, we will generally *select* a capacitor value quasi-arbitrarily, and then work the equation to find a value of R. The equation can be rewritten as follows:

$$R_1 = \frac{t_{sec}}{1.1\ C_1}$$

Where:

All factors are the same as above

Example

Suppose we need to make the pulse width 1 milliseconds. Let's find a resistor and capacitor value that will generate this time duration. First, let's select a capacitor value. A "gut feeling" tells us that 0.001 μF is about right. So let's plug that value into the equation and see what falls out:

(Note: 1 mS = 0.001 s.)

$$R_1 = \frac{0.001 \text{ s}}{(1.1)(1 \times 10^{-9} \text{ farads})}$$

$$= 0.001 / 1.1 \times 10^{-9} \text{ ohms}$$

$$R_1 = 909{,}090 \text{ ohms}$$

This value is possible, and could be used. But it is also at the high end of the range permissible for the 555 device. So our "gut-feeling" was only partially correct. If we increase the value of the capacitor, then we can decrease the value of the resistor. We can use 0.01 μF, for example, and then a 90.1-kohm resistor is useful. The timing of the pulse is only occasionally important in this type of application, so use the nearest "standard" value (e.g., either 820 K or 1 megohm).

The sensor circuit consists of a 4049 CMOS inverter circuit with its output connected to the trigger input of the 555 device. When the output of this inverter drops LOW, the 555 will trigger and issue one output pulse. Normally, the light falls on the photoresistor (PC1), so its resistance is very LOW. We select a value for R2 that is HIHG with respect to the resistance of PC1 when light is falling on the light-sensitive element. The voltage at V1 is always:

$$V1 = \frac{(V+)R_2}{R_2 + R_{PC1}}$$

Where:

$V1$ is the voltage at the input of the 4049 inverter

R_2 is the resistance of R2

R_{PC1} is the resistance of the photocell

(Note: R_2 and R_{PC1} are expressed in the same units.)

The idea is to select value of R2 such that V1 will be in the range considered a logical-LOW by 4049. If the circuit is built as shown, then this potential is ½(V+). In this case, we will need a value for R2 that will be greater than twice the resistance of the photoresistor when light is falling on it. If the "lighted" resistance is 10 kohms, for example (a common value), then this resistance should be more than 20 kohms. A reasonable margin of safety is to make R2 a 50 kohm potentiometer, and then adjust the sensitivity to trigger the circuit at the light intensity demands by the application. In most cases, the potentiometer will be either a trimmer potentiometer for printed circuit mounting, or a screwdriver adjusted panel-mounted potentiometer.

A second version of this circuit is shown in Fig. 12-3B. This circuit will produce a constant-level output anytime it is triggered. The "heart" of the circuit is an operational amplifier (741-family device) that is operated as a *voltage comparator.* So what's a "comparator?" A comparator was once described as an ***amplifier with too much gain and differential inputs.*** If the voltages applied to the two inputs are equal, then the output will be zero. But, if the two input voltages are not equal, then the output will be either very high or very low (close to the maximum permitted output voltage, which is a little below the V+ and V− voltages). Thus, if V1 is ***greater*** than V2, then the output of the operational amplifier is negative. If the situation is reversed, i.e., V1 is less than V2, then the output will be very positive. In this case, we have three series-connected diodes at the output circuit in order to prevent the negative output from getting to the V_o terminal. In this case, therefore, the output will be zero when V1 $\doteq$ V2, or when V1 is greater than V2, and positive when V1 is less than V2. The values of R1, PC1, and the setting of R3 are adjusted to turn on the output (V_o) only when the light beam is interrupted.

We can also adjust the circuit to turn on at some specific light level. This means that the circuit could be used to turn on or turn off some device at either sunup or sundown.

The order of turn on/off can be changed by reversing the roles of R1 and PC1, or, by reversing the isolation diodes in the output circuit. The latter method assumes, of course, that the circuit that Fig. 12-3B drives will recognize the negative output voltage as the alarm condition.

1-SECOND TIMER

Timer circuits fall into several different categories. One is actually clocks because they produce a train of equally-spaced pulses that can be used to time circuits. Others will produce a pair of pulses separated by a fixed interval, while still others will produce a given binary state (either HIGH or LOW) for a period of time. The circuit of Fig. 12-4 is of the latter type. In this circuit, the output will go HIGH for a period of time after the trigger switch is closed and

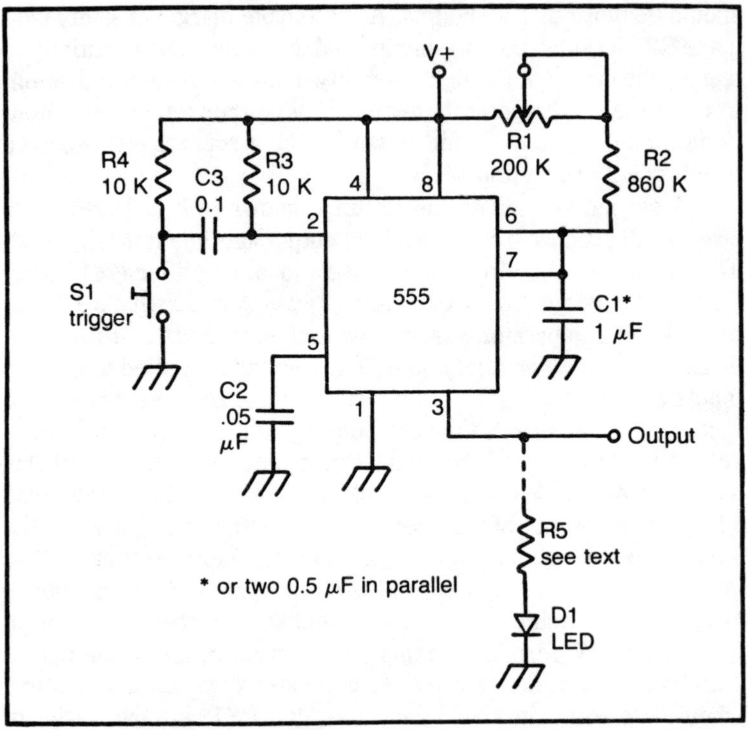

Fig. 12-4. 1-second timer.

will remain HIGH for the time set by the combination RC network R1, R2, and C1.

The duration of the active-HIGH output pulse taken from pin no. 3 of the 555 is set according to the equation:

$$t = 1.1\,(R1 + R2)\,C1$$

Where:

t is the output pulse duration, in seconds
$R1$ is the resistance of R1, in ohms
$R2$ is the resistance of R2, in ohms
$C1$ is the capacitance of C1, in farads

Note in the above equation that the capacitance is specified in farads, but the capacitors which you will use are specified in microfarads or picofarads. The conversion relationship is 10^6 microfarads per farad, and 10^{12} picofarads per farad. Always be sure to convert your units. In the example of Fig. 12-4, the values are: C1 = 1 μF, R1 = 200 kohms and R2 = 860 kohms. In addition, the value of R1 is variable, and can be set to anything between 0 and 200 kohms. With these limits in mind, let's discover the range of timer durations allowable by this circuit.

Condition I: R1 set at zero ohms.

$$t = (1.1)(0 + 860{,}000)(1 \times 10^{-6} \text{ farads})$$
$$t = (1.1)(860{,}000)(1 \times 10^{-6})$$
$$t = (1.1)(0.86)$$
$$t = 0.95 \text{ seconds}$$

Condition II: R1 set at 200 kohms.

$$t = (1.1)(200{,}000 + 860{,}000)(1 \times 10^{-6} \text{ farads})$$
$$t = (1.1)(1{,}060{,}000)(1 \times 10^{-6})$$
$$t = (1.1)(1.06)$$
$$t = 1.17 \text{ seconds}$$

Be sure to know something about the tolerance of resistors selected for R2. The error permitted in timing components is small. A hand-selected resistor for R2 is advisable (ohmmeter selection is accurate enough). Otherwise, if you wish to allow lesser tolerances, reduce the value of R2 to either 750 kohms or 680 kohms, and increase the maximum value of R1 accordingly. If everything were perfect, the formular value of the resistor combination would be 909,090 ohms. You need to select a value for R1 that will permit the total resistance to go past this value, preferable to at least 1.1 megohms. Thus, an 860-kohm resistor at R2 would require a maximum value of at least 1,100,000 − 860,000 ohms, or, 240 kohms (250 kohms is a standard value).

The 555 integrated circuit timer chip used for this project is capable of sourcing 200 mA on HIGH output, and sinking 200 mA on LOW output. In this case, the active condition is HIGH, so we want to turn on an indicator (shown in the schematic as optional) for the period when the output is HIGH. We can do this by connecting the light-emitting diode (LED D1) to the output line through a current-limiting resistor, R5. The value of the resistor used as R5 must be such that it will limit the current to 15 to 20 mA at the V+ voltage selected for this circuit. Since the 555 will operate from supply voltages between +4.5 and +18, this resistance can vary considerably. Below are tabulated values selected for 15 mA at several standard voltages normally used for V+ in small projects

R5 value for 0.015 ampere in LED

5-volts	330 ohms
6-volts	400 ohms
9-volts	600 ohms
12-volts	800 ohms
15-volts	1000 ohms

Of course, it is not necessary that you stick closely to these resistor values. I would select the nearest standard value. For example, for the 6-volt power supply, which requires 400 ohms, I would select 390 ohms.

The circuit used in Fig. 12-4 is called the "monostable multivibrator" or "one-shot" configuration for the 555 timer IC. In a one-shot, the circuit will produce one and only one output pulse of fixed duration and amplitude when the trigger pulse is received. The output is monostable, meaning that it has but one stable state (in this circuit LOW) to the unstable state (i.e., HIGH) for a period of time defined by the equation discussed above. If the one-shot circuit is nonretriggerable, then the input will not respond to further trigger pulses until the output pulse duration is expired. In some older circuits, there was also a short refractory period after the expiration of the output pulse when the trigger pulse would not affect the output. This period was needed to recover the original "waiting to be triggered condition," usually a capacitor discharge operation. The 555 contains an internal discharge transistor which dumps the charge in C1 rapidly when the duration is over, so this refractory problem is minimized, if not zero.

If the one-shot is retriggerable, then the output pulse duration is extended by time *t* every time a trigger pulse is received. We will discuss the retriggerable version of this circuit in Fig. 12-5, momentarily.

Triggering in Fig. 12-4 is applied to pin no. 2. The protocol for the 555 device is to bring pin no. 2 down to a voltage equal to ⅓ (V+) or less. Normally, pin no. 2 is held at V+ by the 10-kohm resistor, R3. This pull-up resistor is attached between pin no. 2 and the V+ power supply. One end of capacitor C3 is also connected to pin no. 2. The other end of the capacitor is connected to another pull-up resistor. R4 and to open switch S1. When pushbutton switch S1 is pressed, the contacts close allowing the end of C3 to become grounded. Previously, the capacitor sides had both been at the same potential, so the capacitor would not charge. When the capacitor is grounded, however, there is a brief instant when the discharged capacitor potential is zero volts and is thus less than ⅓ (V+), so the 555 is triggered. This capacitor charges at a rapid speed determined by the values of R3 and C3. With the values selected for Fig. 12-4, for example, the voltage will reach ⅓ (V+) in 0.0004 seconds, or 400 microseconds.

It is often difficult to obtain accurate capacitors for timer applications, especially in the range shown in Fig. 12-4. Most electrolytic capacitors, which is the type of capacitor normally found in 1-microfarad values, have a tolerance of −20 to +100 percent, which directly affects the error of the timer. We can overcome this problem in any of three ways:

1. Hand-select C1 from a pile of several capacitors to find the one that is closest to 1 μF; or,
2. Use a pair of 0.5 μF capacitors in parallel, selected from 5 percent or 10 percent capacitor types; or,
3. Reduce the value of C1 from 1 μF to 0.1 μF, while increasing the values of R1 and R2 accordingly (e.g., use 2 megohms for R1 and 8.6 megohms for R2), and make the capacitor one of close tolerance.

Figure 12-5 shows the circuit for a retriggerable version of the one-shot timer circuit. A retriggerable monostable multivibrator *will* respond to further input trigger pulses, where the nonretriggerable multivibrator will not. The duration of the output pulse following a trigger pulse is *t*. If a trigger pulse is received, the output will go HIGH for time *t*, and then drop LOW again. If the one-shot receives another trigger pulse before *t* expires, then the timer is retriggered and the output pulse will remain HIGH for a time *t* from the instant that the second pulse is received. The total time, then, will be *t* plus the expired time of the previous trigger period. The total period is not 2*t*, but *t* plus the unexpired part of the previous *t*, which will be a fractional value.

An NPN transistor is shown in Fig. 12-5, and this capacitor is used to dump the charge in capacitor C1, permitting the 555 to respond to the additional trigger pulse. When S1 is closed, the trigger pulse is applied to the bases of Q1 and Q2. This causes both Q1 and Q2 to turn on hard, effectively grounding their collector terminals. For Q2, this effect is identical to the effect of grounding one end of the capacitor with S1 of Fig. 12-4. For capacitor C1, the

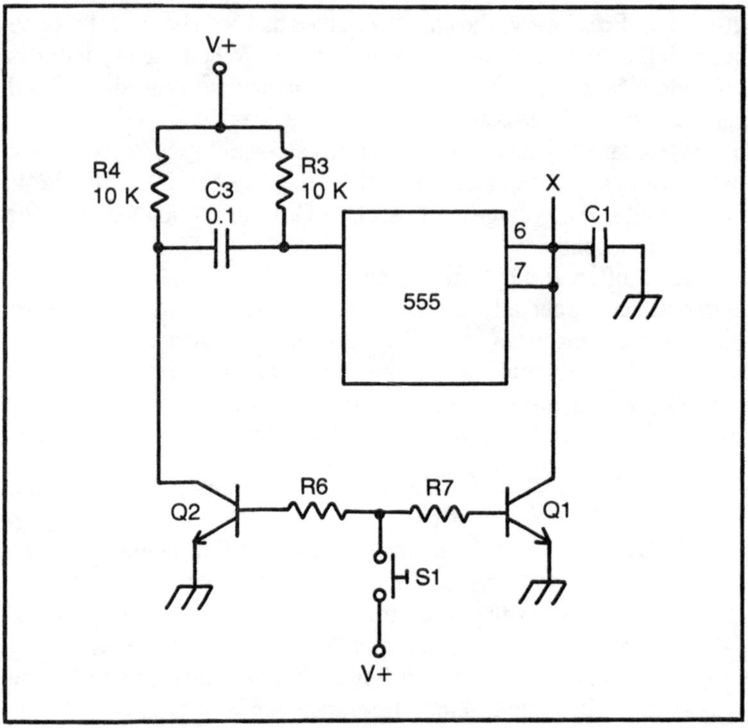

Fig. 12-5. Retriggerable 1-second timer.

effect is to place its "hot" end at ground potential and thereby reseting the one-shot stage.

An application of the retriggerable monostable multivibrator is in alarm circuits. It is often the case that we can build the circuit such that it will alarm when a pulse train disappears. For example, suppose we want to build a burglar/intruder alarm that will not permit an intruder to shine a flashlight into the photocell and thereby "blinding" it. We could pulse the light source at a certain frequency, say 10 Hz. This can be done with a mechanical or rotary shutter, or, electronically in some kinds of light source (e.g., an LED). The output of the photocell, therefore is a pulse of 10 Hz. A zero-crossing detector or Schmitt trigger could be used to clean up the photocell pulse and make it into a square wave. This square wave is then used in place of switch S1 to turn on the bases of Q1 and Q2 in a circuit like Fig. 12-5. As long as the trigger pulses were received, then the retriggerable monostable would remain triggered (i.e., output = HIGH). But, if an intruder passed through, or, if he attempted to fool the alarm by shining a fixed brightness light source into the photocell, then the trigger pulses would go away and the output of the one-shot will go LOW again. The idea is to retrigger the circuit before it "times out," and to let it time out only when an alarming condition exists. This circuit was used by a friend of mine who designed a respirator alarm. If the patient takes a breath on his own, or, if the respirator "breathes him," then a pulse is generated. The duration of the one-shot is selected for the maximum period between breaths that the doctors claim is "safe." If the patient continues breathing, then the one-shot never times out, but if the patient times out, then the alarm times out and lets the staff know that something ickky just happened.

The circuit in Fig. 12-5 is only a partial circuit. All other components which were used in Fig. 12-4 are also used in this circuit.

1000-kHz CRYSTAL MARKER OSCILLATOR

A crystal marker is an oscillator of a fixed, standard frequency that is used in receivers and in radio receiver alignment to find fixed frequency locations. In most cases, we will use either 100-kHz or 1000-kHz marker oscillators. The harmonics of the 1000-kHz marker will appear every 1 MHz up the frequency spectrum. This will permit the operator to align the receiver dials according to the crystal oscillator.

The oscillator circuit in Fig. 12-6A is a transistor Colpitts oscillator in which the LC tuned circuit is replaced with a piezoelectric crystal element. We can tell that the circuit is a Colpitts oscillator because of the tapped capacitor feedback network, C1 and C2. These capacitors are silver mica (DM-15 or equivalent) in order to overcome drift due to temperature changes.

The oscillator element is a transistor. This transistor can be any high-frequency NPN type that will operate into the high-frequency region. In most cases, use a device with a gain-bandwidth product of 100 MHz or more. A likely candidate from the replacement transistor lines (HEP, ECG, etc.) are those intended for VHF or UHF oscillator circuits, as in TV receivers or FM broadcast receivers.

The output signal is taken across resistor R2, by way of capacitor C3. The function of C3 is to block the dc component appearing across resistor R2, while passing the rf component (1000 kHz).

The output signal is applied to the input of a TTL Schmitt trigger that will create a harmonic-rich square wave from the almost-sine wave. Since this integrated circuit is a TTL device, the supply voltage must be +5 volts dc. In order to reduce the 9 to 12 volt dc potential that operates the oscillator to +5 volts for the 7414 device we must use an LM-309H three-terminal IC voltage regulator. This voltage regulator is packaged in the TO-5 transistor case, and will produce an output of +5 Vdc at a current of 100 milliamperes.

Temperature stability is a critical concern in marker oscillators. The frequency will vary somewhat with temperature in any case. The problem is partially overcome by using silver mica capacitors for C1 and C2, and a good quality ceramic for C3. In one circuit, which was popular some years ago, we can use a differential capacitor to select from a pair of temperature dependent capacitors. In normal disc ceramic capacitors, the temperature coefficient will be NPO (for near-zero temperature change of capacitance), Nxxx for a negative temperature coefficient, as expressed in parts per million per degree centigrade—PPM/° C. By using an N750(−750 PPM/° C) and a P1500 (+1500 PPM/° C) capacitor in the circuit of Fig. 12-6B, we can select the temperature coefficient that exactly counterbalances the drift of the crystal oscillator frequency. The trimmer capacitor in this case is a differential capacitor in which the two sections are mechanically linked to make one increasing in capacitance as the other decreases in capacitance. The total capacitance of the pair in parallel will remain constant, so this circuit will permit us to "crank in" the degree of temperature coefficient required.

For the circuit of Fig. 12-6, the alignment can be performed in one of two ways. Both require the temperature to stabilize as much as possible, so the oscillator should be allowed to operate for ten minutes or so before the adjustment is made. The first method uses a digital frequency counter. These instruments are now very low in cost, so are found in more shops than previously was the case. Adjust trimmer C3 with a nonmetallic screwdriver until the output of the oscillator is as close as possible to 1000 kHz.

The second method of alignment is to zero-beat the output signal against the standard frequency station of WWV or WWVH. These stations operate on 5, 10, and 15 MHz, and have a very accurately known frequency. Adjust the trimmer for zero-beat on the highest WWV/WWVH frequency which will produce a strong signal in your area.

The higher the WWV/WWVH frequency that you select for comparison with your crystal marker oscillator, the more accu-

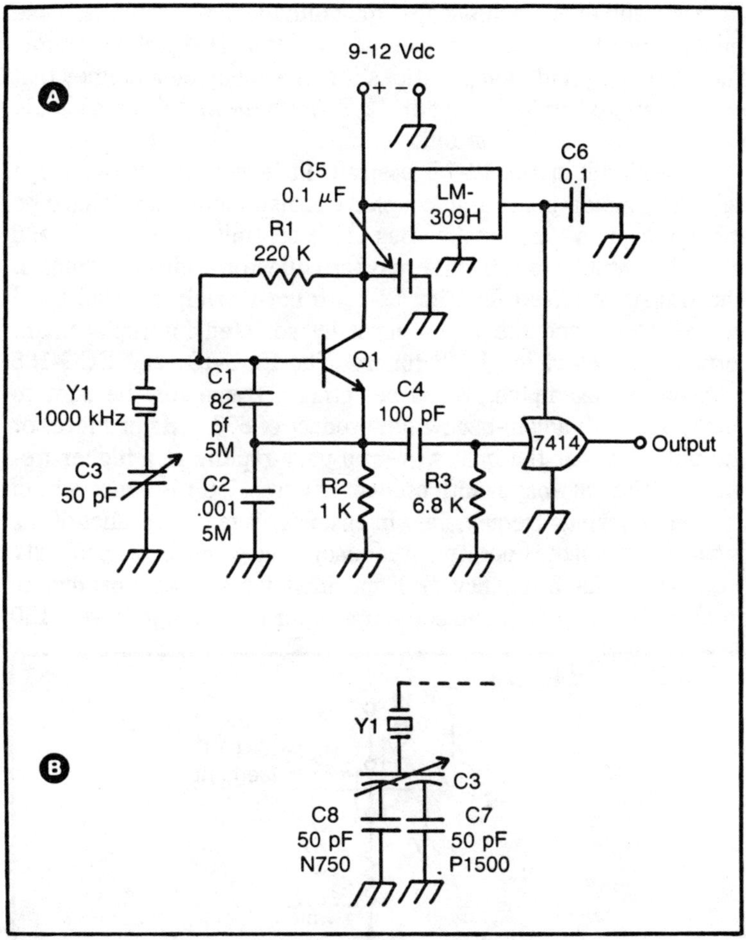

Fig. 12-6. 1000 kHz crystal calibrator.

rately you will be able to set the marker oscillator. The reason for this is that the WWV/WWVH signal will zero-beat against the 5th, 10th, or 20th harmonic of a 1 MHz signal, depending upon which frequency is selected. The human ear can permit us to set the frequency by the zero-beat method to as close as 40 Hz (plus or minus your hearing acuity). If you set the frequency at 15 MHz, then this error is reduced by 40/15, or to 2.7 Hz.

WIDEBAND AMPLIFIER

There are occasional uses for a wideband amplifier that will provide gain as high as 150 MHz. Such an amplifier could be used as the front-end of an oscilloscope, or, voltmeter. We could also use the amplifier to increase the output level of an rf signal generator. There are a myriad of applications for an ac-coupled amplifier that will operate to 150 MHz. Figure 12-7 shows the circuit for a simple amplifier that will fill the bill.

The circuit in Fig. 12-7 is basically little more than a two-stage cascade common-emitter ac-coupled transistor amplifier. This type of stage applies the signal across the base-emitter terminals, and takes the output signal from across the collector-emitter terminals. The transistors used for Fig. 12-7 are not terribly critical, but I would recommend the types normally selected for replacement purposes in television UHF tuners. The ECG-107 and ECG-108 sylvania are examples. When selecting a transistor, be sure to obtain one with a gain-bandwidth product of 800 MHz or more, or you will find that the gain will drop very rapidly at a higher frequency. The gain-bandwidth product of a transistor is a measure of its gain at various frequencies. In essence, it is the product of the current gain and the operating frequency. The gain will drop to unity (i.e., 1) at the frequency (F_t) specified for the gain-bandwidth product. Hence, when you see a transistor specified to have a 150

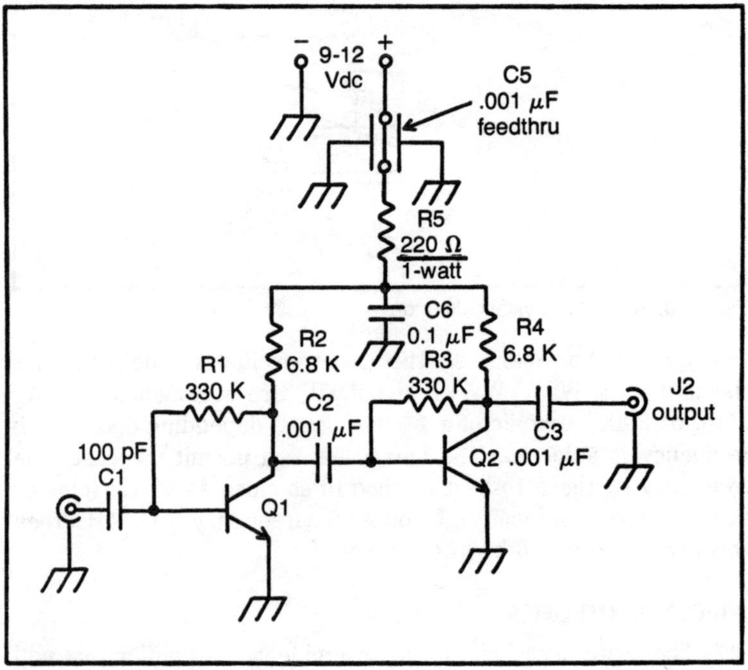

Fig. 12-7. Wideband amplifier.

MHz "F_t," you cannot amplify anything at a frequency for the current gain dropped to one. At higher frequencies, the gain drops off rapidly to zero. In order to have any substantial gain at 150 MHz, therefore, we need a transistor that will operate at 150 MHz with some current gain remaining. That is why we specify a gain-bandwidth product in the 800 to 1000 MHz region for this project.

In both stages of the circuit, the transistor bases are biased from the collector, which is a cheap 'n' dirty method of obtaining a little negative feedback to stabilize the circuit.

The input, output, and coupling capacitors are all disc ceramic types. On these capacitors, be sure to break off the ceramic that dripped onto the capacitor leads when the capacitor was made. This will allow you to place the capacitor as near the printed board or wiring board as possible. This amplifier will operate well into the VHF region, so VHF techniques must be used. If the capacitor leads are too long, then it is possible that lead inductance will combine to either force the circuit into an unwanted resonance, or, to add a series inductive reactance that will interfere with proper operation of the circuit.

"Good VHF techniques" also means keeping all lead lengths as short as possible, and using tight layout in order to reduce the effects of stray inductance and stray capacitance on the circuit at VHF.

The decoupling used in this circuit is also consistent with good VHF practices. There are two power supply decoupling capacitors used here. The 0.1 μF capacitor (C6) in the V+ power supply line is a disc ceramic and is mounted inside the metal box that houses the amplifier. This capacitor must be used in the same manner as described above for the coupling capacitors. The other capacitor, C5, is a *feedthrough* capacitor. This coaxial form of capacitor is in the form of a small threaded stud with connectors on either end. The feedthrough capacitor decouples anything trying to get into or out of the metal box. The 220-ohm resistor in series with the V+ line not only serves to current limit the power supply, but also to increase the effectiveness of the decoupling capacitors by forming what are, in effect, RC low-pass filters.

Some builders prefer to further improve the decoupling by placing small ferrite beads on the power supply lead external to the metal box. The function of these beads is to act as if they were rf chokes in the VHF region, while not affecting the dc at all.

The input and output jacks should be some form of rf-compatible jack, but "RCA" phono-jacks are sufficient. The important thing is to make the system compatible with your other instruments. If you want to use this device with an oscilloscope that uses BNC connectors, then by all means use BNC connectors in order to keep the patch-cord system easy to understand. In certain ham uses, you will want to use SO-239 "UHF Connectors," while in some commercial applications you will want Triax or N-series coaxial connectors.

RF VOLTMETER PROBES

It is difficult to measure rf voltage directly, but we can use a dc voltmeter in order to measure the peak rf voltage. If we know anything about the rf waveform, especially if it is a sine wave, then we can *compute* the average peak or peak-to-peak values.

Figure 12-8 shows two methods of making an rf voltmeter probe. Both of these methods require a fairly high-level signal in order to overcome the diode junction potential (i.e., more than 0.7 volts peak), but can be made more sensitive by connecting them at the output of the amplifier shown in Fig. 12-7. In that case, the demodulator probe is rectifying the output signal from the rf amplifier, so will measure a lot weaker signal. It is, of course, necessary for you to run a frequency response curve for the amplifier in order to know how much scale factor is involved at the measured frequency. Unless, of course, you are not looking for a specific voltage, only some *maxima* or *minima*.

That last bit in the above paragraph leads us to one possible use for the demodulator probe. I used it in order to measure things like the resonant frequency of an antenna, or, the resonant frequency of an LC tank circuit by applying the signal from a generator to the circuit, and then measuring on a dc voltmeter the signal level across the circuit. The parallel resonant tank circuit will show maximum voltage at resonance, while the series resonant type will show minimum signal at resonance. You can also use the method for measuring the electrical length of a piece of coaxial cable. In many antenna measurements or other applications we need to use a half wavelength transmission line. Impedance of the antenna, for example, can only be measured at the antenna end of the coaxial cable, or, at half wavelength multiples reflected back down the line. This is because the reflected impedance will repeat itself at half wavelength intervals. In order to cut a piece of cable we might want to use the old formula

$$L_{ft} = 492/F_{MHz}$$

Where:

L_{ft} is the length in feet for half wavelength

F_{MHz} is the frequency in MHz

But this formula is illusory for the transmission line application because of a little problem called *Velocity Factor* In all wave situations, the frequency and wavelength product is always a *constant*. The "492" constant in this equation is derived from the velocity of light (300,000,000 meters per second) expressed in terms of feet. But in a transmission line, the actual velocity will be less than this free-space velocity by a fractional factor known as the velocity factor (V). The actual equation is:

$$L_{feet} = \frac{492\ V}{F_{MHz}}$$

Where:

L_{ft} is the length in feet

F_{MHz} is the frequency in megahertz

V is the velocity factor, 0 to 1

Typical values for "V" range from about 0.6 for old-style "regular" coaxial cable, to 0.8 or so for some of the newer dielectrics. This factor, however, appears to vary somewhat from one lot of coax to another, and is nowhere near standardized, despite the proclamations of the specifications sheets. We can, however, measure the velocity factor on a sample piece of coaxial cable from a roll, and then use that value for the same roll, or, in some cases where the manufacturer is known to have good quality assurance, for all of that manufacturers product was bought under the same stock number. This latter isn't guaranteed, of course, because the manufacturer may well elect to change some part of the process and thereby change the actual velocity factor experienced.

Start the measurement with a piece of coaxial cable cut according to the formula above, plus about 5 percent. Use the velocity factor indicated by the manufacturer, and then add 5 percent to the total. Then connect in parallel together at the input end of the cable the voltmeter probe and an rf signal generator; the other end of the coaxial cable is shorted. Adjust the signal generator frequency up and down to find the null. When the null frequency is known accurately, then you can measure the physical length of the coax in feet, and find the velocity factor by rearranging the formula above to the form following:

$$V = L_{ft} F_{MHz}/492$$

Alternatively, if you have some frequency in mind, and don't accurately know the velocity factor, then we can use the voltmeter probe to find the exact length required for half wavelength. The procedure is the same as above, except that we preset the desired frequency, and then use a small finishing nail or large needle to short the coax every few inches (marking where we do the dirty deed) until the null is located. The circuit action here is that the zero-ohms short circuit impedance is reflected back down the line every half wavelength so will cause the null in rf voltage at the input end when the coax is exactly half wavelength.

The simple circuit in Fig. 12-8A uses a single UHF-type diode (1N914, 1N4148, etc.) in a half-wave rectifier circuit. A 10 pF capacitor couples the input signal to the rectifier. This small capacitance is needed in order to prevent loading of the circuit being measured by the probe. Following the rectifier diode is a ripple filter, similar to those used in dc power supplies, except that the component values are selected for higher frequencies.

The circuit in Fig. 12-8B uses a half-wave bridge circuit for the rectifier. This circuit is essentially similar to the previous circuit, except for the extra diode, and a little rearrangement of the filter components.

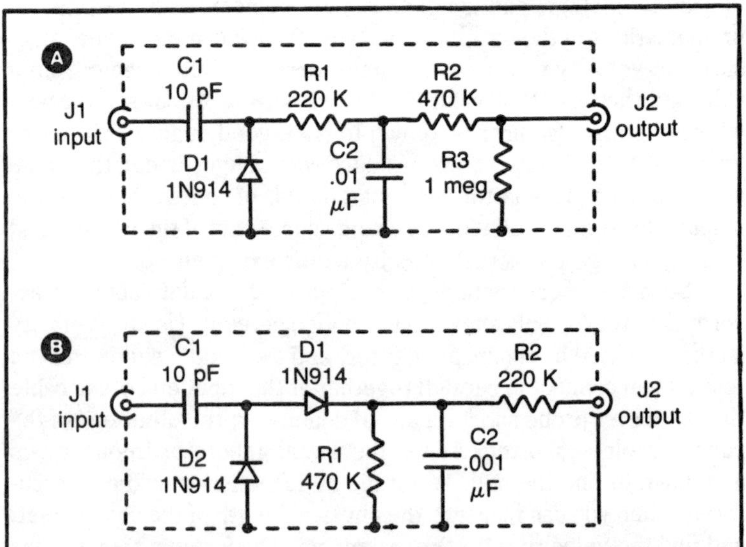

Fig. 12-8. Rf demodulator probes A) half-wave circuit, B) full-wave circuit.

In both cases, the probe circuitry must be built inside of a shielded box, with connectors on both ends. The connectors should be of the same type as is used by other coaxial connectors in your systems. If the equipment that this probe will be used with uses BNC connectors, then BNC should be used for this product. Similarly, if SO-239 UHF connectors are used on the equipment, then use the same for here.

Some people build this circuit inside of a "universal" voltmeter or oscilloscope housing. These housings used to come with a small skinny piece of perfboard so that the user could build his or her own oscilloscope probe for special applications. Eico and Heath both used to sell the probes.

NOISE GENERATOR CIRCUIT

A "noise generator" circuit will produce a semi-gaussian rf signal that contains many thousands of frequency and phase components. It can be used to test many different electronic circuits. In tuned circuits, for example, we can spot resonances with the simple facility to find the frequencies at which noise nulls occur.

There are several ways to build noise generators, some of them are quite complex digital circuits, while others are simple analog circuits such as Fig. 12-9.

The main noise generator element in Fig. 12-9 is the zener diode, D1. All zener diodes operate in the avalanche mode, which is inherently noisy. In fact, this noise is a problem in other circuits. For us, however, it is the root source of our signal. The noise created by the avalanching process in the zener diode is semi-gaussian (in other words, nearly *white noise*) in nature, so can be used for this purpose. The noise signal from the diode is coupled into the base of transistor Q1.

All three transistors in this circuit are in the common-emitter class of amplifier, and are ac coupled. You may use MPS-6531 Motorola transistors or their equivalent. If you plan to use a transistor from a replacement line, then let me recommend that you find out the currently-recommended Motorola HEP replacement type number. Because of the economics of semiconductor making, they don't set up a separate assembly line for the replacement types as opposed to the regular types, but merely take the regulars off the assembly line and mark them with the replacement number. If you use a Motorola replacement for a Motorola transistor, therefore, you are more likely to obtain the *exact* transistor rather than some near-placement.

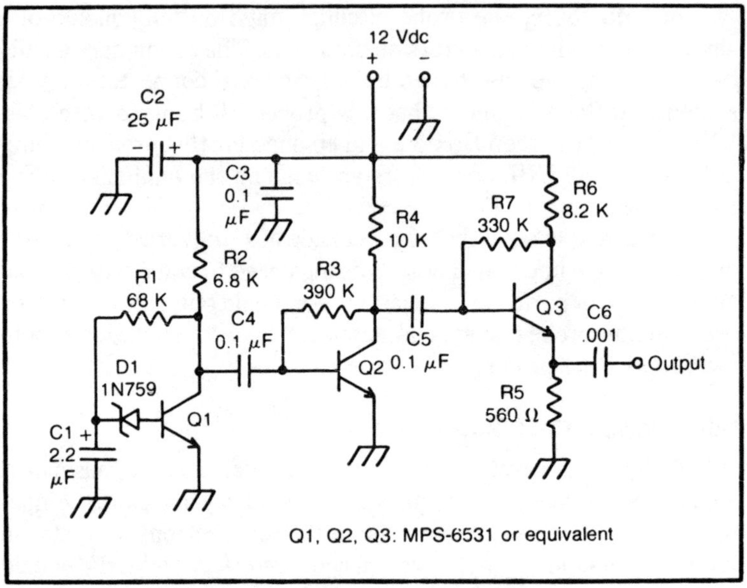

Fig. 12-9. Noise generator.

The amplifier circuit used in this project, can, incidentally, also be used without the noise as a wideband amplifier in a manner similar to Fig. 12-7.

TUNABLE NOTCH FILTER

Earlier in this book we discussed the uses of the 60 Hz notch filter. This filter is used to remove 60 Hz interference from an analog waveform. There are, however, other frequencies that may at one time or another cause interference with another analog signal. A Morse code operator, for example, will often hear two signals in the passband of his or her receiver that are separated by only a small amount. We can use a tunable notch filter to remove one of these signals so that the operator can listen to only one at a time. Another radio application for short wave listeners is to remove heterodynes, which are whistles caused by beating two AM or CW signals against each other (this problem doesn't exist to such a great extent in Single Sideband operations). When, for example, there are two AM stations operating on frequencies that are, say, 5 kHz apart, the resultant beatnote will be 5 kHz—which is well within the passband of the audio amplifier. Normal AM broadcast station channels assigned for use in the United States are spaced 10 kHz apart on the 550 to 1650 kHz band. Thus, if we have two adjacent channel signals in the passband at the same time, the resultant heterodyne is 10 kHz. For example, if a 1390 kHz and a 1400 kHz signal are in the receiver's passband at the same time, then a beat note of (1400 – 1390 kHz) = 10 kHz will exist. This situation is not likely during daylight hours because the Federal Communications Commission takes pains to keep from assigning adjacent channels in the same market. But at night, we find that out of town stations will "skip" in from a great distance, and the heterodyne interference mounts. Normal AM receivers include an RC network called a "Tweet filter" in the detector circuit. This filter consists of resistors and capacitors that form a low-pass audio filter with a −3 dB point of 5 kHz or so, and a −6 dB/octave attenuation slope. Thus, the attenuation at 10 kHz will be approximately −9 dB relative to the mid-band (usually 1 kHz) properties of the circuit.

The notch filter transfer characteristic is shown in Fig. 12-10A. The frequency is graphed along the horizontal axis, while the attenuation is along the vertical axis. Attenuation is the opposite of amplification. An attenuation of 0 dB is none at all, while a degree of attenuation preceded with a "+" sign is amplification. The loss is indicated by a "−" sign.

At frequencies close to the design frequencies of F_n the attenuation is on the order of −25 dB, while at points removed from F_n the attenuation is 0 dB, indicating that no attenuation at all takes place. In actual fact, there might be a small degree of attentuation at frequencies away from F_n and that is called *insertion loss*. Thus, the notch filter will take out only those frequencies that are close to F_n and will transmit all others. In the case of the 60 Hz interference filter which we built earlier, the purpose was to excise 60 Hz from the signal and leave all other frequencies.

The "Q" of the filter determines how wide is the band around F_n that will be removed, i.e., the sharpness of the filter. Ideally, we would want as high a "Q" as possible so that no other information is lost except the interference. But in the real world, we have to recognize that high-Q filters tend to *ring*, which means that they will oscillate for a brief period after they are hit with a fast rise-time input signal. Thus, if we intend to apply fast rise-time signals, then we would want to avoid the high-Q filter in order to not create more interference than we take out!

The basis for most popular notch filters for the audio range is the twin-tee RC filter, as shown in Fig. 12-10B. This is a variation the circuit shown in the earlier 60 Hz circuit, and involves replacement of the two resistors ("R") with a single potentiometer with a

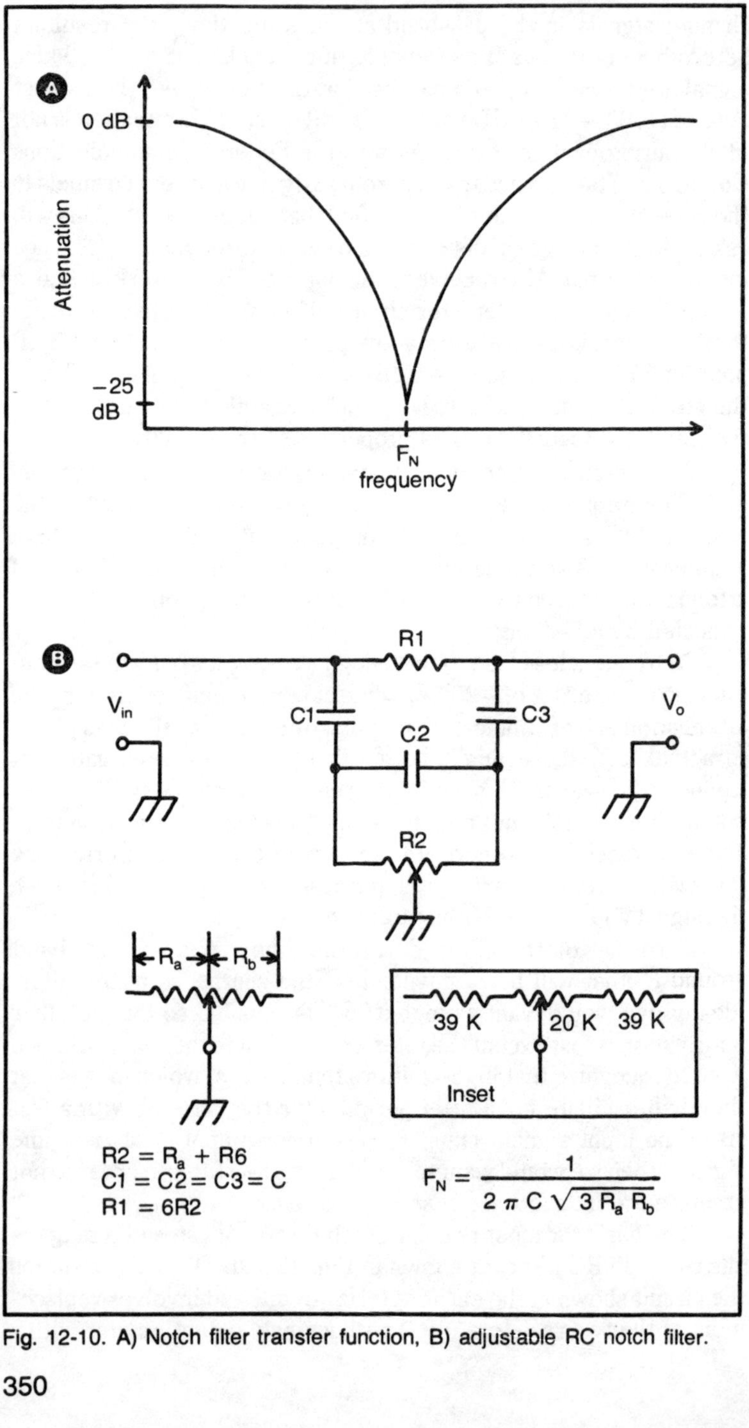

Fig. 12-10. A) Notch filter transfer function, B) adjustable RC notch filter.

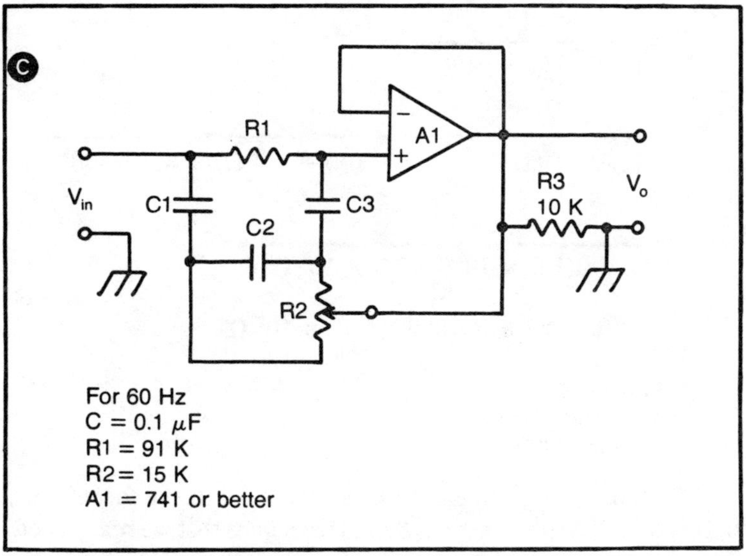

Fig. 12-10. C) active adjustable notch filter.

value of $2R$. We can vary the notch frequency by varying the resistance of $R2$.

The resistance of $R2$ is broken into two portions, labeled R_a and R_b in Fig. 12-10B, which are used in the following equation to find the resonant frequency:

$$F_n \quad \frac{1}{2\pi C\sqrt{3 R_a R_b}}$$

Where:

C is the capacitance of C1, C2, and C3, in farads
R_a is a portion of $R2$, in ohms
R_b is the rest of $R2$ not represented by R_a, in ohms

Assuming:

$R2 = R_a + R_b$
$R1 = 6R2$
$C = C1 = C2 = C3$

Example:

Find the resonant frequency of a notch filter as in Fig. 12-10B, assuming that the capacitors are all 0.0015 μF, and the resistance of R2 is 100 kohms, when the potentiometer wiper is at dead center (i.e., $R_a = R_b$ = 50 kohms).

$$F_n = \frac{1}{(2)(3.14)(C)(3R_aR_b)^{1/2}}$$

$$F_n = \frac{1}{(2)(3.14)(1.5 \times 10^{-9}F)((3)(5 \times 10^5\ \Omega)(5 \times 10^4\ \Omega))^{1/2}}$$

$$F_n = \frac{1}{(6.28)(1.5 \times 10^{-9}F)(7.5 \times 10^9\ \Omega^2)^{1/2}}$$

$$F_n = 1/(6.28)(1.5 \times 10^{-9}F)(8.66 \times 10^4\ \Omega)$$

$$F_n = 1/(8.16 \times 10^{-4})$$

$$F_n = 1226 \text{ hertz}$$

The equation given earlier is not the most useful form for us, and must be rewritten in order to find the values that we really need. For example, it is rare that a designer will want to know the resonant frequency of a circuit such as Fig. 12-10B, but instead, he will know the resonant frequency from the application! Also, it is a lot easier to obtain "oddball" values of resistance than capacitance, so we will probably select a capacitor value and then plug it into the equation to find which value of capacitance will yield suitable resistance values for the frequency that we want to take out. In order to make the circuit most useful, therefore, one must rewrite the expression above in the form:

$$R_aR_b = \frac{1}{118\ F_n^2\ C^2}$$

Where:

R_a and R_b are the resistances, in ohms, as defined above
F_n is the frequency of the notch, in hertz
C is the capacitance, in farads, of C1, C2, and C3
118 is the composite of all constants

Notice that the square root radical is missing, so we are left with a simple algebraic expression. The F and C terms, however, must be squared in order to account for removal of the radical. The constant 118, therefore, is a combination of pi-squared, 2^2 (i.e., 4) and the 3 that was under the radical. With this new equation, we can select a design notch frequency, and then find the R_aR_b product which will yield this frequency at some selected value of C.

Example.

Design a notch filter of the type shown in Fig. 12-10B where the notch frequency is 1000 Hz.

1. Select a value of C. Let's pick 0.002 μF (it's arbitrary, but based on experience).
2. Calculate R_aR_b

$$R_aR_b = \frac{1}{118\ F^2\ C^2}$$

$$R_aR_b = \frac{1}{(118)(1000)^2(2 \times 10^{-9})^2}$$

$$R_aR_b = \frac{1}{(118)(10^6)(4 \times 10^{-18})}$$

$$R_aR_b = \frac{1}{4.72 \times 10^{-10}}$$

$$R_aR_b = 2.12 \times 10^9$$

3. Find combinations of resistance that will yield 2.12×10^9. We can do this by selecting a value, and then dividing it into this figure. If we call the selected resistance R_a, then the result will be R_b. For example, let's make R_a equal to 47 kohms (a standard value), R_b is then 45.6 kohms. One way to achieve these resistances is to select precision resistors of exactly the correct values. Another way is to use a pair of 39-kohm resistors at the ends of the potentiometer, and then make the potentiometer resistance 20 kohms. A typical circuit is shown in the inset to Fig. 12-10B.

There is another reason for using the circuit shown in the inset. Note that both versions of our equation use the *product* R_aR_b. This means that the resultant frequency will be zero if either R_a or R_b is zero ohms—which is what happens when the potentiometer is at one extreme or the other. Use of the fixed resistors at the ends of the potentiometer limit the low-end value that will be accommodated, thus limits the frequency range.

An active version of the filter is shown in Fig. 12-10C. In this circuit, the output end of the twin-T filter is applied to the noninverting input of an operational amplifier. The op amp is connected in the unity gain, noninverting follower configuration. The wiper of the potentiometer is connected to the output of the operational amplifier, which is normally at ground potential when the output is zero. The values for a variable 60 Hz filter are shown in Fig. 12-10C. The resistance of the potentiometer is 15 kohms (a standard value).

Some amateur radio operators and short-wave listeners will use a circuit such as Fig. 12-10C to form a preamplifier, and then follow it with a power amplifier that will drive a pair of headphones or a small communications-grade loudspeaker.

OPERATIONAL AMPLIFIER MANIFOLD

An operational amplifier manifold is a device that will permit any user to do wonderful things with operational amplifiers, without having to first know something about electronic construction techniques. We will include the power supply connections and all other connections, except those used by the "customer," inside of the box. The connections that are needed are the output, inverting input, noninverting input and the common. These connections are brought to the outside world by way of binding posts (Fig. 12-11, *J*1 - *J*6).

The case for the operational amplifier is any form of metal case or box. If you have some sort of mainframe, such as those made for the purpose by *Vector Electronics*, then the modules that fit the mainframe can be used for the operational amplifier manifold. In that case, the card-edge connector that fits the modules can bring a common power supply to all modules. In other cases, we will want to include internal batteries (9 to 12 Vdc) for power supplies, as shown in Fig. 12-11. In either case, it is a good idea to include an on-off switch.

The operational amplifier used for the manifold can be selected according to the range of applications. If only the lowest criticality applications are anticipated, then we can use a 741 operational amplifier. If a higher criticality is anticipated, however, a more premium-grade device is needed. The RCA CA3140 is recommended, as are BiFET equivalents.

1-MILLIVOLT CALIBRATOR

There are many applications for a voltage calibrator source. Almost all analog instruments, especially those which depend upon low-cost oscilloscopes or strip-chart/X-Y recorders, require some

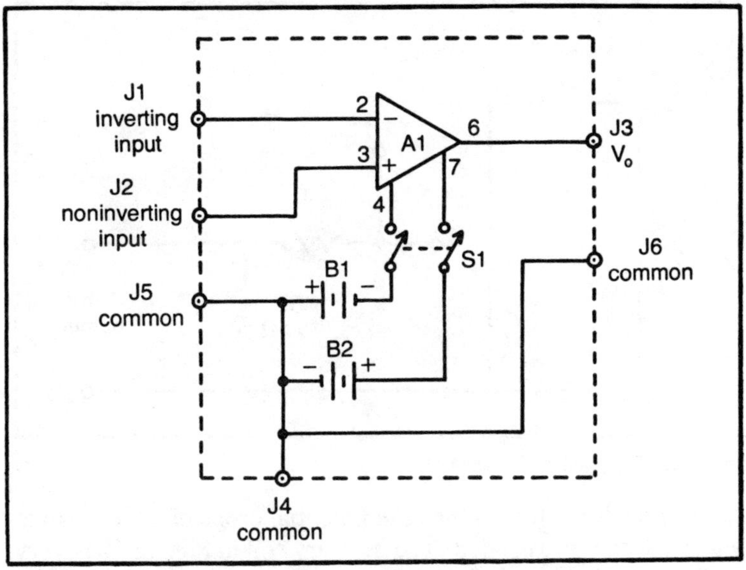

Fig. 12-11. Operational amplifier manifold.

form of amplitude or "span" adjustment. In most cases, the signal calibrator for these devices will be on the order of 1 volt. In a lot of instruments, however, a 1-millivolt source is needed. Most strip-chart and X-Y paper recorders have 1- or 10-millivolt inputs rather than 1 volt. Also, biopotentials amplifiers (e.g., the electrocardiograph and electroencephalograph amplifiers) used in medical instrumentation typically have very low-level signal inputs. The ECG waveform, which is generated by the beating of the heart, has a peak amplitude in the 1-mV region. In testing ECG machines, ECG amplifiers, biopotentials amplifiers and/or hospital "bedside monitors," therefore, we need a 1-millivolt source.

Figure 12-12 shows a simple circuit which is used by a lot of biomedical equipment technicians to test ECG amplifiers. It is simple, very low cost, and can be easily built into the smallest "Pomona Box" for easy portability. The power source is a small mercury instrumentation battery—either HG1 or HG2 will do nicely. Alternatively, one could also use hearing-aid/digital-watch batteries, although the life expectancy will be less. The standard mercury battery (B1) has a potential when new of 1.36 volts dc. The nice thing about mercury cells, which is why they are used for instrumentation purposes, is that the output potential will remain nearly constant over most of its life, and then drops off rapidly as the battery is finally depleted. This is in contrast to the ordinary

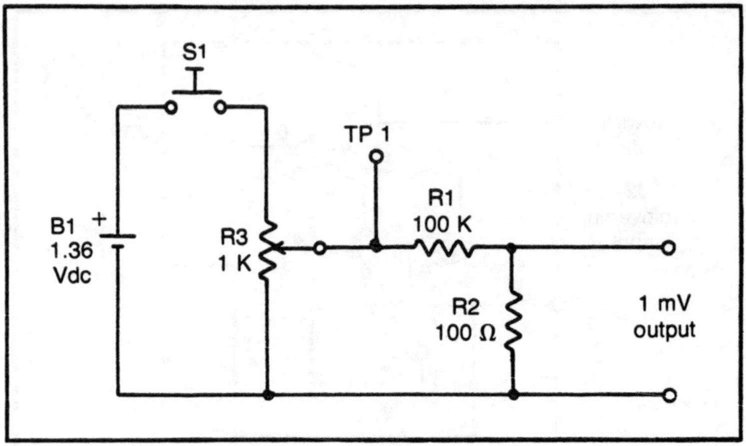

Fig. 12-12. Simple 1-mV calibrator.

carbon-zinc dry-cell, in which the potential drops off at a constant rate as energy is expended. The mercury cell in Fig. 12-12 is very small, on the order of ⅝-inch long and half-inch diameter. You will be able to buy battery holders for these cells at almost any well-equipped parts house/wholesaler.

The "guts" of the calibrator is the resistor voltage divider consisting of R1 and R2. Provided that the output is connected to an impedance of 10 kohms or more (which it will be if used for ECG or other forms of amplifier calibration) the output voltage will be determined solely by the voltage at test point *TP1* and the ratio of R1 and R2. The actual output voltage will always be:

$$V_o = \frac{R2}{R1 + R2} \times V_{TP1}$$

or, since the values of R1 and R2 are fixed:

$$V_o = \frac{100}{100{,}100} \times V_{TP1}$$

$$V_o = 0.000999\ V_{TP1}$$

Since we really want V_{TP1} (because V_o is known to be 0.001 volt—i.e., 1 mV), we can rewrite this equation in the form:

$$V_{TP1} = V_o/0.000999$$
$$V_{TP1} = 0.001\ V/0.000999$$

$$V_{TP1} = 1.001$$

In most cases, we would not adjust the voltage at test point *TP1*, but rather to be exactly 1.00 volts. Of course, if we have the resolution on the adjustment potentiometer (R3) and a meter capable of measuring 1.001 volts, then we would use the capability. But for most readers, only a small and inconsequential error results through the use of 1.00 volts instead of 1.001 volts. The error due to resistor tolerances will be somewhat larger!

The adjustment device is a potentiometer, shown as R3 in Fig. 12-12. This 1 K, linear taper potentiometer should be at least a 10-turn model, and use of a 20-turn model is well advised. For most, the potentiometer will be a screwdriver adjustable type—very few readers will have a use for a panel-mounted adjustment.

Operation of the voltage calibrator is simplicity itself. When you need an output signal, just depress push buttom switch S1. This action will connect the 1.36-volt battery (B1) to the top end of the potentiometer. In the case of biopotentials amplifiers, which are ac coupled, the switch will be depressed and released in a rhythmic action that produces a rough square wave output. Otherwise, the signal will make the ECG amplifier output snap once and then gradually decay. Although that method is proper for testing the low-frequency response of an ECG recorder, it is not useful for examining the signal or setting/testing the gain of the amplifier. In normal ECG practice, the operator will adjust the amplifier gain for a two-division amplitude on a 1-mV square wave.

The resistors used in the voltage divider section of the calibrator must be 1-percent precision types if accuracy is to be maintained. For many applications, including a cheapie checker for amplifiers, ordinary 5-percent carbon resistors will suffice. These should be hand selected for tolerance as close as possible to 100 kohms and 100 ohms (R1 and R2, respectively). Also, be aware that the temperature coefficient of carbon resistors is terrible, so the voltage divider will never be considered precision even if you hand select resistors for precise values.

The ECG amplifiers (as well as most other amplifiers that use a 1-mV input signal level) will succumb to testing with a "single-ended" tester as shown in Fig. 12-12, but only if the unused amplifier input is grounded. If you want to test the amplifier as a differential amplifier, then build the version shown in Fig. 12-13. This device is a pseudo-differential 1-mV signal source. The principle of operation is the same, so will not be expanded here. The big difference is that there is no adjustment for the output voltage. The values are selected so that 1-percent tolerance resistors will yield a differential output signal close to 1 millivolt.

The ground shown in Fig. 12-13 is optional. Most ECG equipment uses a floating common in lieu of a chassis ground common. This is done for safety reasons and should never be compromised.

1-HERTZ, 1-MILLIVOLT CALIBRATOR

The two calibrators shown earlier were manual types. When a push buttom switch is pressed, the 1-millivolt signal appears at the output terminal, and when the button is released the signal disappears. In order to make a square wave, one must alternatively press and release the push buttom switch. Since there are numerous types of amplifiers which have ac coupling (even though frequency responses go down to 0.05 Hz for many biopotential amplifiers!). The most convenient method for testing these amplifiers is to produce a square wave with a 1-millivolt amplitude. For best application, the amplitude and the frequency must be well-known . . . precisely. Figure 12-14 shows a candidate circuit that is based on the same voltage divider that was used in Fig. 12-12.

The power source for this circuit is any battery from 8 to 15 volts dc. Particularly recommended are the 8.2 Vdc mercury batteries that are sometimes used in place of 9-volt transistor radio batteries. For this application, incidentally, the current drain is

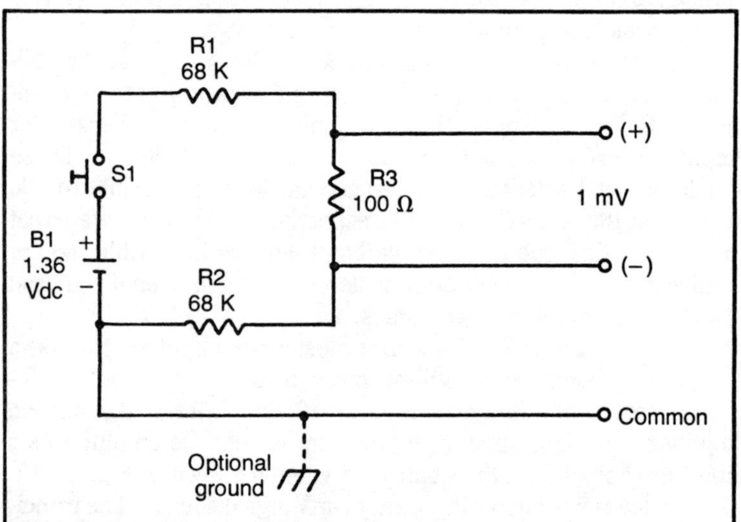

Fig. 12-13. Differential 1-mV calibrator.

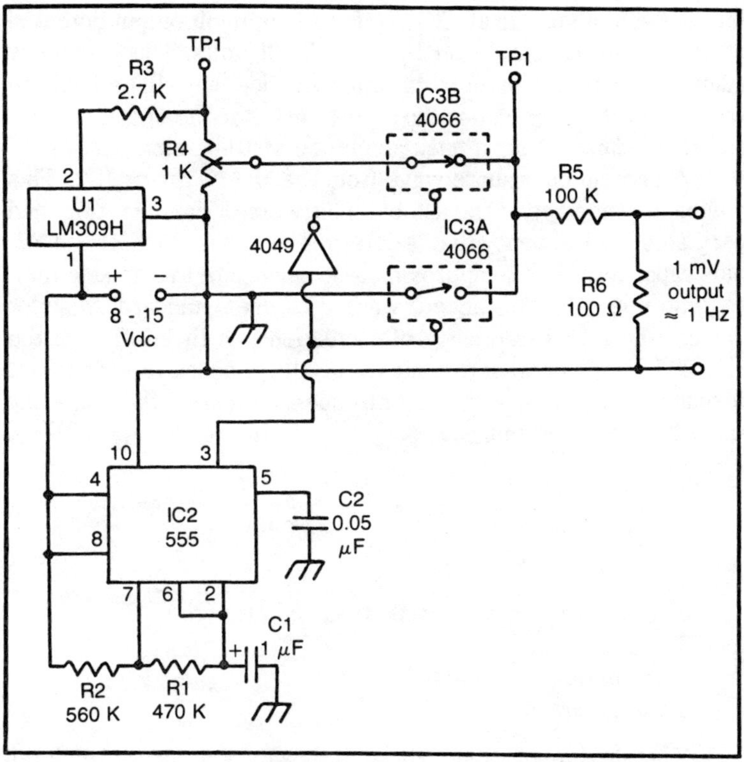

Fig. 12-14. 1-Hz, 1-mV calibrator.

small enough that almost any premium-grade 9 Vdc battery will give satisfactory service. The power supply voltage is reduced to +5 volts ±25 mV by the action of U1, and LM-309H device. The LM-309H is a 5-volt three-terminal IC voltage regulator rated at 100 milliamperes. The "H" suffix indicates that the regulator comes in the "similar-to-TO-5" transistor can, or, a black plastic case. There are only three terminals on the LM-309H: input, output, and common (1, 2, and 3, respectively). The approximately 5-volt dc output of the LM-309H device (U1) is applied to one end of an adjustable voltage divider consisting of R2 and R4. The potentiometer will be used to set the input potential to the precision 1000:1 voltage divider at the output.

In order to simplify our initial discussion, lets assume that the only parts of the circuit which exist are R3, R4, R5, and R6. Assume that the electronic switch IC3B is permanently wired closed, connecting the output of the potentiometer to *TP1*. Like the similar circuit in Fig. 12-12, the output voltage is very nearly 1000 times lower than the voltage at TP1, so for a 1-millivolt output potential we will want to make the TP1 potential 1.00 volts. This is done by adjusting the potentiometer. In actual practice, we will probably be satisfied with using an oscilloscope to monitor the signal at TP1, and then adjust R4 for a peak amplitude of 1.00 volts.

We obtain our square wave from the 555 IC timer, IC2. This device is connected in the astable multivibrator configuration, and oscillates at a frequency that is determined by resistors R1 and R2, plus capacitor C1. The approximate frequency of this astable multivibrator is 1 hertz. Since I wanted to use "standard" resistor values, the actual frequency of oscillation is a little off from the "ideal" 1 Hz. If you want to make the oscillator exact, or, if you want to make it oscillate at some other frequency, then set the frequency according to the formula:

$$F = \frac{1.44}{(R1 + 2R2)C1}$$

Where:

F is the frequency of oscillation, in hertz (Hz)
*R*1 is in ohms
*R*2 is in ohms
*C*1 is in farads

The design of timer circuits is covered in more detail elsewhere in this book, so will not be covered here. Also, for those who feel the need for even more information I can heartily recommend one of my other books: ***IC Timer Handbook . . . with 100 projects & experiments*** (TAB book No. 1290). Just send $9.95 to TAB Books Inc., Blue Ridge Summit, PA 17214.

The output of a 555 IC timer is pin no. 3. This output is zero volts when LOW, and will approach V+ when HIGH. Thus, we have an output that will oscillate back and forth between 0 volts and about 8 volts.

The electronic switches are part of a quad electronic switch, the CMOS 4066 device. Two sections are used (IC3A and IC3B). A 4066 device will exhibit an open switch when the control terminal is HIGH, and a closed switch when the control terminal is LOW. We drive the control terminal of IC3A directly from the output of the 555 timer, while the control terminal of IC3B is driven from the output of a 4049 CMOS inverter that is connected also to the output of the 555. Thus, IC3A and IC3B are driven 180 degrees out of phase with each other—they are complementary outputs.

Since IC3A and IC3B are driven by complementary signals, one will be open and the other closed at all times. Thus, test point *TP*1 will be alternately grounded and then connected to 1 volt. The result is to make the output alternately ground (0 volt) and 1 millivolt. The output is, then, a 1-millivolt square wave.

The square wave calibrator can be easily built into a small aluminum box and is therefore portable. Many biomedical equipment technicians use "test equipment" exactly like Fig. 12-14 in order to "quick check" ECG equipment and other forms of low-level biopotentials amplifiers/recorders.

LIQUID-LEVEL MONITOR

Figure 12-15 shows the basic circuit for a monitor that will give an output voltage proportional to the liquid level in a vessel. The particular scheme shown uses a toilet-bowl float and mechanism,

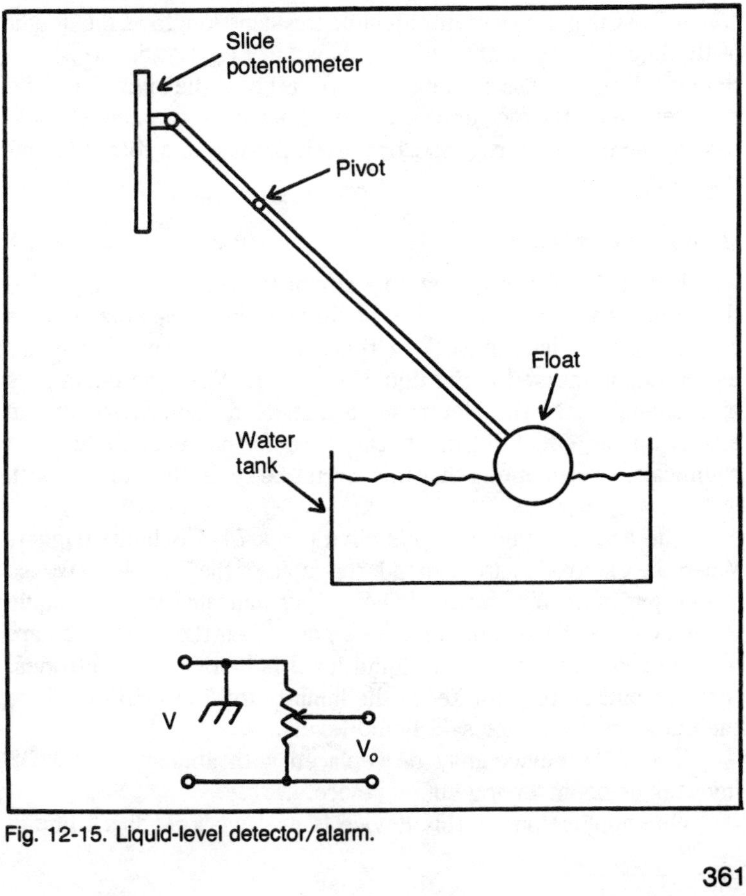

Fig. 12-15. Liquid-level detector/alarm.

but other methods are equally useful. The transducer is a slide potentiometer. Select one with a resistance of 1 kohms (or 10 kohms), and with a linear taper. The longer the "throw" of the potentiometer, the better the result will be.

The potentiometer is connected to the float via a rod that is secured at a pivot. As the float ball rises (i.e., when the liquid level rises), the arm will rotate about the pivot point, so the end connected to the potentiometer will drop . . . changing the resistance. If we can connect the ends of the potentiometer across a voltage V, the output voltage V_o will vary from 0 to V as the float rises from minimum to maximum level.

There should be a large number of alternate methods for making a float vary the resistance of the potentiometer. One alternative is to make the float rise up and down on a fixed rod as the liquid level varies. A second rod permanently affixed to the float will vary the potentiometer. The disadvantage of this system is that the "throw" of the potentiometer must be the same length as the height of the liquid level. In the system shown in Fig. 12-15, the potentiometer throw is less because the pivot is not at the midpoint of the rod. Perhaps in the former case a system of wheels or pulleys would reduce the needed throw and/or permit the use of a rotary potentiometer.

LIQUID-LEVEL ALARM

Figure 12-16 shows a simple circuit for creating a logic level that tells us whether or not the liquid level in a vessel is above a certain point. The "sensor" in this circuit is a pair of ordinary electrodes immersed in the liquid container. When the liquid gets high enough, then the electrodes are shorted. Unless you are an expert on the effects of electrical currents, however minute, on chemicals, I recommend that this circuit be restricted for use with water.

The active element of this circuit is a 7414 Schmitt trigger. When the electrodes are shorted, the input of the 7414 is LOW, so the output is LOW. Thus, a LOW output indicates that the liquid level exceeds the recommended height. When the electrodes are not shorted, e.g., when the liquid level is below the electrodes, then the pull-up resistor keeps the input of the 7414 HIGH. Thus, the output is HIGH for safe liquid levels.

The 7414 device may be replaced with almost any CMOS inverter or noninverting buffer device.

One application for this device is to drive a feedback liquid-

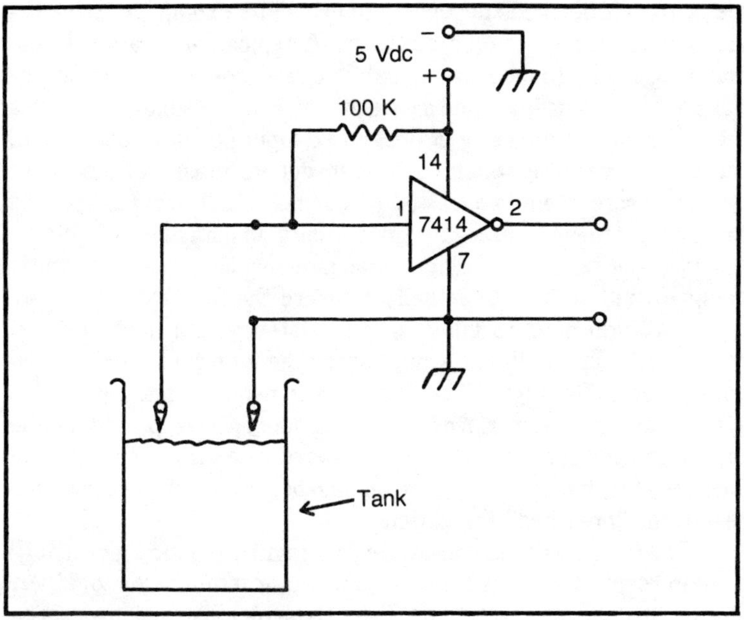

Fig. 12-16. Liquid-level alarm.

filling system. The output of the 7414 can be interfaced through a power transistor to a solenoid-operated water valve. When the liquid level is below the desired point, the electrodes will be uncovered, so the solenoid valve would be turned on. When the water pours into the vessel sufficiently to raise the level to a point where the electrodes are covered, then the solenoid will shut off.

In other cases, we might be satisfied with a simple on/off light to indicate whether or not the liquid level is at the electrode level. This feature can be accomplished with a simple transistor to turn on and off a light-emitting diode (LED), or, an open-collector TTL inverter to do the same job.

One has to be aware, incidentally, that really pure water (e.g., double distilled water) used in some scientific laboratories will not trigger some circuits such as Fig. 12-16. The problem is that water is nearly an insulator, and it is normally the impurities that cause electron conduction.

GAS/AIR FLOW DETECTOR

Detecting the flow of a gas such as air or carbon dioxide is not the easy chore that it seems. If you want to quantify the measurement, then it becomes necessary to use some expensive transducers and related electronic circuitry. A typical respiration transducer, one used to measure actual flow rate, consists of a differential pressure transducer and a wire mesh. The transducer measures the difference in pressure between two input ports, as opposed to the gauge pressure which more transducers measure. The wire mesh is a screen with about 400 wires/inch. As flowing gas passes through the mesh, a pressure drop occurs which is measurable on the transducer. By applying a standard formula, the instrument designer will be able to actually measure the flow rate.

If all you need to know is the existence of a flow rate, or, possibly, its approximate waveshape, then a lot less costly transducer is possible. Figure 12-17 shows a transducer that was used in a medical respirator alarm. This transducer was placed in the exhalation line of the respirator, so will indicate a flow regardless of whether the patient breathed on his/her own, or whether the respirator "breathed" the patient.

The housing of the transducer is a regular plastic "tee-fitting" used in respirator connections to join two or more hoses together. These fittings have a 25-mm (i.e., 1-inch) bore diameter. The

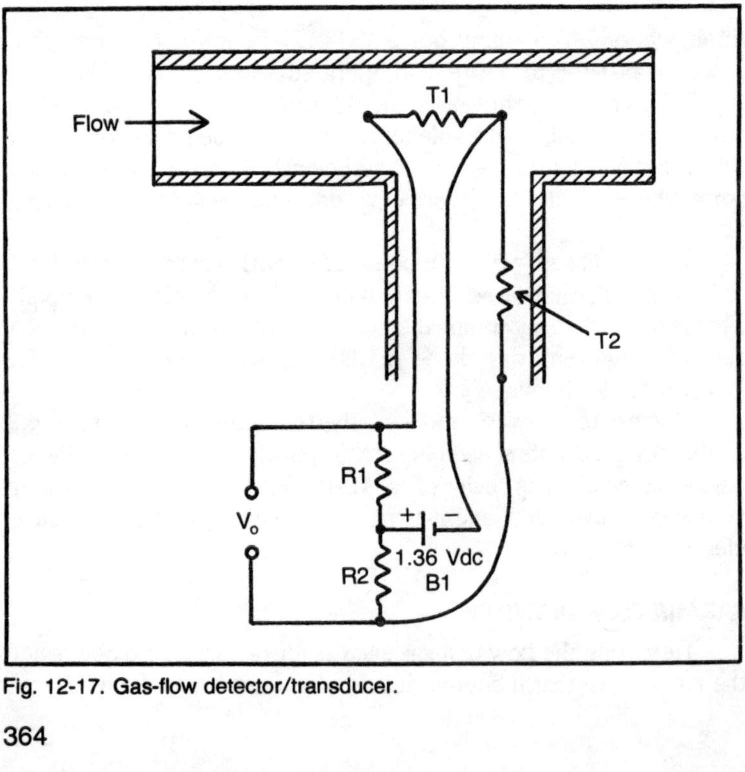

Fig. 12-17. Gas-flow detector/transducer.

transducer elements are fast rise-time thermistors, *T*1 and *T*2. One thermistor, T1, is placed in the airflow path, while the other, T2, is placed in a dead space that is approximately at ambient temperature. The battery voltage is selected to provide a current flow that is barely at the point of self-heating, i.e., the power dissipation that will just barely prevent the internal temperature of the thermistor from rising.

The thermistors are resistors that will change resistance when the temperature changes. These resistances, T1 and T2 are placed in a Wheatstone-bridge circuit with two fixed resistors, R1 and R2. The values of the resistors are selected to exactly balance the bridge when no gas is flowing in the channel. Under this condition, the output voltage (V_0) is zero. As gas flows in the channel, however, the resistance of T1 will change and that tends to unbalance the bridge and produces an output voltage. Of course, if the flow is pulsatile, then the output voltage will also be pulsatile.

The transducer shown in Fig. 12-17 is based on a design in a thermistor handbook, and was used in at least one commercial respirator alarm used in hospitals. It is also useful anywhere that a nonflammable gas is used (the design has not been tested with flammable gases, so don't try it!). It is also necessary that you consider what effects "out-gassing" of the materials used in the transducer design. In the respirator alarm, for example, the epoxy formulation used to hold the thermistors in place had to be a special high-priced material designed for medical applications. Such a material has a very low out-gassing rate, and is non-toxic to humans.

VOLTAGE-TO-FREQUENCY CONVERTER (VFC)

A voltage-to-frequency converter is a device which will produce an analog signal at a frequency that is proportional to an applied voltage. These devices are essentially forms of analog-to-digital converters. They are essentially frequency modulated signal generators which produce a signal that can be converted to a digital signal for use in a converter, or, for other instrumentation applications. An example of a voltage-to-frequency converter, or VFC, is shown in Fig. 12-18. This particular example is an integrated circuit VFC manufactured by Analog Devices, Inc.

The AD537 device used in Fig. 12-18 can operate on dc power supply voltages from +5 to +36 volts dc. It is able to accept input voltages up to ±30 volts, and will output frequencies up to 100 kHz. The linearity of the device (for 10 kHz full scale) is 0.05 percent, and

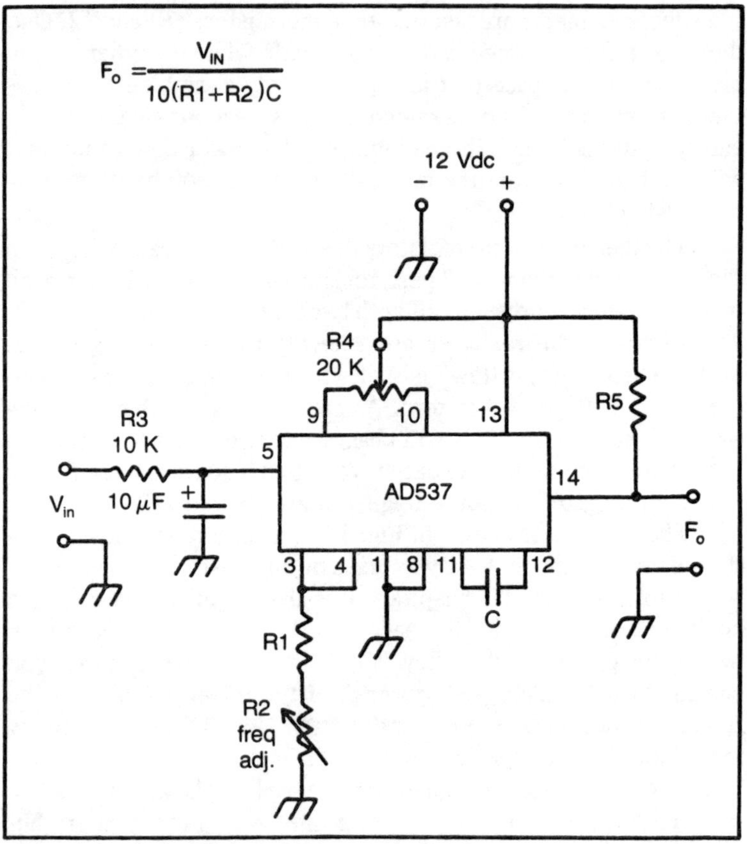

Fig. 12-18. Voltage-to-frequency converter circuit.

the device will operate over at least an 80 dB dynamic range.

The input circuitry is a simple low-pass filter consisting of a 10-kohm resistor (R3) and a 10 μF capacitor. This particular circuit is designed for a positive input voltage. The output frequency is set by the timing capacitor, C, and the series combination of R1 + R2. The following equation is used:

$$F_o = \frac{V_{in}}{10(R1 + R2)C}$$

Where:

F_o is the output frequency, in hertz (Hz)
*R*1 is in ohms
*R*2 is in ohms
C is in farads

In general, the input voltage range and the maximum frequency range will be known to the designers, and will be set by external factors. The input voltage will be a function of the transducer or amplifier used to acquire the analog signal, and can be scaled by the designer where necessary with VFC input amplifiers. The frequency can be up to 100 kHz where the signal is counted on a digital frequency counter or input to a computer that is capable of measuring this frequency. But, if the signal is to be transmitted over a radio channel, telephone line, or into a tape recorder, then the bandwidth may be so limited that only 10 kHz is allowable. Thus, we will know the frequency range, and can then select resistor and capacitor values that will perform this job.

Perhaps it is better to rearrange the above equation in order to produce a version that will be more useful in this type of designing:

$$(R1 + R2)C = \frac{V_{in}}{10\ F_o}$$

Example:

A VFC is required that will represent the voltage range 0 to 10 volts with frequencies up to 10 kHz. Find a combination of R1, R2, and C that will provide this range.

Solution

$$(R1 + R2)C = V_{in}/10F_o$$
$$(R1 + R2)C = (10\ V)/(10)(10{,}000\ Hz)$$
$$(R1 + R2)C = 1/10{,}000$$
$$(R1 + R2)C = 0.0001 = 10^{-4}$$

The next step will be to select a capacitor value and try plugging it into the formula to see whether or not a reasonable resistor combination is available. If we insert 0.001 μF (i.e., 10^{-9}), for example, we find the following:

$$(R1 + R2)C = 10^{-4}$$
$$(R1 + R2)(10^{-9}) = 10^{-4}$$
$$(R1 + R2) = 10^{-4}/10^{-9}$$
$$(R1 + R2) = 10^{5}\ \text{ohms} = 100{,}000\ \text{ohms}$$

Thus, if we select a combination of R1 and R2 that total 100,000 ohms, then the VFC will work. In order to have a range above and below the design ideal value, we will select 82 kohms for R1, and 50 kohms for R2. The potentiometer is used to set the oscillating frequency to 10 kHz when the input voltage is set to exactly 10 volts dc.

The stability of the VFC will be enhanced when the capacitor is a low-drift type, such as polystyrene or silver mica. Also, the resistors should be low temperature coefficient precision types. The potentiometer is best selected as a multi-turn (10-turn or more) trimmer type. Potentiometer R4 is to cancel offset voltages.

One use of the VFC is to serve as an analog-to-digital converter in situations where it is difficult or expensive to transmit binary data. We could, for example, find a situation where the VFC is part of an instrument in an inaccessible location that must communicate with an instrumentation package over a single conductor pair or single radio channel. One place where this might be done is in oceanographic work. Suppose the job at hand is to measure the light intensity of sunlight at the bottom of a body of water, such as the Chesapeake Bay. This job was done by a friend of mine at Old Dominion University. The photoresistor circuit that measured the sunlight (through a set of filters) produced an output of 0 to 10 volts. The voltage is then converted into a frequency (0 to 10 kHz) that can be transmitted over a telephone line to the surface, and tape recorded. If it is necessary, we can also count the frequency and input the resultant data word into a computer.

There are two approaches to frequency counting. First, and possibly most obvious is the standard digital frequency counter. The binary (BCD) outputs can be read by a computer and stored in memory. Alternatively, we can use the microcomputer to count the frequency. There are two approaches to this job. One is to measure the time between two successive pulses of the input frequency, while the other is to make the input frequency operate a counter that will cause a computer program to "jump" when it times out at 00H.

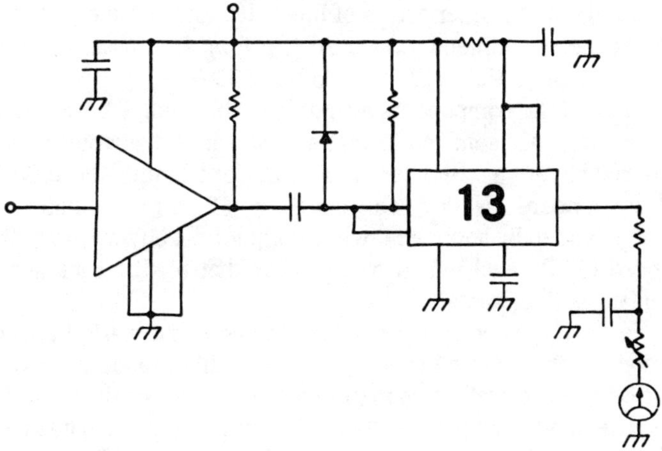

Construction and Components

Although the design aspects are terribly important to the success of any electronic construction project, one must not overlook some of the practical aspects as well. It is simply foolish to spend a large amount of time in the designing and planning of a project, only to "blow it" in the final implementation. If the mechanics of electronic construction are now understood, and if you are unwilling to invest in a little time to learn the skills, then all of the planning in the world will come to naught. There are several matters to be considered: e.g., tools, soldering and wire-wrapping, component selection and so forth. In this final chapter we will consider some of these matters and will at least shower you with my experience.

COMPONENT SELECTION

Perhaps the most significant components to be selected by the hobby project builder will be the semiconductors. The first decision to be made is the type of technology which will be used. Choices for a simple amplifier include bipolar transistors (PNP/NPN), certain linear IC devices, operational amplifiers, current amplifiers (CDAs or "Norton" amplifiers), operational transconductance amplifiers and so forth. For most users, I suppose, the old-fashioned operational amplifier is the most easily applied. The rules of operation are simple, and the devices are well-behaved. For those who might like to learn about the other forms of linear IC, and some of the tradeoffs, let me recommend *Linear IC/Op-Amp Handbook–2nd Edition* by Joseph J. Carr (TAB Book No. 1550).

Most of the components available to the hobbyist are ordinary commercial grade semiconductors sold at prices that would make an industrial buyer gag. Standard rectifiers, for example the 1000-volt PIV a 1-ampere 1N4007, sell for $1 or $2 in most blister pack displays tended by hobbyists, while commercial buyers pay $22 per hundred ($0.22 each). You will pay for that blister pack and the attendant selling costs!

Some of the components sold in the hobbyist market are commercial surplus, and might be of considerably higher quality than those devices normally sold in this market. I hint that this is the case is the use of odd-ball part numbers. The blister pack will be marked with the generic part number (74xx for 74xxx for TTL or 4xxx or 45xx for CMOS), but the markings on the device are strange. These numbers that look so strange might be some companies drawing number or standard house part number. You will also see some TTL devices marked with 54xx and 54xxx numbers. The 54-series are merely military spec TTL devices (e.g., 5490 is merely a 7490 in uniform). Generally these are higher quality units than the 74xx/74xxx devices. They usually have ceramic packages rather than plastic in order to accommodate the −55 to +125° C military specification temperature range. Also, the 54-series devices have been burned-in for more hours than the commercial devices (in fact, on low cost devices the 74xx/74xxx are not burned in at all . . . they are tested and shipped green).

Components marked with numbers beginning with "38510/xxxx" are also military specification devices. This designation indicates that the device is constructed and screened according to military specification MIL-M-38510. While most such devices are too expensive to buy for ordinary projects, they often show up in blister packs at commercial device prices, so are affordable. Unfortunately, the numbers don't correlate to standard type numbers. A guide to many of the 38510/xxxx numbers is found in the annually issued *IC Master*.

Be careful of deals that seem too good to be true . . . they probably are! While the existence of good deals cannot be denied, there is also a lot of ripping off going on. One major electronic retailer routinely buys factory seconds and passes them off as good parts. I recall buying "5 741 op-amps" for 98¢. It turned out that all of them were out of specification, most of them because of a 1 to

2-volt dc output offsetting existing when the input terminals are shorted (output voltage should be zero under that circumstance). In general (which means the following advice cannot be trusted 100 percent!), if the manufacturer's logo or name appears on the IC or transistor, then it will most likely be a quality unit. It will be at least the lowest grade sold by that manufacturer. Most companies do not put the logo/name on until after the initial testing phase, so the seconds are unbranded. If the only branding is the type number, and the deal seemed too good to be true then test the device thoroughly. Notice that I didn't say not buy the part, only test it properly before using. I have seen perfectly wonderful op-amps, for example, marked only with "741" or "1458," which are type numbers.

Perhaps the most shabby of the cheapie parts are switches. I have found that special deal switches are either made so cheaply to begin with that they are useless, or, are seconds that failed test at a major manufacturer. If the action feels a little erratic, or perhaps a little too soft, then beware of the device . . . it might not operate properly or for long. The most common defect is either no connection. Junk switches, even if obtained for a "good" price, are no fun at all . . . they cost more in the long run than a properly made switch.

HAND TOOLS

My first job after high school graduation, several years before I went to engineering school, was working in a radio-TV repair shop near Washington, D.C. One of the two shop owners was a retired Navy chief electronics technician. He had a slogan emblazoned on the wall of the shop that still rings in the ears of all apprentice techs who got their start under his rather loud tutelage: USE THE RIGHT TOOL, USE THE TOOL RIGHT. Oh, so true—even if it was a lesson painfully learned by a teenager who knew everything (it's amazing how much ol' Moodie has learned in the last 20 years!).

Many people who do not make their living with hand tools tend to either misuse them or abuse them (usually misuse and abuse are synonyms!). If you are the kind of turkey who would pick up a screwdriver and pound nails with it, then shame on you—go read another book. But, if you care about tools and their proper use, then read on.

Each tool has a single purpose or at most, a limited range of purposes. In radio and electronics, there are quite a few items that would be classified as the "right tool" under different circumstances. Obviously, an antenna party would require different tools than building a Heathkit.

It is essential that anyone who wants to have an electronics workbench have a small collection of "basics." While most of us have our own preferences (and prejudices), there are a few items that belong in every tool box:

1. Long-nose pliers—jeweler's size
2. Long-nose pliers—4-6 inch size
3. Diagonal side cutters ("dikes")—jeweler's size
4. Diagonal side cutters—4-6 inch size
5. At least three Phillips screwdriver sizes—Nos. 1, 2, and 3. Note I also recommend owning Nos. 1 and 2 in all three lengths: *Long, regular* (6 inch), and *stubby.*
6. Set of straight-blade screwdrivers, from 1/16 inch to 3/8 inch width.
7. Complete set of nutdrivers from 1/16 inch to 5/8 inch size (3/4 and 7/8 inch are also available, but I prefer to use a socket set for these sizes).
8. Socket set, ¼-inch drive
9. Socket set, ⅜-inch drive (if you want to do beam antenna work and fix cars).
10. Crescent wrench, miniature
11. Crescent wrench, regular
12. Crescent wrench, long
13. Gas, or slip-joint, pliers (large and small)
14. *Channel Lock®* pliers
15. Vise-Grips®
16. Soldering apparatus
17. Hack saw
18. Wood saw
19. Hand drill
20. Milling machine (only kidding)
21. All three sizes of miniature Phillips and regular screwdrivers

You will see the wisdom of owning both long-john and stubby sizes of all types of screwdrivers, especially if you decide to work on television sets or car radios! If you prefer (I don't), then buy one of the *Xcelite®* tool rolls, such as the Master Servicer set. These have universal handles, and inserts for the tools. They also have extensions.

When you pick out a jeweler's-size pair of side cutters, select one that has straight sides rather than curved sides. This feature allows you to get in close to pins of items like ICs and sockets on a printed circuit board. The types with curved sides look good in the tool store, but work less well. Before you buy a pair of side cutters, *examine* them for proper fitting of the jaws. A good pair of side cutters will have jaws that just touch together at the tip when the jaws are closed—and only at the very tip. If the jaws show up as irregular, or they meet other than at the tip, then reject it out of hand! The way to see this is to hold the "dikes" up to the light, and view them from the hollow side of the jaws. Dikes that do not meet this criterion are pure junk and should be despised!

Over the years, because of my own ignorance and sometimes stupidity, I have learned some harsh lessons about tools—and their acquisition. The one thing that I can say with certainty is: DON'T BUY CHEAP TOOLS! They are a waste of money. While the "thrifty" might view the cheaper tool as a way to save money, it is really false economy. I paid $15 for a pair of heavy duty dikes in 1963, and they lasted for the twenty years that I was in the service business—and I still use them as a hobbyist now that I am "white collar." Of course you can overspend, but a good tool is a wonder. I like *Craftsman* (Sears) and *Channel Lock* brand (yes, Channel Lock makes more than oversized slip-joint pliers!) although those are only preferences. In a pinch, I would buy *Xcelite* brand.

Let me interject another word of wisdom here: *wire cutters* ("dikes") and *wire strippers* are different tools and have different purposes. If you attempt to strip insulation from hook-up wire with a pair of dikes, you might nick the wire and that could lead to a break in the possibly near future. Also, that pair of dikes will tend to stretch the insulation (it tears it, not cuts it), and the end result will not look like you expected.

Interestingly enough, it is possible to buy low-cost wire strippers of high quality from *Miller, Xcelite* and others. These are of simple design and made of what looks like sheet metal stampings. These strippers are remarkably alike and not at all complex. Funnily enough, the fancy $25 type wire strippers rarely work all of the time—they get out of adjustment. It's nice to have all those whistles and bells, but they are of little importance when the thing stops stripping wire!

The quality of tools can often be deduced from their "feel" as you work them. This is why I hate to see tools inside of blisterpacks—I want to feel them. I know that "feel" is terribly unscientific, but it is a valid way to evaluate tools. Once you get used to the feel of good tools, it will be all but impossible to satisfy you with the junk. Another indication of quality on any tool that has a cutting edge, is the appearance of that edge: it should look sharp and seem to be blued (like the blade of a fine pocketknife).

One of the basic metalworking tools required in most workshops is a hand drill. In addition, many also find a power "saber saw" also of immense use. Both tools can be bought in local Harry and Harriet Homeowner stores—the do-it-yourself emporium—for only a few bucks. The cost of good tools is a lot less when they go on sale, which seems to happen every three or four months. Make sure that your drill has a ¼-inch or ⅜-inch chuck (the latter seems most universal). Do not buy a drill that is equipped with a ½-inch chuck unless you really need it: if that half-incher breaks loose, it can cripple and injure the unwary. All good hardware stores sell drill bits up to half inch that have a shank reduced to either ¼ inch or ⅜ inch. For larger holes, up to 2½ inches, you can also buy a hole saw. These require a *mandrel* (or two if certain large sizes are needed) and various saw blades. But these require a bit of warning! The mandrel and the saw blade are separate pieces. If you do not tighten them properly, they may disassemble at high speed—with you in the way. I prefer the *Black & Decker* hole saws intended for metalworking. The woodworking types are not useful on metal.

If you tend to handle drills with the finesses of Atilla the Hun, then don't try to use hole saws or the larger sizes of drill bit. These devices can be dangerous in the wrong hands. Instead, invest in a set of chassis punches. The punches can be used only if you can reach both sides of the metal being worked. The *Greenlee* brand hole punches use a fixed mandrel, a cutting edge or block, and a bolt that drives the two together under the influence of a large wrench. To use the punch, drill a hole in the metal that is just large enough to accept the pilot-drive bolt. Thread the cutting edge onto the bolt, and begin to tighten the bolt. You will soon feel and hear a snap sound—but you are not finished yet. This snap indicates that one tip of the cutting edge has broken through. Keep turning the wrench until you hear the second snap. At that point, you should be able to lift the punch out of the freshly cut hole. The cost of hole punches is high, so only buy those that you need.

Files are another metalworking tool that can be used to very good advantage by the skilled worker. The results of cheapie files are so poor compared with the results of costlier files, that there is simply no excuse for buying the cheapie! Especially since the cost difference is a lot smaller than the quality difference. You will need several sizes of files. If you do small work, or work on printed circuit cards, then a set of jeweler's files are needed. For working sheet metal chassis, equipment cabinets, and small brackets use a larger file. The wrong size will do one of two things: it will cut off too much metal (sigh), or not enough. In the latter case, you work yourself to death, in the former you work the work to death. Keep at least three different sizes of all the main types of files, in addition to the jeweler's files: flat bastard files, rat-tail files and triangular files.

One of the most clever metalworking tools for use on sheet-metal chassis is the so-called "nibbling tool." The name sounds silly, but it is very descriptive of the way that the tool works: it munches out (nibbles) little bits of metal every time you squeeze the handle. The form of the thing looks like those hand exercisers that one finds in sporting goods stores. You can cut out almost any shape over ⅜-inch at the expense of a well-exercised hand.

Before drilling holes in sheetmetal chassis, it is necessary to use a center-punch to guide the drill point. The punch will make a small dimple in the metal that is used as the drilling guide. Be careful, though, most of the types of punch which require a hammer blow tend to dent the metal and make it hard to work into the correct shape. The *automatic center punch* is a clever little device that works by hand. You center the punch point at the correct place on the metal, and then press down on its body by hand. You will hear a *snap!* when the punch punches. Most larger hardware stores (plus Sears' tool department) sell these wonder punches. Another neat little thing for drilling, especially when you want to drill cylindrical things like antenna tubing, is a *Black and Decker* (plastic) drilling guide. When drilling larger holes, start with the smaller sizes (⅛-inch or smaller), and work up. This is also the method (mandatory method!) for drilling into plastics. If you try to drill too large a hole into plastic at once, then you will chip your way through instead of drilling—and that's *uncool*, to quote the Fonz.

CONSTRUCTION

The final phase of any amateur radio construction project is the actual construction. After the circuit is designed, tested, and debugged—which is often the hard part—it is then time to commit the design to hardware.

The actual approach to construct taken depends a lot upon the type of project being built. Obviously, a VHF converter that is stuck inside an hf communications receiver would take a different approach from that, say, of an outboard audio cw filter. Any time the project is external to another piece of equipment, you want to consider buying a good quality painted chassis. Otherwise, use an unpainted chassis box.

Enclosures and Cabinets

The cabinet or enclosure that you use for your project is both a practical matter, and an artistic concern. After all, there are times when we build things for the pleasure of building as much as any thing else. These projects should be housed in as good a cabinet as we can afford.

Be fully awake and aware when reading catalogs of enclosures. There are few bargains in this market! The best that you can hope for is to get what you pay for. If you buy for price, then expect a junker cabinet. If you demand a well-constructed painted (decorator colors, of course), then expect to pay a pretty penny for it. You *will* pay dearly for these professional-grade cabinets. If, however, you are satisfied with an ordinary metal chassis or a gray hammertone snap-together sheetmetal box, then you will gain low cost in exchange for less ruggedness and looks. Of course, if the project does not require ruggedness, then don't buy it. Be honest about this matter. There is little reason to buy a "MIL-specced" cabinet for a project that will sit on top of your operating desk for a few years!

There are several different types of low-cost and medium-cost cabinets available. Where you desire pleasant and professional appearance, but not lots of mechanical strength, look for the metal plastic and plastic-only by *Ten-Tec* and *Hammond*. Both of these cabinet lines are available in several colors and color combinations. Although these cabinets are not very strong, I have used them in scientific and medical instruments built as part of my job and found both brands reasonably rugged if they are not overly abused. *LBM* also offers some nice low-cost cabinets. Their series in white on black-crackle are particularly attractive if you like a stark "electronicky" look. These are among the best buys in the cabinet market because they are all aluminum (where shielding is important, the plastic cabinets are useless) and look nice.

All of the cheap cabinets can be stiffened and with little effort. I have used aluminum "home handyman" stock from the hardware and plastic rods for this purpose. A little cross-bracing goes a long way, so, in the interest of low-cost cabinetry, you might want to buy a Ten-Tec, Hammond, or LMB cabinet and stiffen it yourself.

The Pomona blue-box series is particularly useful; and these boxes are very tough. There are several useful sizes, painted and unpainted. The painted types are a pleasant blue shade. These rugged boxes come "with or without" card guides milled into the sides; and "with or without" assorted forms of rf and audio connector, as you require (and are willing to pay for!). My major complaint

Fig. 13-1. Modules and strot cage (courtesy Vector).

with the Pomona line is that they seem to lack some of the very sizes that I personally would find useful. But that does not diminish the quality of the products that I can use!

Figure 13-1 shows rack-mountable Vector modules in a cage. The individual modules are aluminum (hence shielded) and will accept either a collection of DIP cards or perfboard. You can also mount DIP cards directly to the mainframe by placing the cardedge connectors along the rear apron of the rack. If you are building a system of instruments, then use the cage as the mainframe and build the sections as needed in EFP module cases.

Vector actually offers at least two different styles of module rack. One has a closed-back, and you must provide appropriate connectors for power and whatever control or signal lines are needed. The other type is an open-back rack and is designed so that the Vector 3677-2DP DIP card edge will protrude from the read end. The card edge connector can then be attached permanently to the cage's rear strut.

The closed-back modules can also be pressed into service as mini-equipment cabinets that are not altogether unattractive. Place the case on its side and install little rubber feet along the edges of the rear sliding cover. I am sure that the person at Vector who designed these would choke if this were known, but I have found that the EFP modules are a low cost alternative to providing a sturdy (key word, that!) enclosure for small projects—including some that have been beat half to death by the "customer."

Finishing Touches

The final touch to any professional looking project is the finish. If you are going to paint the project yourself, then perform *all* of the machining prior to painting. I once had a machinist tell me that he preferred to deburr the holes after it was painted (!?!?!?!), but I didn't believe him! If it is an aluminum panel, then use a base primer (for aluminum surfaces) before painting. The paint should be enameled and be baked on in the oven.

My friend Chuck, WB4DVR, settled onto a method that resulted in some of the most professional-looking panels that I have seen. He had several projects in which a metal-finishing company black (or, in one case, blue) anodized the aluminum panel (again, after all machining was finished). The lettering was then let into the anodized surface intaglio-style by a signmaker who had one of those name tag engraving machines (the kind you see at hamfests making name tags for a bunch of bucks).

At the very least, buy several packages of dry transfer Datamark labels. After the paint is finished and cured, you can apply these labels with ease. Note that a crayonlike pencil (the red or blue pencils teachers use to correct papers) work a lot better for this operation than regular writing pencils or ballpoint pens.

SOLDERING AND WIRE-WRAP

The first thing that you should learn in practical electronics is how to solder! No, I'm not being a smart-aleck, I said *learn to solder.* Soldering is a skill that eludes most newcomers and some old timers as well. Once upon a time, I worked in a local radio shop that was a Heathkit dealer. (Before they had the Heath Electronic Centers all over the country they had little local radio shop dealers.) My job in the service shop often required me to troubleshoot the projects that customers could not get working. By far, the biggest problem was soldering: too much solder, too little solder, or incorrectly done soldering.

Soldering looks so bloody easy that is receives too little of the attention that it deserves until it is too late. Please pay attention to this section, where we will discuss some of the implements used in soldering, and don't let your own cocky arrogance fool you into thinking that you know all about it—unless, of course, you can show me a decade of industrial, aerospace, phone company, or military experience; or, a certificate from a NASA soldering school. (Yes, NASA *did* once have soldering schools).

There are a lot of different ways to learn the delicate art of ordinary soldering, and all of them can be summarized by two words: DO IT! One method favored by some technical institute instructors who have a flair for the unexpected is to give the students exactly one foot of #12 or #14 solid bus wire. The student is to cut this wire into twelve 1-inch lengths—all exactly the same length. Why do you suppose that this malicious instructor wants *exactly* twelve pieces of the same length? Why not five, or thirteen, or ten? Well, you see, there happen to be twelve sides to a *cube*. The students are required to take the twelve pieces of wire, and construct a cube using only the soldering iron, solder and—there's a catch—a single pair of long nose pliers. No vises, no "third hands," no helpers, no nothing else. The idea is to teach the student the importance of keeping the work from moving before the solder has cooled. Once the student has finished the cube and let it cool (about 15 to 30 minutes), the instructor would walk over and pick up the cube "for examination." He would then crush the cube in his hand, and if any of the joints broke the student was given the pleasure of doing it again. The student would make cubes until Hell froze over, or until he/she made a cube that survived that crushing experience.

Another alternative is to go to a hobbyist-type electronics parts store and buy a couple of fistfulls of terminal strips, and a few dozen rejected printed circuit boards. The national mail-order chains usually offer two- or three-dollar assortments of terminal strips, and old PC boards for 50¢ apiece, that can be used in your practice. Practice the delicate arts of: a) wrapping a wire around a terminal strip and soldering it to obtain a strong mechanical joint and an electrically sound joint; and b) soldering to printed circuits without causing solder breaks or solder bridges.

For the type of soldering that you will do in typical amateur radio and electronic hobbyist projects there are only a few basic rules to worry about:

1. Always use quality solder; never reuse old solder. New solder contains a resin flux that prevents oxidation on the surface of the work from ruining the job. Low grade solders (off-brands) often have insufficient flux. I prefer Kester and Ersin solders in a 60/40 or 50/50 lead/tin mixture. NEVER USE ACID-CORE, or plumber's solder. If in doubt, do not use it; buy a new roll of resin (or "rosin") or "radio-TV-electronic" solder.
2. Make the *joint* hot enought to melt the solder, and then apply

the solder to it. Let the heat of the joint do the melting. A frequent error is to let the iron melt the solder, and then hope that it will flow down onto, and adhere to, the joint. This is always a vicious and poor practice, but seems to persist no matter what people like me keep saying.

3. Make absolutely sure that the work is clean. Use a pencil eraser, sandpaper, solvent, or whatever it takes to make the joint bright and clean. Even the tip of the screwdriver is better than nothing!
4. Keep the joint still until the molten solder is again solidified. If the joint moves, a cold solder joint will result (not *may, will!*). Such a joint has a dull, brittle, and porous look.
5. Do not overheat the joint. Too much heat will use up the flux, and will often damage the component.
6. Don't use too little solder.
7. Don't use too much solder.

These last two admonitions cover what are probably the most frequent errors made by the new solderer. Use just enough to cover the joint, without overloading it. If two adjacent terminals are overloaded with solder they will short out, and create damage to your hard built project. I have seen printed circuit boards in which an unskilled assembler doubled the weight of the work just by improper solder technique! No wonder NASA specified the weight of solder on PC boards. A few milligrams of extra solder on each of over a million joints in a space craft or satellite, and the thing simply could not get off the ground!

Soldering Apparatus

Soldering and desoldering are primary skills for the electronics hobbyist. If your amateur radio career makes you more than an appliance operator, then it is a fair bet that you will eventually wind up soldering something sometime.

A basic mistake made by amateurs is to use incorrect, or insufficient tools in the process of soldering. At *least* two types of soldering apparatus are needed in most amateur work. For printed circuit soldering or light "hand-wired circuit" wiring, a 25-50 watt soldering pencil is needed. For heavier grades of work, a 150-250 watt soldering iron is needed.

Pencil-type soldering irons are so numerous that selection is often very difficult, if only for sheer numbers. The only *general* advice is to select a model that uses a number of different interchangeable tips. My personal preferences are the *Weller SP-28*, and the *Weller W-TCP Soldering Station*. I keep the SP-28 in my portable tool case (a vinyl "leatherette" affair that looks like a ladies clutch handbag), and the W-TCP on the permanent workbench. In both cases, I keep a number of well-selected tips for the soldering irons. Several thin conical and blade tips are needed, and make them all at least 700° F types.

A good selection for the 150-250 soldering iron is the *Weller D-440* soldering gun. It is usually wise to buy the kit that comes along with the soldering gun, as it will contain several little widgets and whatchamacallits that are disgustingly clever.

As important as proper soldering is *desoldering*. Almost any fool can desolder a joint, especially on a terminal strip (PC boards are a little harder), but they will splatter all over the place. Components and printed wiring may be summarily destroyed by the desoldering process and solder bridges may be created half-way across the board. Of course, Finagle's Law tends to hide all of the really dangerous solder bridges until 1.12 milliseconds after irrevocable damage has been done!

Finagle's Law: *"The perversity of the universe tends to a local maximum whenever and wherever it will cause the greatest harm!"* Believe it!

There are two general classes of desoldering implement: *solder suckers* and *solder wick*. Both types have their fans, but I maintain that both have their place and should be obtained.

There are two different forms of solder sucker available to the amateur (discounting the professional kind that uses a vacuum pump and "plumbing system"), and both use a form of "manual vacuum pump" to do the job. The cheapest kind uses a squeeze ball pump that looks like an ear irrigation syringe (or the squeeze ball thing that gas station attendants use to check battery acid or antifreeze). The idea is to squeeze the ball, melt the solder joint, and then suck the solder into the device by suddenly releasing the ball.

My own experience with this grade of desoldering device is that the tips soon deteriorate. While they work well when in good shape, they will soon require a replacement tip. As a result of this experience over 20 years, let me suggest rather pointedly that you obtain two or more replacement tips.

The other type of solder sucker is a long cylinder that contains a piston. A nozzle at the working end of the cylinder is used to suck the solder. When you cock the device, the piston is driven all the way down the cylinder to the nozzle face. A spring and latch device is used to operate the sucker. When you touch the trigger, the piston smacks back along its track, pulling a vacuum as it goes. To use the device, cock it, melt the solder, and then place the tip against the melted solder and press the trigger. When the piston goes slamming back up the cylinder the partial vacuum pulled will suck the solder up into the device.

There is only one real problem with this device: it loves to choke on solder. It must be cleaned of solder every now and then, and the "oh-rings" that seal the vacuum must be re-lubricated (silicon heat-transfer grease for transistors works well!). There are two different mechanisms on the market, and both are made by at least one company. In one type, the cocking rod and the piston are not attached to each other. They make contact only when you are in the act of cocking the spring. At all other times the rod is in its resting position, and is harmless. In the other type, however, the rod and piston are of one-piece construction. When the device is cocked, the rod stays down until the trigger is pressed. The only hang-up is that when you are bending over the work trying to see what you are doing, you may fail to recognize that the rod is pointed at your eye and then *WHACCKKKKO!* it smacks you good. A word to the wise should be sufficient.

Solder wick is one of the more clever methods of desoldering, and is probably the method of choice for the person who works on printed circuits. Solder wick leaves a remarkably clean joint. This "wick" consists of copper braid that is filled with resin. When you lay it on a joint, and apply heat, it will cause the solder to be lifted off the printed circuit track or terminal strip by capillary action. Solder wick is a little bit expensive, however, so use it wisely.

Selection of Solder

Solder can be specified according to several parameters: composition, size, form, and so forth. Electronic solder should be a composition of lead and tin. This composition, incidentally, will melt at a lower temperature than either metal alone. Some formulations will melt as low as 275° F, while most melt between 300° F and 450° F. There are a few premium solders available that contain 2 percent silver, which raises the melting point while simultaneously reducing the resistivity.

WORD OF CAUTION: DON'T EVER USE PLUMBER'S SOLDER FOR ELECTRONICS APPLICATIONS!

The proper solder for radio and electronics applications is

50/50 or 60/40 lead/tin formulation solder and a *resin* (or *rosin*) core. For small work on IC-laden printed circuit boards use 22, 24, 26 or 28 gauge solder. For larger work, use 14, 16 or 18 gauge solder. Solder is terribly expensive, so you might have to compromise. If you can afford two rolls, then select a #14 or #16 for larger work and a #24 or #26 for the smaller stuff. If, however, your operation is a one-roll concern, select #22 as the best trade-off.

Most solder used by amateur radio and electronic hobbyists comes in the form of wire, is stored on rolls, and has a resin core. But there are at least two other types of use: cordless wire solder and bar solder. Both of these are sometimes used in amateur antenna work to connect the heavy ground wire from the transmitter to the ground rod driven into the ground. Do not try to use a soldering iron for this application! The ground makes a good heat sink, so will not allow the joint to get hot enough to melt the solder. I use a *Bernz-o-matic* propane torch for this application. Neither of these solders have a flux, so buy a can of resin flux before starting.

Wire-Wrap

Wire-wrap is a powerful electronic construction technique in which the connecting wires are wrapping tightly without solder around square terminals. The only real drawback to this method is that genuine wire-wrap boards tend to be expensive. This is a difficulty that is easy to overcome with the native cleverness of amateurs. Buy wire-wrap sockets and then solder them to a DIP plug board or glue them to perfboard. Make sure that the socket is a real wire-wrap socket with square terminals. Most electronic stores that carry either Vector or Jameco products will have these sockets.

The interterminal connections on a wire-wrap board are made using a special tool. You can buy a nice electrically powered wire-wrap tool for something like $60, but there is a cheaper and easier method. You could also buy a $6 wire-wrap tool, but these are a pain in the *tschush* to use. One of the best alternatives combines the ease of the electrical type (almost) and the lower cost of the cheapie (almost): the *Vector P180 Slit'n'Wrap* tool shown in Fig. 13-2, or, its more recent cousins from the same line of products. The special wire used with this tool has a special insulation that will scrape away when the P180 tool scratches it, and will melt if a soldering iron is applied. This feature allows you to solder errant joints (or if you don't really trust (!?!) wire-wrap).

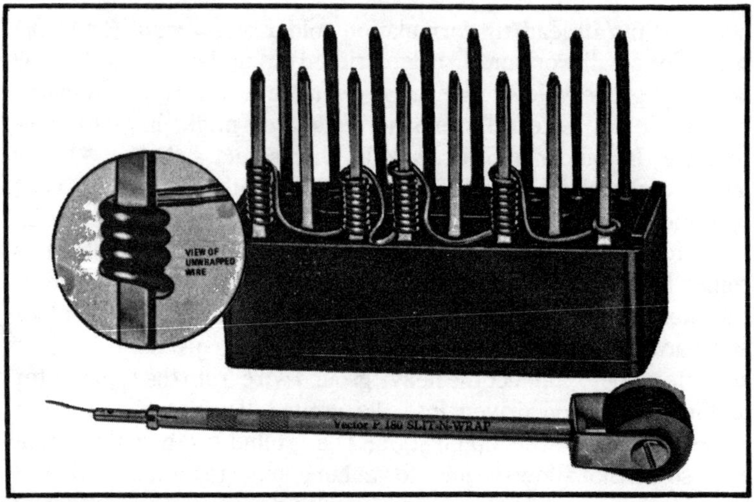

Fig. 13-2. Vector P180 Slit-n-Wrap tool.

PRINTED CIRCUIT AND WIRE BOARDS

With the possible exception of high-power linear amplifiers, the hand-wired method of construction is all but dead. Even the well-known manufacturer of a particular consumer electronic product that advertised "hand-wired chassis" has now opted for printed circuit boards.

Amateurs can make their own printed circuit boards, buy them from a company that makes the project in kit form, or use wire boards. The wire boards come in two basic forms: solder-type and wire-wrap. Almost any of your electronic construction projects can be given a really professional appearance if it is housed in a properly designed cabinet, but the use of a printed circuit card will make it an elegant project even if you are the only one that ever sees it!

Professional printed circuit layout can be expensive if done commercially. Of course, the PC board industry is loaded with moonlighting craftsmen that will do the job at reasonable cost on a part-time basis. There are also some amateur-oriented PC craftsmen around, such as O.W. Stafford in North Carolina (see *QST* ads).

Of course, you can also try to make your own printed circuit board. This job has to be done at least by all electronic hobbyists, so go ahead and get it out of your system. You will find that most local electronic supplies stores also sell printed circuit materials, including PC kits, direct transfer kits, rubylith film, supplies, and PC layout graphics aids. There are several brands that are generally available: Vector, Calectro, Kepro, and others.

The layout graphics are particularly good for small production efforts. You can buy small quantities of dry transfer (rub-on) PC patterns in 1×, 2× and 4× size. Work with the 2× size. These dry transfer PC patterns include tape for making straight lines, corners and edges of various types, and patterns for all of the different IC packages and transistor cases. You will work on mylar film, rubbing the dry transfer patterns onto the surface.

I personally prefer to make a "rough draft" on ordinary white paper. I draw the outlines in 2× scale, and then try (in pencil!) to make all of the interconnections. Where it is impossible to jump over some feature of the printed circuit, use wire jumpers on the component side of the board. This is "unelegant," but is often the only way around the problem short of making a double-sided printed circuit board.

Once the rough draft is completed, overlay a sheet of transparent mylar or acetate sheet (mylar works best because it doesn't change dimensions over time!) and lay down the dry transfer patterns. PC tape is used to make the interconnections. Alternatively, you may use an India ink pen such as a *Rapidograph,* or *Technic* pen. In some cases, I have been successful using an ordinary marker pencil, although the ordinary felt-tip sort of marker pencil is generally too light for this application.

Once the positive master (black on clear film) is made, then it is time to make a negative. Go to a print or lithography shop—or any other place that has a litho camera—and have the negative made. Remember that you made the master 2× normal size, so ask the printer for a *50 percent reduction.* Most printers only offer nominal accuracy unless you specify closer tolerances (remember, those darn ICs might not fit!). I recommend making a line exactly 4 inches long (precisely) on the master, or, alternatively, make the card edges a precise dimension such as 4 inches, and tell the printer to carefully adjust the camera so that this line is reduced to exactly 2 inches. This will ensure that the printed board will fit!

There are two basic types of PC stock: photosensitive and nonphotosensitive. For small runs, I prefer to use the photosensitive stuff because with the other you also have to buy a photosensitive resist coating (often in spray cans) that will deteriorate over time. Lay the negative on top of the photosensitive layer, and expose it to a bright white light. There are different specifications for this process, but I can tell you that room light is terrible. I prefer buying a photoflood light, or a movie light (i.e., "sun gun") and using it to expose the board. The manufacturer of the board will usually provide exposure information for their products.

The printed circuit board will require developing after exposure (remember, this is a photographic process). When the board is developed, those areas where we want the copper cladding to remain will be resistive to acid or other corrosives.

There are two basic forms of etchant used in amateur printed circuit work: *ferric chloride* and *ammonium bichloride*. The ferric chloride stains everything it contacts, and the ammonium bichloride eats everything it contacts—BE CAREFUL! I personally prefer the ferrous chloride because it is easier to tame.

Place the exposed and developed board into a tray of ferrous chloride, and wait for the process to complete. It takes from 20 minutes to an hour. The process is facilitated by heating the bath with an incandescent lamp placed a few inches above the surface, and by occasional agitation of the bath.

Figure 13-3 shows a PC bath container designed by a friend of mine. It is made of plexiglass sheets that are cemented together. Wire hooks (plastic is preferred—metal is eaten away) suspend the PC card being etched in the chamber (which is 1 inch wide). Ferric chloride is then added to the chamber and the etching begins. Again, an incandescent lamp can be used to heat the solution to make it work faster.

We can also use a DIP board by Vector (See Fig. 13-4). These boards are basically perforated stock, but some have printed circuit pads for ICs and transistors (a universal pattern), and have power distribution paths etched onto the board. These can be used either as wire-wrap or solder-boards. The type shown is available in 4.5 × 6 and 4.5 × 9 inch sizes, and uses the standard 22/44 pin card edge connectors. The pads on the DIP board for the card edge connector are gold plated to resist corrosion. The plating cost a little more, but works so much better a year later than do the thinned types. Both the DIP boards, and the correct type of perfboard (also a Vector product) are available with 0.042 inch holes in 0.1 × 0.1 inch centers.

You can "work" printed circuit board, and the printed sections of DIP boards, using an Xacto knife or razor blade (the knife is safer in my opinion, having butchered myself once on a razor blade!). We can, for example, break up the two major PC tracks in the power distribution system for several uses.

Wire-wrap perf boards may or may not have any printing, but I

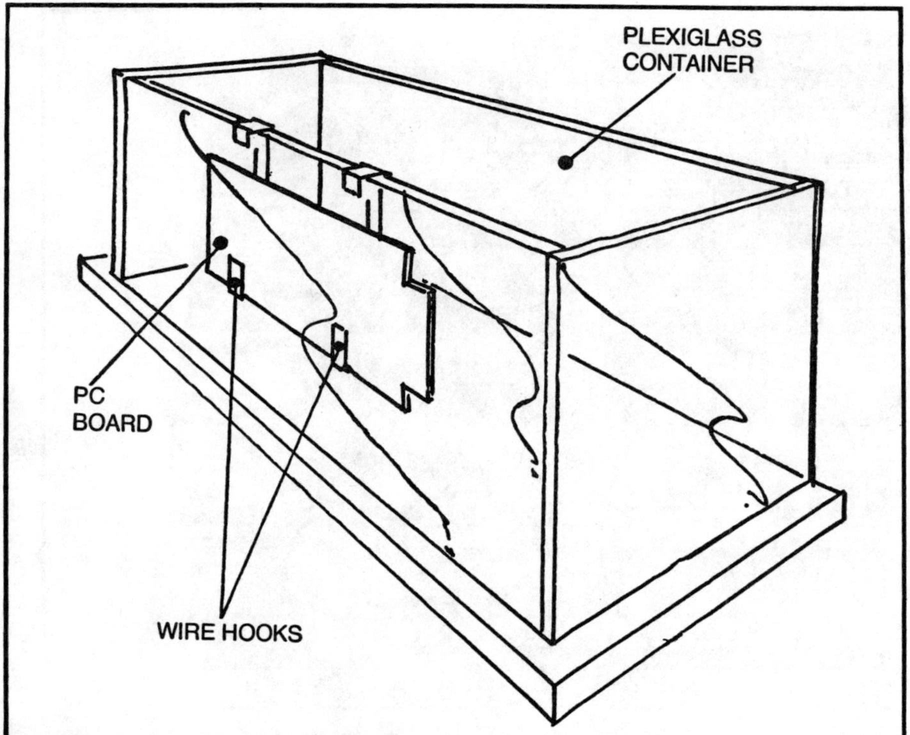

Fig. 13-3. Pc bath container.

Fig. 13-4. Vector Electronics DIP board.

prefer the types that have at least the power distribution tracks printed. It is also nice to have the IC pads printed onto the board because rare is the amateur wire-wrap board that cannot benefit from a few soldered connections—not all components are amenable to wire-wrap! You may be surprised to find out that ordinary DIP IC sockets (14 and 16 pins) are suitable for mounting diodes and quarter-watt resistors. Bend the leads on these components so that they are at right angles to the body of the component, in the same direction, and are 0.3 inch apart (approximately). Cut off the leads to ¼-inch and insert them into the holes of the IC socket. Neat, Huh?

Index

Index

U

V

W

Z